INEQUALITIES IN
ANALYSIS AND
PROBABILITY

INEQUALITIES IN
ANALYSIS AND
PROBABILITY

Odile Pons
INRA, France

World Scientific

NEW JERSEY • LONDON • SINGAPORE • BEIJING • SHANGHAI • HONG KONG • TAIPEI • CHENNAI

Published by

World Scientific Publishing Co. Pte. Ltd.
5 Toh Tuck Link, Singapore 596224
USA office: 27 Warren Street, Suite 401-402, Hackensack, NJ 07601
UK office: 57 Shelton Street, Covent Garden, London WC2H 9HE

British Library Cataloguing-in-Publication Data
A catalogue record for this book is available from the British Library.

ISBN 978-981-4412-57-5

Printed in Singapore.

Preface

The inequalities in vector spaces and functional Hilbert spaces are naturally transposed to random variables, martingales and time indexed stochastic processes with values in Banach spaces. The inequalities for transforms by convex functions are examples of the diffusion of simple arithmetic results to a wide range of domains in mathematics. New inequalities have been developed independently in these fields. This book aims to give an account of inequalities in analysis and probability and to complete and extend on them.

The introduction gives a survey of classical inequalities in several fields with the main ideas of their proofs and applications of the analytic inequalities to probability. This is not an exhaustive list. They are compared and sometimes improved with simple proofs. Further developments in the literature are mentioned. The book is organized according to the main concepts and it provides new inequalities for sums of random variables, their maximum, martingales, Brownian motions and diffusion processes, point processes and their suprema.

The emphasis on the inequalities is aimed at graduate students and researchers having the basic knowledge of courses in Analysis and Probability. The concepts of integration theory and of probabilities are supposed to be known, so the fundamental inequalities in these domains are acquired and references to other publications are added to complete the topic whenever possible. The book contains many proofs, in particular, basic inequalities for martingales with discrete or continuous parameters are detailed and the progress in several directions are easily accessible to the readers. They are illustrated by applications in probability.

I undertook this work in order to simplify the approach of uniform bounds for stochastic processes in functional classes which are presented in

Chapter 5. In the statistical applications, the assumptions for most results of this kind are specific to another distance rather than the uniform distance. Here, the results use inequalities of Chapter 4 between the moments of martingales and those of their predictable variations, then the conditions and the constants of the probabitistic bound differ from those of the other authors. During the preparation of the book, I added other inequalities while reading papers and books containing errors and unproved assertions; it should therefore fill some gaps. It does not cover the convex optimization problems and the properties of their solutions. It can be used as an introduction to more specific domains of the functional analysis or probability theory and as a reference for new applications to the asymptotic behaviour of non-standard empirical processes in statistics. Several applications to the tail behaviour of processes are developed in Chapters 5 and 6.

Odile M.-T. Pons
April 2012

Contents

Preface   v

1. Preliminaries   1

    1.1   Introduction   1
    1.2   Cauchy and Hölder inequalities   2
    1.3   Inequalities for transformed series and functions   6
    1.4   Applications in probability   9
    1.5   Hardy's inequality 13
    1.6   Inequalities for discrete martingales 15
    1.7   Martingales indexed by continuous parameters 19
    1.8   Large deviations and exponential inequalities 23
    1.9   Functional inequalities 27
    1.10   Content of the book 28

2. Inequalities for Means and Integrals   31

    2.1   Introduction 31
    2.2   Inequalities for means in real vector spaces 31
    2.3   Hölder and Hilbert inequalities 35
    2.4   Generalizations of Hardy's inequality 38
    2.5   Carleman's inequality and generalizations 46
    2.6   Minkowski's inequality and generalizations 48
    2.7   Inequalities for the Laplace transform 52
    2.8   Inequalities for multivariate functions 54

3. Analytic Inequalities   59

    3.1   Introduction 59

3.2	Bounds for series	61
3.3	Cauchy's inequalities and convex mappings	64
3.4	Inequalities for the mode and the median	68
3.5	Mean residual time	72
3.6	Functional equations	74
3.7	Carlson's inequality	80
3.8	Functional means	83
3.9	Young's inequalities	86
3.10	Entropy and information	88

4. Inequalities for Martingales   **91**

4.1	Introduction	91
4.2	Inequalities for sums of independent random variables	92
4.3	Inequalities for discrete martingales	99
4.4	Inequalities for martingales indexed by $\mathbb{R}_+$	104
4.5	Poisson processes	108
4.6	Brownian motion	111
4.7	Diffusion processes	116
4.8	Level crossing probabilities	120
4.9	Martingales in the plane	124

5. Functional Inequalities   **127**

5.1	Introduction	127
5.2	Exponential inequalities for functional empirical processes	128
5.3	Exponential inequalities for functional martingales	135
5.4	Weak convergence of functional processes	139
5.5	Differentiable functionals of empirical processes	142
5.6	Regression functions and biased length	146
5.7	Regression functions for processes	151

6. Inequalities for Processes   **153**

6.1	Introduction	153
6.2	Stationary processes	154
6.3	Ruin models	156
6.4	Comparison of models	162
6.5	Moments of the processes at T_a	164
6.6	Empirical process in mixture distributions	166
6.7	Integral inequalities in the plane	169

6.8   Spatial point processes 170

7. Inequalities in Complex Spaces   179

7.1   Introduction 179
7.2   Polynomials 182
7.3   Fourier and Hermite transforms 183
7.4   Inequalities for the transforms 190
7.5   Inequalities in $\mathbb{C}$ 192
7.6   Complex spaces of higher dimensions 193
7.7   Stochastic integrals 197

Appendix A   Probability   201

A.1   Definitions and convergences in probability spaces 201
A.2   Boundary-crossing probabilities 206
A.3   Distances between probabilities 207
A.4   Expansions in $L_2(\mathbb{R})$ 210

Bibliography   213

Index   219

Chapter 1

Preliminaries

1.1 Introduction

The origin of the inequalities for convex functions are the inequalities in real vector spaces which have been extended to functional spaces by limits in Lebesgue integrals. They are generalized to inequalities for the tail distribution of sums of independent or dependent variables, under conditions for the convergence of their variance, and to inequalities for the distribution of martingales indexed by discrete or continuous sets. These inequalities are the decisive arguments for bounding series, integrals or moments of transformed variables and for proving other inequalities.

The convergence rate of sums of variables with mean zero is determined by probability inequalities which prove that a sum of variables normalized by the exact convergence rate satisfies a compactness property. If the normalization has a smaller order than its convergence rate, the upper bound of the inequality is one and it tends to zero if the order of the normalization is larger.

Many probability results are related to the Laplace transform, such as Chernov's large deviations theorem, Bennet's inequalities and other exponential inequalities for sums of independent variables. This subject has been widely explored since the review papers of the Sixth Berkeley Symposium in Mathematical Statistics and Probability (1972) which covers many inequalities for martingales, Gaussian and Markov processes and the related passage problems and sojourn times. Some of them are revisited and extended after a brief review in this chapter. The upper bounds for the tail probability of the maximum of n variables depend on n, in the same way, the tail probability of the supremum of functional sums have upper bounds depending on the dimension of the functional classes.

1.2 Cauchy and Hölder inequalities

Inequalities for finite series were first written as inequalities in a vector space V provided with an Euclidean norm $\|x\|$, for x in V. The scalar product of x and y in V is defined from the norm by

$$< x, y > = \frac{1}{2}\{\|x + y\|^2 - \|x\|^2 - \|y\|^2\} \qquad (1.1)$$

and, conversely, an Euclidean norm is a ℓ_2-norm related to the scalar product as

$$\|x\| = < x, x >^{\frac{1}{2}}.$$

From the definition (1.1), the norms of vectors x and y of an Euclidean vector space V satisfy the geometric equalities

$$\|x + y\| + \|x - y\| = 2(\|x\| + \|y\|) \qquad (1.2)$$
$$\|x + y\| - \|x - y\| = 4 < x, y >.$$

The space $\ell_2(V)$ is the space of series of V with a finite Euclidean norm. An orthonormal basis $(e_i)_{1 \leq i}$ of V is defined by the orthogonality property $< e_i, e_j > = 0$ for every $i \neq j$ and by the normalization $\|e_i\| = 1$ for every $i \geq 1$. Let V_n be a vector space of dimension n, for example $V_n = \mathbb{R}^n$, for an integer n and V_∞ be its limit as n tends to infinity. Every vector x of $\ell_2(V_n)$, $n \geq 1$, is the sum of its projections in the orthonormal basis

$$x = \sum_{i=1}^{n} < x, e_i > e_i,$$

its coordinates in the basis are $x_i = < x, e_i >$, $i = 1, \ldots, n$, and its norm is $\|x\|_2 = (\sum_{i=1}^{n} x_i^2)^{\frac{1}{2}}$. In $\ell_2(V_\infty)$, a vector x is the limit as n tends to infinity of $\sum_{i=1}^{n} < x, e_i > e_i$ and its norm is the finite limit of $(\sum_{i=1}^{n} x_i^2)^{\frac{1}{2}}$ as n tends to infinity. The space $\ell_p(V_n)$, $1 \leq p < \infty$, is defined with respect to the norm

$$\|x\|_p = (\sum_{i=1}^{n} |x_i|^p)^{\frac{1}{p}}$$

and the space $\ell_\infty(V_n)$ is the space of vector with a finite uniform norm $\|x\|_\infty = \max_{1 \leq i \leq n} |x_i|$. In $\ell_p(V_\infty)$ and $\ell_\infty(V_\infty)$, the norms are defined as the limits of the norms of $\ell_p(V_n)$ as n tends to infinity. The norms $(\|x\|_p)_{0 < p \leq \infty}$ are an increasing sequence for every x in a vector space V. The triangular inequality is $\|x + y\| \leq \|x\| + \|y\|$. Consequently, for all x and y in a vector space $|\|x\| - \|y\|| \leq \|x - y\|$.

The Cauchy inequality (1821) in the vector space $V_n = \mathbb{R}^n$, for an integer n, or V_∞ is

$$< x, y > \ \leq \|x\|_2 \|y\|_2$$

for every x and y in V_n, with equality if and only if x and y are proportional. It is established recursively from the triangular inequality in V_2.

All norms are equivalent in an n-dimensional vector space V_n: For every x in $\mathbb{R}^n$ and for $1 \leq p, q \leq \infty$, there exist constants $c_{p,q,n}$ and $C_{p,q,n}$ depending only on n, p and q, such that $c_{p,q,n}\|x\|_p \leq \|x\|_q \leq C_{p,q,n}\|x\|_p$

$$\|x\|_\infty \leq \|x\|_p \leq n^{\frac{1}{p}}\|x\|_\infty,$$
$$n^{-1}\|x\|_1 \leq \|x\|_\infty \leq \|x\|_1,$$
$$n^{-\frac{1}{p}}\|x\|_p \leq \|x\|_1 \leq n^{\frac{1}{p'}}\|x\|_p, \qquad (1.3)$$
$$n^{-\frac{1}{p'}}\|x\|_1 \leq \|x\|_p \leq n^{\frac{1}{p}}\|x\|_1,$$
$$n^{-\frac{1}{p'}-\frac{1}{q}}\|x\|_q \leq \|x\|_p \leq n^{\frac{1}{p}+\frac{1}{q'}}\|x\|_q,$$

with conjugate integers $p^{-1} + p'^{-1} = 1$ and $q^{-1} + q'^{-1} = 1$.

Extensions of Cauchy's inequality to bilinear series have been studied by Hilbert, who proved that for positive real series $(x_n)_{n \geq 1}$ in ℓ_p and $(y_m)_{m > 1}$ in ℓ_q

$$\sum_{n \geq 1} \sum_{m \geq 1} \frac{x_n \, y_m}{n + m} \leq \frac{\pi}{\sin(p^{-1}\pi)} \| \sum_{i \leq n} x_i \|_p \| \sum_{j \leq m} y_j \|_{p'}$$

where p and p' are conjugate integers. Other examples are given by Hardy, Littlewood and Pólya (1952).

The Cauchy inequality is extended to an inequality for integrals with respect to the Lebesgue measure on $\mathbb{R}$. Let f and g be square-integrable functions in $L_2(\mathbb{R})$, the Cauchy-Schwarz inequality is

$$| \int_\mathbb{R} f(x)g(x) \, dx| \leq (\int_\mathbb{R} f^2(x) \, dx)^{\frac{1}{2}} (\int_\mathbb{R} g^2(x) \, dx)^{\frac{1}{2}},$$

with equality if and only if f and g are proportional. Let μ be a positive measure on $\mathbb{R}$ and w be a positive weighting function, the Cauchy-Schwarz weighted inequality is

$$| \int_\mathbb{R} wfg \, d\mu| \leq (\int_\mathbb{R} wf^2 \, d\mu)^{\frac{1}{2}} (\int_\mathbb{R} wg^2 \, d\mu)^{\frac{1}{2}},$$

with equality under the same condition. A simple proof for both inequalities relies on the inequality $\int_\mathbb{R} (tf(x) - g(x))^2 w(x) \, d\mu(x) \geq 0$ which develops as an equation of the second order with respect to t, with a negative discriminant $(\int wfg \, d\mu)^2 - (\int wf^2 \, d\mu)(\int wg^2 \, d\mu)$.

It is extended to the Hölder inequalities for L_p-integrable real functions with a finite or infinite support $[a, b]$, where $-\infty \leq a < b \leq \infty$. Let p be an integer $L_{p,\mu}(a, b) = \{f : (\int_a^b |f|^p \, d\mu)^{\frac{1}{p}} < \infty\}$, if $1 \leq p < \infty$, and $L_\infty(a, b) = \{f : \sup_{(a,b)} |f| < \infty\}$. Let p and q be conjugate integers such that $p^{-1} + q^{-1} = r^{-1}$. For all f in $L_{p,\mu}(a, b)$ and g in $L_{q,\mu}(a, b)$, Hölder inequality is

$$\left(\int_{\mathbb{R}} |fg|^r \, d\mu \right)^{\frac{1}{r}} \leq \left(\int_{\mathbb{R}} |f|^p \, d\mu \right)^{\frac{1}{p}} \left(\int_{\mathbb{R}} |g|^q \, d\mu \right)^{\frac{1}{q}}. \tag{1.4}$$

The L_p norms are increasing. This implies Khintchine's inequality for $p \geq 1$. Let $(r_k)_{k \geq 0}$ be the Rademacher functions on $[0, 1]$, for every measurable function $f = \sum_{k \geq 0} a_k r_k$ on $[0, 1]$ satisfying $\sum_{k \geq 0} a_k^2 < \infty$, there exist constants $A_p > 0$ and B_p such that

$$A_p \left(\sum_{k \geq 0} a_k^2 \right)^{\frac{1}{2}} \leq \|f\|_p \leq B_p \sum_{k \geq 0} (a_k^2)^{\frac{1}{2}}. \tag{1.5}$$

Since the constants do not depend on the dimension of the projection of the function onto the basis, (1.5) implies the equivalence between the norms of the function, $A_p \|f\|_2 \leq \|f\|_p \leq B_p \|f\|_2$.

Inequalities for a countable sequence $x = (x_i)_{i \geq 0}$ in the space $\ell_p(\mathbb{R}^\infty)$ are deduced from functional inequalities for piecewise constant integrable functions of a space $L_{p,\mu}$ or by setting $x_i = \int_{a_{i-1}}^{a_i} f \, d\mu$ for some function f of $L_{p,\mu}$ and $\cup_{i \geq 0} (a_{i-1}, a_i) = (a, b)$. Let $r < p$ be integers and q be the conjugate of p such that $r^{-1} = p^{-1} + q^{-1}$, then $\|xy\|_r \leq \|x\|_p \|y\|_q$.

The Cauchy and Hölder inequalities are recursively generalized to more than two series or functions. Let $k \geq 2$ be an integer, let $p_1, \ldots, p_k \geq 1$ be integers such that $\sum_{i=1}^k p_k^{-1} = r^{-1}$ and let $f_1, \ldots, f_k$ be functions in $L_{p_1,\mu}(a, b), \ldots, L_{p_k,\mu}(a, b)$ respectively, the Hölder inequality for the functions $f_1, \ldots, f_k$ is

$$\left(\int_a^b |\prod_{i=1}^k f_i|^r \, d\mu \right)^{\frac{1}{r}} \leq \prod_{i=1}^k \|f_i\|_{p_i,\mu} \tag{1.6}$$

with equality if and only if the functions f_i are proportional.

The analogous Cauchy inequality for a set of k series $(x_{ij})_{1 \leq i \leq n, 1 \leq j \leq k}$ of L_{p_k}, respectively, is obtained with piecewise constant functions

$$\left(\sum_{i=1}^n |\prod_{j=1}^k x_{ij}|^r \right)^{\frac{1}{r}} \leq \prod_{j=1}^k \|x_j\|_{p_j}. \tag{1.7}$$

It is extended to infinite arrays $(x_{ij})_{1 \leq i < \infty, 1 \leq j < \infty}$.

The Hölder inequality is a consequence of the concavity of the function $\varphi_\alpha(u,v) = u^\alpha v^{1-\alpha}$ with $\alpha = p^{-1}$ (Neveu, 1970) and the same argument applies for the other inequalities. More generally, let f be a real function of $L_1(a,b)$, then for every convex function φ from $\mathbb{R}$ to $\mathbb{R}$, Jensen's inequality for integrals is

$$\varphi\left(\int_a^b f\, d\mu\right) \leq \int_a^b \varphi(f)\, d\mu$$

and it extends the property $\varphi(\sum_{i=1}^n \alpha_i a_i) \leq \sum_{i=1}^n \alpha_i \varphi(a_i)$ for convex functions in vector spaces, with $\sum_{i=1}^n \alpha_i = 1$, and the inequality is an equality if and only if the a_i are proportional. The Minkowski inequality is the additivity property of the norms in the spaces $\ell_p(\mathbb{R}^n)$, $1 \leq n \leq \infty$

$$\|x+y\|_p \leq \|x\|_p + \|y\|_p, 1 \leq p \leq \infty$$

and it is extended to $L_p(\mathbb{R})$

$$\|f+g\|_p \leq \|f\|_p + \|g\|_p, 1 \leq p \leq \infty,$$

with equality if and only if x and y and, respectively, f and g are proportional. In $\ell_p(\mathbb{R}^n)$, it is a consequence of the concavity and the homogeneity of the function $\varphi_p(x) = \|x\|_p$ and it generalizes to $L_p(\mathbb{R})$.

In the complex space $\mathbb{C}$, the Euclidean norm of $x = x_1 + ix_2$ is

$$\|x\| = (x\bar{x})^{\frac{1}{2}} = (x_1^2 + x_2^2)^{\frac{1}{2}},$$

with the complex conjugate $\bar{x} = x_1 - ix_2$, and the scalar product $< x, y >$ of x and y is still defined by (1.1). In the product spaces $\mathbb{C}^n$ and $\mathbb{C}^\infty$, let $x = (x_i)_{i \in I}$ and $y = (y_i)_{i \in I}$ be vectors of finite or countable dimension, where I is a subset of integers $I = \{1, \ldots, n\}$, $I = \mathbb{N}$ or $I = \mathbb{Z}$, its Euclidean norms is $\|x\| = (\sum_{i \in I} x_i \bar{x}_i)^{\frac{1}{2}}$ and $< x, y >$ follows by (1.1). The inequalities of Cauchy and Minkowski are written in the same form as in $\mathbb{R}^2$.

The Fourier transform of a variable X with density function f_X is defined as $\widehat{f}_X(\omega) = Ee^{2\pi\omega X} = \int_{-\infty}^\infty e^{2\pi\omega x} f_X(x)\, dx$. Its norms $L_p([0,1])$ satisfy the properties

$$\|\widehat{f}_X\|_p \leq \|f_X\|_p,$$
$$\|\widehat{f}_X\|_1 \leq \|\widehat{f}_X\|_p^p \|\widehat{f}_X\|_{p'}^{p'}, \ p^{-1} + p'^{-1} = 1.$$

It is generalized to functions of $L_p(\mathbb{R}^d)$. The Fourier transform of the sum of independent variables is the product of their transforms, hence by Hölder's inequality $\|\widehat{f}_{X+Y}\|_r \leq \|\widehat{f}_X\|_p \|\widehat{f}_Y\|_{p'}$, $r^{-1} = p^{-1} + p'^{-1}$.

1.3 Inequalities for transformed series and functions

The Cauchy inequality has been extended by convex transformations of the series. A convex real function f on $\mathbb{R}$ satisfies

$$f\left(\frac{\sum_{i=1}^{k} \alpha_i x_i}{\sum_{i=1}^{k} \alpha_i}\right) \leq \frac{\sum_{i=1}^{k} \alpha_i f(x_i)}{\sum_{i=1}^{k} \alpha_i}$$

for every linear combination $\sum_{i=1}^{k} \alpha_i x_i$ with all $\alpha_i > 0$, or equivalently $f(\sum_{i=1}^{k} \alpha_i x_i) \leq \sum_{i=1}^{k} \alpha_i f(x_i)$ for every linear combination such that $\alpha_i > 0$ for every i and $\sum_{i=1}^{k} \alpha_i = 1$. With $k = 2$ and $\alpha > 0$, consider $x > y$ and f increasing at y then

$$0 \leq f(y + \alpha(x - y)) - f(y) \leq \alpha\{f(x) - f(y)\} \to 0$$

as α tends to zero. Otherwise, let f be decreasing at x, $0 \leq f(x - (1 - \alpha)(x - y)) - f(x) \leq (1 - \alpha)\{f(y) - f(x)\}$ which tends to zero as α tends to 1, so every convex function is continuous.

Let f be increasing and belong to $C_2(\mathbb{R})$ and let y satisfy $f'(y) = 0$. Let α in $[0, 1]$ and $x > y$. As α tends to zero, a Taylor expansion of f in a neighbourhood of y is written $f(y + \alpha(x - y)) - f(y) = \frac{\alpha^2}{2}(x - y)^2 f''(y) + (\alpha^2)$ therefore $f''(y) > 0$. If f is a decreasing convex function of $C_2(\mathbb{R})$, its second derivative at y such that $f'(y) = 0$ is also strictly positive, so the second derivative of a convex function is strictly positive where its first derivative is zero and it is a minimum of the function. Conversely, a function f is concave if $-f$ is convex, its second derivative is strictly negative where its first derivative is zero and it is a maximum of the function. The polygons generate the space of convex functions.

The Hadamard inequality for a convex function f on a finite interval $[a, b]$ is

$$f\left(\frac{a + b}{2}\right) \leq \frac{1}{b - a} \int_a^b f(x)\, dx \leq \frac{f(a) + f(b)}{2}.$$

Cauchy (1821) proved other inequalities for convex functions of a countable number of variables, for example let $(x_i)_{i \geq 1}$ be a sequence of real numbers, then

$$\prod_{i \geq 1}(1 + x_i) < \exp\left\{\sum_{i \geq 1} x_i\right\}.$$

Let f be a positive and concave function on a subinterval $[a, b]$ of $\mathbb{R}_+$ and $(x_i)_{i=1,\ldots,k}$ be k points in $[a, b]$ and let $\bar{x}_\alpha = \sum_{i=1}^{k} \alpha_i x_i$ be a weighted mean of these points, with $\sum_{i=1}^{k} \alpha_i = 1$. By concavity

$$\prod_{i=1}^{k}\{f(x_i)\}^{\alpha_i} \leq f(\bar{x}_\alpha), \tag{1.8}$$

with equality if $\alpha_1 x_1 = \cdots = \alpha_k x_k$. The exponential function is convex and the equality is satisfied for all sequences $(x_i)_{i=1,\dots,k}$ and $(\alpha_i)_{i=1,\dots,k}$ then conditions for equality in (1.8) depend on the function f. The inequality (1.8) is also written $\sum_{i=1}^{k} \alpha_i \log f(x_i) \leq \log f(\bar{x}_\alpha)$ and it is satisfied by Jensen's inequality, due to the concavity of the logarithm function.

With $f(x) = 1 + x$, (1.8) implies

$$\prod_{i=1}^{k}(1 + x_i) \leq (1 + \bar{x}_k)^k,$$

with $\bar{x}_k = k^{-1}\sum_{i=1}^{k} x_i$. Applying this inequality with $x_i = 2i$ provides a bound for the $2k$-th moments $m_{2k} = 1.3\ldots(2k-1) = \{2^k(k)!\}^{-1}(2k)!$ of the normal distribution, $\prod_{i=0}^{k-1}(1 + 2i) \leq k^k$. This is a mean value bound but it is not very sharp. By the convexity of the exponential function, the converse inequality at points $(t_1, \dots, t_k)$ yields $(1 + e^{\bar{t}_k})^k \leq \prod_{i=1}^{k}(1 + e^{t_i})$, which it is equivalent to

$$1 + \prod_{i=1}^{k} x_i^{\frac{1}{k}} \leq \prod_{i=1}^{k}(1 + x_i)^{\frac{1}{k}}.$$

Replacing x_i by $x_i y_i^{-1}$ and multiplying both sides by $\prod_{i=1}^{k} y_i^{\frac{1}{k}}$ implies

$$\prod_{i=1}^{k} x_i^{\frac{1}{k}} + \prod_{i=1}^{k} y_i^{\frac{1}{k}} \leq \prod_{i=1}^{k}(x_i + y_i)^{\frac{1}{k}}$$

for all sequences $(x_i)_{i=1,\dots,k}$ and $(y_i)_{i=1,\dots,k}$. More generally, let $(\alpha_1, \dots, \alpha_k)$ be positive numbers such that $\sum_{i=1}^{k} \alpha_i = 1$, then

$$\prod_{i=1}^{k} x_i^{\alpha_i} + \prod_{i=1}^{k} y_i^{\alpha_i} \leq \prod_{i=1}^{k}(x_i + y_i)^{\alpha_i}.$$

For an array $(x_{ij})_{i=1,\dots,k,j=1,\dots,n}$, a recursive argument implies

$$n^{-1}\sum_{j=1}^{n}\prod_{i=1}^{k} x_{ij}^{\alpha_i} \leq \prod_{i=1}^{k} \bar{x}_i^{\alpha_i} \tag{1.9}$$

with the partial means $\bar{x}_i = n^{-1}\sum_{j=1}^{n} x_{ij}$. This inequality differs from the inequality (1.7) and from the Minkowski inequality for a sum of n series in $\mathbb{R}^k$, $\|\bar{x}\|_p \leq n^{-1}\sum_{j=1}^{n}\|x_j\|_p$, for $p \geq 1$.

With the convex function $f(t) = \log(1 + e^t)$, Equation (1.8) implies that for every $x_i > 0$, for $i = 1, \dots, k$

$$\log(1 + \prod_{i=1}^{k} x_i^{\alpha_i}) \leq \prod_{i=1}^{k}\{\log(1 + x_i)\}^{\alpha_i}.$$

Replacing x_i by $x_i y_i^{-1}$, with $y_i > 0$ for $i = 1, \ldots, k$, and adding to the left member $\log \prod_{i=1}^{k} y_i^{\alpha_i} \leq \log \bar{y}_\alpha$ yields

$$\prod_{i=1}^{k} \log\{(x_i + y_i)^{\alpha_i}\} \leq \log \prod_{i=1}^{k} (x_i + y_i)^{\alpha_i} - \log \prod_{i=1}^{k} y_i^{\alpha_i} + \log \bar{y}_\alpha.$$

For every convex function φ from $\mathbb{R}$ to $\mathbb{R}$, Jensen's inequality for the integral of a function f with respect to the positive measure μ on $[a, b]$ is

$$\varphi\{\frac{1}{M(b) - M(a)} \int_a^b f \, d\mu\} \leq \int_a^b \varphi \circ f \, d\mu,$$

with $M(a) = \mu(] - \infty, a])$ and $M(b) = \mu(] - \infty, b])$. As a consequence, for every real function f on $[a, b]$

$$\exp\{\frac{1}{M(b) - M(a)} \int_a^b \log f \, d\mu\} \leq \frac{1}{M(b) - M(a)} \int_a^b f \, d\mu.$$

Other integral inequalities for convex function can be found in Maurey (2004).

An integral equation similar to Equation (1.8) is obtained from Jensen's inequality for every concave function $f > 0$

$$\exp\{y^{-1} \int_0^y \log f \, d\mu\} \leq f\{y^{-1} \int_0^y x \, d\mu(x)\}, \, y > 0. \qquad (1.10)$$

The bound

$$y^{-1} \int_0^y f \, d\mu \leq f\{y^{-1} \int_0^y x \, d\mu(x)\}$$

is sharper and it is an equality for affine functions $f > 0$.

For example, with $f(x) = 1 + x$, Equation (1.10) is written

$$\int_0^y \log(1 + x) \, dx \leq y \log(1 + \frac{y}{2})$$

for every $y > 0$. With $f(x) = x^\alpha$, with $x > 0$ and $0 < \alpha < 1$, (1.9) implies $\int_0^y \log x \, dx \leq y \log \frac{y}{2}$ and Jensen's inequality yields $y^\alpha \leq \alpha(\alpha + 1) \log \frac{y}{2}$.

Most inequalities for bounding finite series are intervals for the error of their approximation by a Taylor expansion or they are obtained by induction. For example, Cauchy (1833) defined the exponential function as

$$e^x = \lim_{n \to \infty, n\alpha \to x} (1 + \alpha)^n = \lim_{n \to \infty, n\alpha \to x} (1 - \alpha)^{-n}$$

which provided him the following interval

$$(1 + \alpha)^{\frac{x}{\alpha}} < e^x < (1 - \alpha)^{-\frac{x}{\alpha}},$$

for every real $\alpha > 0$. This definition entails immediately the main properties of the exponential. Reciprocally, the unique function ϕ which satisfies

$$\phi(x + y) = \phi(x)\phi(y),$$

for all x and y, is the exponential function $\phi(x) = \{\phi(1)\}^x \equiv e^x$.

Other classical bounds for functions have been established for the first n terms of their Taylor expansions. Alzer (1990b) studied a lower bound of the form $I_{n-1}(x)I_{n+1}(x) > c_n I_n^2(x)$ for the sum $I_n(x)$ of the first n terms in the Taylor expansion of the exponential function. More generally, Darboux's sums for a function f on an interval $[a, b]$ are defined using a partition $\pi_n = (a_k)_{k=0,\ldots,a_n}$ of increasing numbers of this interval such that $a = a_0$ and $a_n = b$

$$S(f, \pi_n) = \sum_{k=1}^{n} m_k(a_k - a_{k-1}),$$

$$T(f, \pi_n) = \sum_{k=1}^{n} M_k(a_k - a_{k-1}),$$

with $m_i = \inf_{a_{k-1} \leq x < a_k} f(x)$ and $M_i = \sup_{a_{k-1} \leq x < a_k} f(x)$ and the convergence of the sum $T(f, \pi_n)$ implies the convergence of the integral $I - \int_a^b f(x)\,dx$, conversely the convergence of I implies the convergence of the sum $S(f, \pi_n)$. The sum $1 + 3 + \cdots + (2k - 1)$ has then the bounds $k^2 - k - 1 \leq \sum_{i=0}^{k-1}(1 + 2i) \leq k^2 + k - 1$ and the product $m_{2k} = 1.3\ldots(2k-1) = \exp\{\sum_{i=0}^{k-1} \log(1 + 2i)\}$ satisfies

$$\prod_{i=0}^{k-1}(1 + 2i) \leq (1 + 2k)^{1+2k}e^{-2k} \leq e^{(2k)^2} = 2ke^{2k}, \qquad (1.11)$$

the inequality $m_{2k} \leq k^k$ obtained from (1.8) is therefore better than the upper bound of (1.11) for m_{2k}. Close approximations by Darboux's sums require a thin partition π_n.

1.4 Applications in probability

Let $(\Omega, \mathcal{F}, P)$ be a probability space and let X be a real variable defined on $(\Omega, \mathcal{F}, P)$, with distribution function F respectively. The norms $L_p(\mathbb{R})$ define norms for variables

$$\|X\|_p = (E|X|^p)^{\frac{1}{p}} = \{\int_{\mathbb{R}} |x|^p dF(x)\}^{\frac{1}{p}},$$

$$\|X\|_\infty = \sup\{x : P(|X| > x) > 0.$$

Let X be a random vector of $\mathbb{R}^n_+$, by concavity of the norm, we have $E\|X\|_p \leq \|EX\|_p$ for every $p \geq 1$. The sequence $(\|X\|_p)_{0<p\leq\infty}$ is increasing and the additive and multiplicative properties of the norms are consequences of the Hölder and Minkowski inequalities. For dependent variables X and Y, the Cauchy-Schwarz inequality entails

$$E(XY) \leq \|X\|_p\|Y\|_q$$

for all conjugate integers $p, q \geq 2$. Let $(X_i)_{i\geq 1}$ be a sequence of random variables in a vector space V, $\|\max_{i=1,...,n} X_i\| \leq \max_{i=1,...,n} \|X_i\|$ and the inequality extends as n tends to infinity.

The density of the sum $X+Y$ of independent variables with distribution functions F and G and densities f and g is the convolution

$$f * g(t) = \int_\mathbb{R} f(t-s)\,dG(s) = \int_\mathbb{R} g(t-s)\,dF(s).$$

The density of the ratio XY^{-1} of independent variables X and $Y > 0$ is the multiplicative convolution $h(t) = \int_\mathbb{R} f(s^{-1}t)\,dG(s)$, its norms has the bound $\|h\|_r \leq \|f\|_p\|g\|_q$, for all integers such that $p^{-1} + q^{-1} = r^{-1}$, $1 \leq p \leq \infty$. If the variables have no densities, the convolution is $P(X + Y \leq t) = \int_\mathbb{R} G(t-s)\,dF(s) = \int_\mathbb{R} F(t-s)\,dG(s)$ and it is sufficient that F or G is continuous (respectively, has a density) to ensure the continuity (respectively, the derivability) of the distribution function of their sum. Their ratio has the distribution function $\int_\mathbb{R} F(s^{-1}t)\,dG(s)$.

The inequalities of Section 1.3 imply the following inequalities for the sequence of variables. The Cauchy inequality implies that for every n, finite or not

$$E\{n^{-1}\sum_{i=1}^n \log(1+X_i)\} < E\{\bar{X}_n\}.$$

Let $(\alpha_i)_{i=1,...,n}$ be a real sequence such that $\sum_{i=1}^n \alpha_i = 1$. For every positive concave function f on $\mathbb{R}_+$, inequality (1.8) entails

$$E\{\sum_{i=1}^n \alpha_i \log f(X_i)\} \leq E \log f(\sum_{i=1}^k \alpha_i X_i)$$

in particular $E\{n^{-1}\sum_{i=1}^n \log f(X_i)\} \leq E \log f(\bar{X}_n)$.

Bienaymé-Chebychev's inequality has been formulated for a variable X of $L_2(\mathbb{R})$ as $P(|X| > a) \leq a^{-2}E|X|^2$, $a > 0$. For a variable X in $L_p(\mathbb{R})$, $1 \leq p < \infty$, and such that $EX = m$, it is

$$P(|X - m| > a) \leq a^{-2}E|X - m|^2, \text{ for every } a > 0, \tag{1.12}$$

where $varX = E(X - EX)^2$ is the variance of X. It extends to $p \geq 2$

$$P(X - m > a) \leq a^{-p} E|X - m|^p, \text{ for every } a > 0.$$

Another Chebychev's inequality for the mean of n variables with mean m and variance σ^2 is

$$P(|\bar{X} - m| > a) \leq \frac{1}{1 + \sigma^{-2} n a^2}, \ a > 0.$$

The Laplace transform of a variable X is $L_X(t) = Ee^{tX}$ and it satisfies $L_X(t) \geq e^{tEX}$ by the Jensen inequality. Its generating function is defined as $G_X(t) = Et^X$ and it satisfies $G_X(t) = L_X(\log t)$. The moments of the variable are obtained as the value at zero of the derivatives of L_X or G_X, $L_X^{(k)}(0) = EX^k$ and $G_X^{(k)}(0) = E\{X(X - 1) \cdots (X - k + 1)\}$, $k \geq 1$.

Let $(X_i)_{i=1,\ldots,n}$ be a sequence of independent random variables with mean zero and respective variances σ_i^2 and let $S_n = \sum_{i=1}^n X_i$, $\bar{X}_n = n^{-1} S_n$ and $V_n = \sum_{i=1}^n \sigma_i^2$. If $\bar{\sigma}_n^2 = n^{-1} V_n$ converges to a finite limit $\sigma^2 > 0$, then for every $\beta > 0$ and $a > 0$, the inequality (1.12) implies

$$P(n^{\frac{\beta}{2}} |\bar{X}_n| > a) \leq n^\beta a^{-2} E\bar{X}_n^2 = n^{\beta - 1} \bar{\sigma}_n^2.$$

As the upper bound tends to zero for $\beta < 1$, the inequality cannot be satisfied for every a. The upper bound being infinity for every $\beta > 1$, the convergence rate of the mean $\bar{X}_n = n^{-1} \sum_{i=1}^n X_i$ to zero is $\beta = 1$. For independent variables with respective means μ_i, the convergence rate of $n^{-\alpha} S_n$ to a limit μ is determined by the convergence rates of the mean $\mu_n = n^{-\alpha} \sum_{i=1}^n \mu_i$ to μ and of the mean variance $\sigma_n^2 = n^{-\beta} V_n$ to a limit $\sigma^2 > 0$. For every $a > 0$

$$P(n^{\frac{\beta}{2}} |n^{-\alpha} S_n - \mu| > a) \leq n^{2(\beta - \alpha)} a^{-2} n^{-\beta} V_n,$$

therefore the convergence rate α and β are equal.

Chow and Lai (1971) proved the following equivalences for a sequence of independent and identically distributed random variables $(X_i)_{i \geq 1}$ with mean $EX_1 = 0$, for α in $]0, 1[$

$$E \exp(t|X_1|^\alpha) < \infty, \ t > 0,$$

$$\lim_{n \to \infty} (\log n)^{\alpha^{-1}} X_n = 0, \text{ a.s.},$$

$$\lim_{n \to \infty} (\log n)^{\alpha^{-1}} \sum_{i=1}^n c_{n-i} X_i = 0, \text{ a.s.},$$

for a sequence of weighting constants such that $c_n = O(n^{-\nu})$, with $\nu > \frac{1}{2}$. Other equivalences between convergences proved by Chow and Lai (1971) are given in the appendix.

Bienaymé-Chebychev's inequality for the maximum of n positive variables is

$$\limsup_{n \to \infty} P(n^{-\frac{1}{2}} \max_{k=1,\ldots,n} |S_k| > a) \leq \frac{\sigma^2}{a^2}, \ a > 0.$$

Billingsley (1968) proved many inequalities for partial sums and their moments and for the empirical process of independent random variables. In particular $E(|S_n - S_k|^2 |S_k - S_j|^2) = (V_n - V_k)(V_k - V_j)$, $0 \leq j \leq k \leq n$, the variable

$$M'_n = \max_{0 \leq k \leq n} \min\{|S_k|, |S_n - S_k|\}$$

satisfies the inequalities

$$M'_n \leq M_n \leq M'_n + |S_n|, \quad M'_n \leq M_n \leq M'_n + 3 \max_{0 \leq k \leq n} |X_k|$$

and for every $\lambda > 0$, there exists a constant K such that

$$P(M'_n > \lambda) \leq K\lambda^{-4}V_m^4.$$

For dependent variables, the variance of S_n is bounded by a sum which also depends on the means $E|X_i X_j|$ for every $i \neq j$ and a mixing coefficient η_p determines the convergence rate of this sum through the inequality $E|X_i X_j| \leq \eta_{j-i}\sigma_i\sigma_j$. If the exist constants $\alpha > 0$ and $\beta > 0$ such that $\mu_n = n^{-\alpha}\sum_{i=1}^n \mu_i$ convergences to μ and $n^{-\beta}\sum_{i=1,\ldots,n}\sum_{j=1,\ldots,n}\eta_{j-i}\sigma_i\sigma_j$ convergences to a limit $\sigma^2 > 0$, then the convergence rate of $n^{-\alpha}S_n$ to μ is still $n^{-\beta}$.

The Berry-Essen inequality for independent variables concerns the convergence of the distribution of a normalized sum of variables $V_n^{-\frac{1}{2}}S_n$ to the Gaussian distribution Φ. If $E|X_i|^3$ is finite, there exists a constant K such that

$$\sup_{x \in \mathbb{R}} |P(V_n^{-\frac{1}{2}}(S_n - EX) \leq x) - \Phi(x)| \leq K\frac{\sum_{i=1}^n E|X_i - EX|^3}{V_n^{\frac{3}{2}}}.$$

For dependent variables with mixing coefficients such that $\sum_{k>0} k^2\eta_k$ is finite, the convergence rate of S_n is $n^{\frac{1}{2}}$ (Billingsley, 1968) and the Berry-Essen inequality is satisfied.

The Fourier transform of the correlation function $R_X(t) = E\{X_0 X_t\}$ of a stationary process X is defined as its spectral density

$$\widehat{f}_X(\omega) = \int_{-\infty}^{\infty} R_X(t)e^{2\pi\omega t}\,dt.$$

Inverting the mean and the integrals, its norms satisfy the properties

$$\|\widehat{f}_X\|_p \leq \|R_X\|_p,$$

$$\|\widehat{f}_X\|_1 \leq \|\widehat{f}_X\|_p^p\|\widehat{f}_X\|_{p'}^{p'}, \quad \text{where } p^{-1} + p'^{-1} = 1.$$

1.5   Hardy's inequality

Hardy's inequalities have been developed in different forms presented in the monograph by Hardy, Littlewood and Pólya (1952). The functional inequality is a consequence of the Hölder inequality. Let $p > 1$ be an integer and let f be a positive function of $L_p(\mathbb{R}_+)$, with primitive F and such that $\lim_{x \to 0} x^{\frac{1}{p}-1}F(x) = 0$ and $\lim_{x \to \infty} x^{\frac{1}{p}-1}F(x) = 0$. The integral

$$I_p(y) := \int_0^y \{\frac{F(x)}{x}\}^p \, dx = \frac{p}{p-1} \int_0^y \{\frac{F(x)}{x}\}^{p-1} f(x) \, dx - \frac{1}{p-1} \frac{F^p(y)}{y^{p-1}},$$
(1.13)

and $I_p = I_p(\infty)$ have the bounds

$$I_p \leq (\frac{p}{p-1})^p \int_0^\infty f^p(x) \, dx, \ 1 < p < \infty,$$
(1.14)

and the inequality is strict except if f is a constant.

With $p = 1$, for every function $f > 0$ on $\mathbb{R}_+$ having a primitive F such that $\lim_{x \to 0} f(x) = 0$, $\lim_{x \to \infty} x^{-1}F(x) = 0$, the integral cannot be calculated in the same way, however we have

$$I_1 = \int_0^\infty \frac{F(x)}{x^2} \, dx = \int_0^\infty \frac{f(x)}{x} \, dx.$$

If $0 < p < 1$, integrating by parts yields

$$I_p(y) = \frac{1}{1-p}\{p \int_0^y \{\frac{F(x)}{x}\}^{p-1} f(x) \, dx - y^{1-p} F^p(y)\}$$

where

$$\int_0^y \{\frac{F(x)}{x}\}^{p-1} f(x) \, dx \geq I_p^{1-\frac{1}{p}}(y)\|f\|_p.$$

Since $y^{1-p}F^p(y) > 0$ and tends to infinity as y tends to infinity, the inequality (1.14) cannot be inversed for $p < 1$.

For positive real series $(a_i)_{i \geq 1}$, let $A_n = \sum_{i=1}^n a_i$ and let $p > 1$ be an integer such that $(a_i)_{i \geq 1}$ belongs to ℓ_p. Hardy's inequality is

$$\sum_{n \geq 1} (\frac{A_n}{n})^p < (\frac{p}{p-1})^p \sum_{i \geq 1} a_i^p.$$

A special case is Euler's limit of the series $\sum_{k>1} k^{-2} = \frac{\pi^2}{6}$ (Euler, 1735) which was obtained as the coefficient of x^2 in the expansion of the function $x^{-1} \sin x$ as a polynomial with roots $\pm k\pi$. Integrating the approximation of $(1-x)^{-1}$ by $(1-x^n)(1-x)^{-1}$, with normalization, it appears that

$$\int_0^1 \frac{1}{x} (\int_0^x \frac{1-t^n}{1-t} \, dt) \, dx = \frac{\pi^2}{6}$$

(Bradley, d'Antonio and Sandifer, 2007, Dunham, 1999, Sandifer, 2007). It follows that $\sum_{n\geq 1} n^{-2}(\sum_{k=1}^n k^{-1})^2 < \frac{2}{3}\pi^2$.

Bickel *et al.* (1993) mentioned a statistical application of the inequality to the derivative of the density of a randomly right-censored time variable. Let S be a positive real random variable with distribution function H and density h, for every square integrable function a such that $\int a\, dH = 0$

$$E(\frac{\int_S^\infty a\, dH}{H(S)})^2 \leq 4Ea^2(S).$$

Kufner and Persson (2003) and others later proved weighted inequalities of the Hardy type. Pachpatte (2005) also provided a review of the classical inequalities of the same type and several generalizations of the inequality

$$\int_0^x \int_0^y \frac{f(s)g(t)}{s+t}\, ds\, dt \leq pq\frac{\sqrt{xy}}{2}\{\int_0^x (x-s)f'^2(s)\, ds \int_0^y (y-t)g'^2(t)\, dt\}^{\frac{1}{2}}.$$

Let f be a complex function defined on $\mathbb{R}^d$, with the Lebesgue measure λ_d and let $B_r(x)$ be the ball of $\mathbb{R}^d$, centered at x and with radius r. The maximum average value of f is

$$Mf(x) = \sup_{r>0} \frac{1}{\lambda_d(B_r(x))} \int_{B_r(x)} |f|\, d\lambda_d.$$

Hardy-Littlewood's weak inequality for Mf is

$$\lambda_d\{x \in \mathbb{R}^d : Mf(x) > a\} \leq a^{-1}C_d\|f\|_{L_1(\mathbb{R}^d)}, \text{ for every } a > 0,$$

where C_d is a constant. A stronger inequality for Mf is due to Marcinkiewicz for every integer $p > 1$

$$\|Mf\|_{L_p(\mathbb{R}^d)} \leq C_{d,p}\|f\|_{L_p(\mathbb{R}^d)}$$

where $C_{d,p}$ is a constant.

Another type of analytical inequalities for integrals is the Mean Value Theorem (Lagrange) and its extensions to higher order approximations. Hardy, Littlewood and Pólya (1952) proved a bound between the difference between the integral of the square of a continuously differentiable function f on $[0,1]$ and the square of its integral

$$0 \leq \int_0^1 f^2(x)\, dx - \{\int_0^1 f^2(x)\, dx\}^2 \leq \frac{1}{2}\int_0^1 x(1-x)f'^2(x)\, dx.$$

Ostrowski's type inequalities (Sahoo and Riedel, 1998, Mitrinović, Pecarić and Fink, 1957, Dragomir and Sofo, 2000, Dragomir and Rassias, 2002) have been written for the approximation of a continuously differentiable function f defined in an interval (a, b), having a derivative bounded by a constant M, in the form

$$|f(x) - \frac{1}{b-a}\int_a^b f(t)\,dt| \leq (b-a)M\,[\frac{1}{4} + \frac{\{x - \frac{1}{2}(a+b)\}^2}{(b-a)^2}].$$

Gruss's inequalities are bounds for the difference between the integral of a product of bounded functions and the product of their integrals. Let $\varphi \leq f(x) \leq \Phi$ and $\gamma \leq g(x) \leq \Gamma$

$$|\frac{1}{b-a}\int_a^b f(x)g(x)\,dx - \frac{1}{b-a}\{\int_a^b f(x)\,dx\}\frac{1}{b-a}\{\int_a^b g(x)\,dx\}|$$

$$\leq \frac{1}{4}(\Phi - \varphi)(\Gamma - \gamma).$$

These inequalities have been developed for several classes of functions and extended to inequalities for higher order approximations (Barnett and Dragomir, 2001, 2002).

1.6 Inequalities for discrete martingales

On a probability space $(\Omega, \mathcal{F}, P)$, let $(\mathcal{F}_n)_{n \geq 0}$ be a filtration of $\mathcal{F}$, i.e. a nondecreasing sequence of subsigma-algebras of $\mathcal{F}$, and let $(X_n)_{n \geq 0}$ be a sequence of $\mathcal{F}_n$-measurable real variables. It is a *martingale* with respect to $(\mathcal{F}_n)_n$ if $E(X_m|\mathcal{F}_n) = X_n$ for every $m > n \geq 0$. Then $EX_n = EX_0$ and $E\{(X_m - X_n)Y_n\} = 0$, for every $\mathcal{F}_n$-measurable variable Y_n. It follows that the conditional variance of a square integrable $(\mathcal{F}_n)_n$-martingale $(X_n)_{n \geq 0}$ is $E\{(X_m - X_n)^2|\mathcal{F}_n\} = E(X_m^2|\mathcal{F}_n) - X_n^2$.

The sequence $(X_n)_{n \geq 0}$ is a *submartingale* if $E(X_m|\mathcal{F}_n) \geq X_n$ for every $m > n \geq 0$ and it is a *supermartingale* if $E(X_m|\mathcal{F}_n) \leq X_n$. Let $(X_n)_{n \geq 0}$ be a real martingale and let φ be a function defined from $\mathbb{R}$ to $\mathbb{R}$. If φ is convex, $E\{\varphi(X_{n+1})|\mathcal{F}_n\} \geq \varphi(E\{X_{n+1}|\mathcal{F}_n\}) = \varphi(X_n)$ then $(\varphi(X_n))_{n \geq 0}$ is a submartingale. If φ is concave, then $(\varphi(X_n))_{n \geq 0}$ is a supermartingale.

The following examples of discrete martingales are classical.

(1) A random walk $X_n = \sum_{i=1}^n \zeta_i$ is defined by a sequence of independent and identically distributed random variables $(\zeta_i)_i = (X_i - X_{i-1})_{i=1,\ldots,n}$. Let $\mu = E\zeta_i$ be the mean increment of X_n, if $\mu = 0$ then X_n is a martingale, if $\mu > 0$ then X_n is a submartingale and if $\mu < 0$ then X_n is a supermartingale.

(2) For every increasing sequence of random variables $(A_n)_{n \geq 1}$ such that A_n is $\mathcal{F}_n$-measurable and $E(A_\infty|\mathcal{F}_0) < \infty$ a.s., $X_n = E(A_\infty|\mathcal{F}_n) - A_n$ is a supermartingale and $EM_n = (A_\infty|\mathcal{F}_n)$ is a martingale.

(3) Let

$$V_n(X) = X_0^2 + \sum_{i=1}^{n}(X_i - X_{i-1})^2$$

be the quadratic variations of a $L_2(P)$ martingale $(X_n)_{n \geq 1}$ with respect to a filtration $(\mathcal{F}_n)_{n \geq 1}$. For every $n \geq 0$, $EX_n^2 = E\{\sum_{i=1}^{n}(X_i - X_{i-1})\}^2 = EV_n(X)$. Let

$$E\{V_{n+1}(X)|\mathcal{F}_n\} = V_n(X) + E\{(X_{n+1} - X_n)^2|\mathcal{F}_n\} \geq V_n(X),$$

hence $(V_n(X))_{n \geq 1}$ is a submartingale and it converges to a limit $V(X)$ in $L_1(P)$.

Theorem 1.1. *Let $(X_n)_{n \geq 1}$ be a square integrable martingale on a filtered probability space $(\Omega, \mathcal{F}, (\mathcal{F}_n)_{n \geq 0}, P)$ and let $(\widetilde{V}_n(X))_{n \geq 1}$ be the sequence of its predictable quadratic variations, $\widetilde{V}_n(X) = E\{V_{n+1}(X)|\mathcal{F}_n\}$. Then $(X_n^2 - V_n(X))_{n \geq 1}$ is a martingale.*

It is proved by noting that $E(X_n^2 - X_{n-1}^2|\mathcal{F}_{n-1}) = E\{(X_n - X_{n-1})^2|\mathcal{F}_{n-1}\}$. A transformed martingale on a filtered probability space $(\Omega, \mathcal{F}, (\mathcal{F}_n)_{n \geq 0}, P)$ is defined by two sequences of $L_1(P)$ random variables $(X_n)_{n \geq 1}$ and $(A_n)_{n \geq 1}$ by $Y_0 = X_0$ and

$$Y_{n+1} = Y_n + A_n(X_{n+1} - X_n), n \geq 1, \qquad (1.15)$$

where A_n is $\mathcal{F}_{n-1}$-measurable, for every integer n, and X_n is a $(\mathcal{F}_n)$-martingale, then $E\{Y_{n+1}|\mathcal{F}_n\} = Y_n + A_n E(X_{n+1} - X_n|\mathcal{F}_n)$, so Y_n is a $(\mathcal{F}_n)$-martingale. If X_n is a $\mathcal{F}_n$-submartingale (respectively supermartingale), then Y_n is a $\mathcal{F}_n$-submartingale (respectively supermartingale). The quadratic variations $V_n(Y)$ of $(Y_n)_{n \geq 1}$ satisfy

$$E\{V_{n+1}(Y)|\mathcal{F}_n\} - V_n(Y) = A_n^2 E\{(X_{n+1} - X_n)^2|\mathcal{F}_n\} \geq 0, \qquad (1.16)$$

hence the process $V_n(Y) = X_0^2 + \sum_{i=1}^{n-1} A_i^2\{V_{i+1}(X) - V_i(X)\}$ defines a convergent submartingale.

Kolmogorov's inequality for a $(\mathcal{F}_n)$-martingale $(X_n)_{n \geq 1}$ of $L_p(\mathbb{R})$, for an integer $1 \leq p < \infty$, is similar to Bienaymé-Chebychev's inequality for independent variables

$$P(|X_{n+1}| > a|\mathcal{F}_n) \leq a^{-p}|X_n|^p, \text{ for every } a > 0. \qquad (1.17)$$

A stopping time T of a uniformly integrable martingale sequence $(X_n)_{n \geq 1}$ defined on a filtered probability space $(\Omega, \mathcal{F}, (\mathcal{F}_n)_{n \geq 0}, P)$ satisfies the next measurability property: $\{T \leq n\}$ is $\mathcal{F}_n$-measurable, for every integer n. Doob established that $E(X_T|\mathcal{F}_S) = X_S$ for all stopping times S and $T > S$, therefore $E(X_T|\mathcal{F}_S) = X_{S \wedge T}$ and $EX_T = EX_0$, for all S and T.

Let $(X_n)_{n \geq 1}$ be a sequence of adapted random variables with values in $[0, 1]$, for every stopping time τ, the martingale $M_n = (X_{n+1}|\mathcal{F}_n)$, for every n, satisfies the property (Freedman, 1973)

$$P(\sum_{i=1}^{\tau} X_n \leq a, \sum_{i=1}^{\tau} M_n \geq b) \leq (\frac{b}{a})^a e^{a-b} \leq \exp\{-\frac{(a-b)^2}{2c}\}, \; 0 \leq a \leq b,$$

$$P(\sum_{i=1}^{\tau} X_n \geq a, \sum_{i=1}^{\tau} M_n \leq b) \leq (\frac{b}{a})^a e^{a-b}, \; 0 \leq a \leq b,$$

where $c = \max(a, b)$, and the bound reduced to 1 if $a = b = 1$.

The inequalities for the maximum of variables extend to martingales. Let

$$X_n^* = \max_{1 \leq i \leq n} |X_i|$$

be the maximal variable of $(X_1, \ldots, X_n)$ and $X_n^+ = \max(X_n, 0)$.

Theorem 1.2. *Let $\tau_1 = 1$ and for $k > 1$, let*

$$\tau_k = \min\{n > \tau_{k-1} : X_n^* = X_n\}$$

be stopping times for the martingale $(X_n)_n$. For every $k \geq 1$ such that τ_k is finite and for every $\lambda > 0$

$$\lambda P(X_{\tau_k}^* > \lambda) < E(X_{\tau_k} 1_{\{X_{\tau_k}^* > \lambda\}}) < EX_{\tau_k}^+$$

and the inequality holds only at the stopping times τ_k.

Proof. The variables τ_k are stopping times for the martingale $(X_n)_n$, $X_{\tau_k}^* = X_{\tau_k}$ for every $k \geq 1$ and $X_n^* > X_n$ for every n which does not belong to the sequence of stopping times. For every $\lambda > 0$

$$E(X_{\tau_k} 1_{\{X_{\tau_k}^* > \lambda\}}) = E(X_{\tau_k}^* 1_{\{X_{\tau_k}^* > \lambda\}}) > \lambda P(X_{\tau_k}^* > \lambda).$$

Otherwise $E(X_n 1_{\{X_n^* > \lambda \geq |X_n|\}}) \leq \lambda P(X_n^* > \lambda \geq |X_n|) \leq \lambda P(X_n^* > \lambda)$ and $E(|X_n| 1_{\{X_n^* > \lambda\}})$ is smaller than $E(X_n^* 1_{\{X_n^* > \lambda\}}) > \lambda P(X_n^* > \lambda)$. $\square$

Note that the a.s. convergence of the sequence $n^{-\alpha} X_n^*$ to a limit, for some $\alpha > 0$, implies the a.s. convergence of the sequence $n^{-\alpha} |X_n|$.

Bürkholder's inequalities for L_p martingales are

$$c_p E \|V_n\|^p \leq E(X_n)^p \leq C_p E \|V_n\|^p$$

with constants $c_p > 0$ and C_p depending only on p, for $1 \leq p < \infty$. The Bürkholder, Davis and Gundy inequality is similar for the maximum variable $(X_n^*)_{n \geq 0}$. Mona (1994) proved that Bürkholder's inequality is not satisfied for discontinuous martingales and p in $]0, 1[$. Meyer's proof (1969) of the Bürkholder, Davis and Gundy inequality holds in $\mathcal{M}_{0,loc}^p$, for every integer $p > 1$.

As a consequence of the inequality (1.3) between norms $L_p(\Omega, \mathcal{F}, P)$ and $L_\infty(\Omega, \mathcal{F}, P)$ in vector spaces of dimension n, for every random variable $Y = (Y_1, \ldots, Y_n)$

$$E(Y_n^*)^p \leq E\|Y\|_p^p \leq nE(Y_n^*)^p,$$

which is equivalent to $n^{-1}E\|Y\|_{n,p}^p \leq E(Y_n^*)^p \leq E\|Y\|_{n,p}^p$. The previous inequalities and the Kolmogorov inequality imply

$$P(X_n^* > a) \leq a^{-p}E\|X_n\|_{n,p}^p, \tag{1.18}$$

for every $a > 0$. The inequality for the norms has been extended to tranformed martingale by a convex function (Bürkholder, Davis and Gundy, 1972).

Theorem 1.3. *Let ϕ be a non negative convex function with $\phi(0) = 0$ and $\phi(2x) \leq 2\phi(x)$. On a filtered probability space $(\Omega, \mathcal{F}, (\mathcal{F}_n)_{n \geq 0}, P)$, let $(X_n)_n$ be a $(\mathcal{F}_n)_n$-martingale, there exist constants such that*

$$c_\phi E\phi(X_n) \leq E\phi(X_n^*) \leq C_\phi E\phi(X_n). \tag{1.19}$$

Bürkholder (1973) proved that for $p \geq 1$ and for independent variables X_i, there exist constants such that the random walk $S_n = \sum_{i=1,\ldots,n} X_i$ satisfies

$$ES_n^p \leq A_p E(V_n(X)^{\frac{p}{2}}). \tag{1.20}$$

Some results have been extended to submartingales indexed by a multidimensional set of integers (Cairoli and Walsh, 1975). Due to the partial order of the sets of indexes, several notions of martingales are defined and the conditions for the filtration are generally stronger than previously. In $\mathbb{R}^d$, $d \geq 2$, the total order $m \leq n$ means $m_k \leq n_k$ for $k = 1, \ldots, d$ and an assumption of conditional independence of the marginal σ-algebras, given the d-dimensional σ-algebra is usual. Christofides and Serfling (1990) proved that under the condition $E\{E(\cdot|\mathcal{F}_k)|\mathcal{F}_1\} = E(\cdot|\mathcal{F}_{k \wedge 1})$, $k = 2, \ldots, d$, a martingale $(X_k)_{k \in \mathbb{N}^d}$ with respect to a filtration $(\mathcal{F}_k)_{k \in \mathbb{N}^d}$ satisfies

$$P(\max_{k \leq n} |X_k| > a) \leq 4^{d-1}a^{-2}E(X_n^2), \ a > 0, \ n \in \mathbb{N}^d.$$

1.7 Martingales indexed by continuous parameters

A time-continuous martingale $X = (X_t)_{t\geq 0}$ on $(\Omega, \mathcal{F}, (\mathcal{F}_t)_{t\geq 0}, P)$ is defined with respect to a right-continuous and increasing filtration $(\mathcal{F}_t)_{t\geq 0}$ of $\mathcal{F}$ i.e. $\mathcal{F}_s \subset \mathcal{F}_t$ for every $s < t$, $\mathcal{F}_t = \cap_{s>t}\mathcal{F}_s$ and $\mathcal{F} = \mathcal{F}_\infty$. For the natural filtration $(\mathcal{F}_t)_{t\geq 0}$ of $(X_t)_{t\geq 0}$, $\mathcal{F}_t$ is the σ-algebra generated by $\{X_s; s \leq t\}$. A random variable τ is a stopping time with respect to $(\mathcal{F}_t)_{t\geq 0}$ if for every t, the set $\{\tau \leq t\}$ belongs to $\mathcal{F}_t$.

A martingale X is uniformly integrable if

$$\lim_{C\to\infty} \sup_{t\geq 0} E(|X_t|1_{\{|X_t|>C\}}) = 0.$$

Then $\lim_{t\to\infty} E|X_t - X_\infty| = 0$ and $X_t = E(X_\infty|\mathcal{F}_t)$ for every t, with the natural filtration $(\mathcal{F}_t)_{t\geq 0}$ of X. Doob's theorem (1975) for stopping times applies to continuously indexed martingales.

The Brownian motion $(B_t)_{t\geq 0}$ is a martingale with independent increments, defined by Gaussian marginals, a mean zero and the variance $B_t^2 = t$, hence $B_0 = 0$. It satisfies the following properties

(1) $E(B_s B_t) = s \wedge t$,
(2) $B_t - B_s$ and B_{t-s} have the same distribution, for every $0 < s < t$,
(3) $(B_t^2 - t)_{t\geq 0}$ is a martingale with respect to $(\mathcal{F}_t)_{t\geq 0}$,
(4) for every θ, $Y_\theta(t) = \exp\{\theta B_t - \frac{1}{2}\theta^2 t\}$ is a martingale with respect to $(\mathcal{F}_t)_{t\geq 0}$, with mean $EY_\theta(t) = EY_\theta(0) = 1$.

For every $t > 0$, the variable $t^{-\frac{1}{2}}B_t$ is a normal variable and the odd moments of the Brownian motion satisfy $\|B_t\|_{2k} \leq (kt)^{\frac{1}{2}}$, $k \geq 1$. Wiener's construction of the Brownian motion is reported by Itô and McKean (1996) as the limit of the convergent series

$$X_t = \frac{t}{\sqrt{\pi}}g_0 + \sum_{k=2^{n-1}}^{2^n-1} \frac{\sqrt{2}}{\sqrt{\pi}}\frac{\sin(kt)}{k}g_k,$$

where the variables g_k are independent and have the normal distribution. Let $T_a = \inf\{s : B_s = a\}$, it is a stopping time for B and the martingale property of the process Y_θ implies that $E \exp\{\frac{1}{2}\theta^2 T_a\} = e^{a\theta}$, therefore the Laplace transform of T_a is $L_{T_a}(x) = e^{a\sqrt{2x}}$. Its density function is

$$f_{T_a}(x) = \frac{a}{\sqrt{2\pi x^3}} \exp^{-\frac{u^2}{2x}}.$$

Let a and b be strictly positive, the independence of the increments of the process B implies that the times T_a and $T_{a+b} - T_a$ are independent and the density function of T_{a+b} is the convolution of the densities of T_a and T_b, according to the property (2) of the Brownian motion. Let P_x be the probability distribution of $x + B$, for all a and b

$$P_x(T_a < T_b) = \frac{b-x}{b-a}, \quad P_x(T_b < T_a) = \frac{x-a}{b-a}, \quad \text{if } a < x < b,$$

and $E(T_a \wedge T_b) = (x - a)(b - x)$, this is a consequence of the expansion of the mean as $E_x B(T_a \wedge T_b) = aP_x(T_a < T_b) + bP_x(T_b < T_a) = x$. The rescaled Brownian motion B_{at} has the same distribution as $\sqrt{a}B_t$, for every real $a > 0$, hence the variable T_a has the same distribution as $a^2 T_1$.

Let $\sigma_a = \inf\{t : B_t < t - a\}$, it is an a.s. finite stopping time for every finite a and $Ee^{\frac{\sigma_a}{2}} = e^a$. Let $\sigma_{a,b} = \inf\{t : B_t < bt - a\}$, then $Ee^{\frac{1}{2}b^2 \sigma_{a,b}} = e^{ab}$ (Revuz and Yor, 1991). Other bounds are proved by Freedman (1975).

Let $T = \arg\max_{t \in [0,1]} B_t$, it is a stopping time of B and its distribution is determined as follows (Feller, 1966, Krishnapur, 2003, Durrett, 2010)

$$P(T \leq t) = P(\sup_{s \in [0,t]} B_s \geq \sup_{s \in [t,1]} B_s)$$

$$= P(\sup_{u \in [0,t]} B_{t-u} - B_t \geq \sup_{v \in [0,1-t]} B_{t+v} - B_t).$$

Let $X_t = \sup_{u \in [0,t]} B_{t-u} - B_t$ and $Y_t = \sup_{v \in [0,1-t]} B_{t+v} - B_t > 0$ a.s., they are independent and the variables $X = t^{-\frac{1}{2}}X_t$ and $Y = (1-t)^{-\frac{1}{2}}Y_t$ have the same distribution as $\sup_{t \in [0,1]} B_t - B_0$, then

$$P(T \leq t) = P(X_t \geq Y_t) = P\left(\frac{Y}{(X^2 + Y^2)^{\frac{1}{2}}} \leq t^{\frac{1}{2}}\right)$$

where the variable $Y(X^2 + Y^2)^{-\frac{1}{2}}$ is the sine of a uniform variable on $[0, \frac{\pi}{2}]$, hence

$$P(T \leq t) = \frac{2}{\pi}\arcsin(t^{\frac{1}{2}}). \tag{1.21}$$

The arcsine distribution has the density

$$f_{\arcsin}(x) = \frac{1}{\pi}\frac{1}{x(1-x)}, \quad 0 < x < 1,$$

it is a symmetric function with respect to $\frac{1}{2}$ and it tends to infinity at zero or one. Geetor and Sharpe (1979) established that for every x, the time variable $S_{t,\varepsilon} = \int_{G_t}^{D_t} 1_{[0,\varepsilon]}(|B_s|)\,ds$ spent in $[-\varepsilon, \varepsilon]$ between two passages at zero of the Brownian motion has a Laplace transform satisfying $\lim_{\varepsilon \to 0} E\exp\{-\lambda\varepsilon^{-2}S_{t,\varepsilon}\} = (\cosh\sqrt{2\beta})^{-2}$.

The Brownian bridge $(G_t)_{t \in [0,1]}$ is the Gaussian process with mean zero and covariance $E\{G(s)G(t)\} = ts - s \wedge t$, it is the limit in distribution of the sum $S_n(t) = n^{-\frac{1}{2}} \sum_{i=1}^n \{1_{\{X_i \le t\}} - P(X_i \le t)\}$ as a process with paths in the space $C([0,1])$ provided with the uniform norm and the Borel σ-algebra. Doob proved

$$P(\sup_{t \in [0,1]} |G(t)| \le x) = \sum_{i \in \mathcal{Z}} (-1)^j e^{-2j^2 x^2},$$

$$P(\sup_{t \in [0,1]} G(t) \le x) = e^{-2x^2}.$$

Dudley (1973) gave a survey of the sample properties of Gaussian processes, including processes satisfying Hölder conditions with various rates.

On a space $(\Omega, \mathcal{F}, P, \mathbb{F})$, let M be a square integrable $\mathcal{F}$-martingale. There exists a unique increasing and $\mathcal{F}$-predictable process $< M >$ such that

$$M^2 - < M >$$

is a $\mathcal{F}$-martingale. The process $< M >$ is the process of the predictable quadratic variations of M. It satisfies $E\{(M_t - M_s)^2 | \mathcal{F}_s\} = E(M_t^2 | \mathcal{F}_s) - EM_s^2 = E(< M >_t | \mathcal{F}_s) - < M >_s$, for every $0 < s < t$. It defines a scalar product for square integrable martingales M_1 and M_2 with mean zero

$$< M_1, M_2 > = \frac{1}{2}(< M_1 + M_2, M_1 + M_2 >$$
$$- < M_1, M_1 > - < M_2, M_2 >), \qquad (1.22)$$

then $E < M_1, M_2 >_t = EM_{1t}M_{2t}$ for every $t > 0$. Two square integrable martingales M_1 and M_2 are orthogonal if and only if $< M_1, M_2 > = 0$ or, equivalently, if $M_1 M_2$ is a martingale. Let $\mathcal{M}_0^2$ be the space of the right-continuous square integrable martingales with mean zero, provided with the norm $\|M\|_2 = \sup_t (EM_t^2)^{\frac{1}{2}} = \sup_t (E < M >_t)^{\frac{1}{2}}$.

A process $(M_t)_{t \ge 0}$ is a local martingale if there exists an increasing sequence of stopping times $(S_n)_n$ such that $(M(t \wedge S_n))_t$ belongs to $\mathcal{M}^2$ and S_n tends to infinity. The space of local martingales is denoted $\mathcal{M}^{2,loc}$. Let $(M_t)_{t \ge 0}$ be in $\mathcal{M}_{0,loc}^2$, it is written as the sum of a continuous part M^c and a discrete part $M_t^d = \sum_{0 < s \le t} \Delta M_s$, with jumps $\Delta M_s = M_s - M_{s-}$. The increasing process of its quadratic variations is $< M >_t = < M^d >_t + < M^c >_t$ where $M_t^{c2} - < M^c >_t$ and $\sum_{0 < s \le t} (\Delta M_s)^2 - < M^d >_t$ belong to $\mathcal{M}_0^{loc}$.

Let M be a local martingale of $\mathcal{M}_{0,loc}^2$ with bounded jumps, Lepingle (1978) proved that

$$Z_t(\lambda) = \exp\{\lambda M_t - \frac{\lambda^2}{2} < M^c >_t - \int (e^{\lambda x} - 1 - \lambda x) \, dM^d\} \qquad (1.23)$$

is a positive local supermartingale. The process Z_t is called the exponential supermartingale of M, for the Brownian motion it is a martingale.

Proposition 1.1. *Let $\lambda > 0$ be real, let M be a local martingale of $\mathcal{M}_{0,loc}^2$ and let T be a stopping time, then*

$$P(\sup_{t \in [0,T]} |M_t| > \lambda) \leq \lambda^{-2} E < M >_T .$$

For a local supermartingale M such that $E\sup_{t \geq 0} |M_t|$ is finite

$$P(\sup_{t \geq 0} M_t > \lambda) \leq \lambda^{-1} EM_0.$$

Let $\delta > 0$ and $\lambda = T^{\delta + \frac{1}{2}}$ and let M be a local martingale of $\mathcal{M}_{0,loc}^2$ with an increasing process $< M >$ satisfying a strong law of large numbers, then $\lim_{T \to \infty} P(T^{-\frac{1}{2}} \sup_{t \in [0,T]} |M_t| > T^\delta) = 0$. Let M be a bounded local supermartingale, then

$$\lim_{A \to \infty} P(\sup_{t \geq 0} M_t > A) = 0.$$

The Birnbaum and Marshal inequality is extended as an integral inequality for a right-continuous local submartingale $(S_t)_{t > 0}$, with a mean quadratic function $A(t) = ES_t^2$ and a increasing function q

$$P(\sup_{t \in [0,s]} q^{-1}(t)|S_t| \geq 1) \leq \int_0^s q^{-2}(t) \, dA(s).$$

Proposition 1.2 (Lenglart, 1977). *Let M be a local martingale of $\mathcal{M}_{0,loc}^2$ and let T be a stopping time, then for all $\lambda > 0$ and $\eta > 0$*

$$P(\sup_{t \in [0,T]} |M_t| \geq \lambda) \leq \frac{\eta}{\lambda^2} + P(< M >_T \geq \eta).$$

For every stopping time of a local martingale M of $\mathcal{M}_{0,loc}^2$, Doob's theorem entails $E(M_T^2) = E < M >_T$. By monotonicity

$$E\{\sup_{t \in [0,T]} M_t^2\} = E(< M >_T).$$

The Bürkholder, Davis and Gundy inequality (1.20) has been extended to L_p local martingales indexed by $\mathbb{R}$, for every $p \geq 2$.

There exist several notions of martingales in the plane, according to the partial or total order of the two-dimensional indices. Walsh (1974) and Cairoli and Walsh (1975) presented an account of the results in $\mathbb{R}^2$. Cairoli (1970) established maximal inequalities for right-continuous martingales of L_p with parameter set $\mathbb{R}^2$. For $\lambda > 0$ and integers $p > 1$

$$E\sup_{z \in \mathbb{R}^2} |M_z|^p \leq (\frac{p}{p-1})^{2p} \sup_{z \in \mathbb{R}^2} E|M_z|^p.$$

1.8 Large deviations and exponential inequalities

Let $(X_i)_{i=1,\ldots,n}$ be a sequence of independent and identically distributed real random variables defined on a probability space $(\Omega, \mathcal{F}, P)$, with mean zero, and let $S_n = \sum_{i=1}^{n} X_i$. For all $a > 0$ and $t > 0$, the Bienaymé-Chebyshev inequality for the variable $\exp\{n^{-1}tS_n - at\}$ implies

$$P(S_n > a) \leq E\{\exp(tS_n - at)\} \leq e^{-at}L_X^n(t). \qquad (1.24)$$

The maximal variable satisfies $P(X_n^* > a) \geq 1 - \{1 - e^{-at}L_X(t)\}^n$ and, for the minimum $X_{*n} = \min(X_1, \ldots, X_n)$, $P(X_{*n} > a) \leq e^{-ant}L_X^n(t)$.

Theorem 1.4 (Chernov's theorem). *On a probability space $(\Omega, \mathcal{F}, P)$, let $(X_i)_{i=1,\ldots,n}$ be a sequence of independent and identically distributed real random variables with mean zero, having a finite Laplace transform, and let $S_n = \sum_{i=1}^{n} X_i$. For all $a > 0$ and $n > 0$*

$$\log P(S_n > a) = \inf_{t>0}\{n \log L_X(t) - at\}.$$

Proof. It is a direct consequence of (1.24) and of the concavity of the logarithm, which implies $\log E e^{S_n t} \geq t E X_1$, so that the inequality is an equality and $\log P(S_n > a) = \inf_{t>0}\{n \log L_X(t) - at\}$. $\qquad \square$

This equality entails

$$\lim_{n\to\infty} n^{-1} \log P(n^{-1}S_n > a) = \inf_{t>0}\{\log L_X(t) - at\}.$$

The function

$$\psi_a(t) = \{\log L_X(t) - at\} \qquad (1.25)$$

is minimum at $t_a = \arg\min_{t>0} \psi_a(t)$, then for every integer n, $n^{-1}\log P(n^{-1}S_n > a) = \{\log L_X(t_a) - at_a\}$ and t_a is solution of the equation $\psi_a'(t) = 0$, i.e. $E(Xe^{Xt_a}) = aE(e^{Xt_a})$.

With the norm L_1 of S_n, the equality in Chernov's theorem is replaced by an upper bound. For all $a > 0$ and $n > 0$

$$\lim_{n\to\infty} n^{-1} \log P(n^{-1}\|S_n\|_1 > a) \leq \inf_{t>0}\{\log L_{|X|}(t) - at\}.$$

The Laplace transform of a Gaussian variable X with mean zero and variance σ^2 is $L_X(t) = e^{\frac{1}{2}t^2\sigma^2}$ and $t_a = \sigma^{-2}a$, then for every $a > 0$

$$P(X > a) = \exp\{-\frac{a^2}{2\sigma^2}\}$$

and by symmetry

$$P(|X| > a) = 2\exp\{-\frac{a^2}{2\sigma^2}\}.$$

For the minimum of n independent Gaussian variables

$$P(n^{\frac{1}{2}}X_{*n} > a) = e^{-a^2(2\sigma^2)^{-1}}.$$

The standard Brownian motion on $[0,1]$ has the Laplace transform $L_{X_t}(x) = e^{\frac{x^2 t}{2}}$ and, for every $0 < s < t$, $L_{X_t - X_s}(x) = e^{\frac{1}{2}x^2(t-s)}$ and

$$P(X_t > a) = e^{-\frac{a^2}{2t}},$$

$$P(X_t - X_s > a) = e^{-\frac{a^2}{2(t-s)}},$$

$$P(X_t > a, X_t - X_s > b) = e^{-a^2\{(2t)^{-1} + (2(t-s))^{-1}\}}.$$

In particular, the probability of the event $\{X_t \leq a\sqrt{t}\}$ is $1 - e^{-\frac{a^2}{2}}$ and it tends to one as a tends to infinity.

The Laplace transform at t of an exponential variable X with parameter α is $L_\alpha(t) = \alpha(\alpha - t)^{-1}$ for every $0 < t < \alpha$ and $L_\alpha(t)$ is infinite if $\alpha < t$, therefore Chernov's theorem does not apply. The probability of the event $(n^{-1}S_n > a)$ can however be bounded if a is sufficiently large

$$\log P(n^{-1}S_n > a) \leq \inf_{0 < t < \alpha} n\{\log L_X(t) - at\},$$

where the function ψ_a defined by (1.25) is minimum at $t = \alpha - \sqrt{a\alpha^{-1}}$ and the bound is negative if $a\alpha > 4$.

Chernov's theorem applies to a sequence of independent and identically distributed random variables with values in $\mathbb{R}^d$. For a sequence of Gaussian vectors with mean zero and a nonsingular covariance matrix Σ, a diagonalization of the covariance allows us to write the quadratic form $X^t\Sigma^{-1}X$ as a sum of independent variables $X^t\Sigma^{-1}X = \sum_{k=1}^d \lambda_k Y_k^2$, where $\lambda_1, \ldots, \lambda_d$ are the strictly positive eigenvalues of Σ^{-1} and $Y_1, \ldots, Y_d$ are independent normal variables, then the Laplace transform of the variable X at $t = (t_1, \ldots, t_d)$ equals

$$L_X(t) = \exp\{-\sum_{i=1,\ldots,d} t_i^2 (4\lambda_i)^{-1}\}.$$

Let $a = (a_1, \ldots, a_d)$ and a^t be its transpose, the minimum in $\mathbb{R}^d$ of the function $\psi(t) = \log L_X(t) - a^t t$ is reached at $t_i = 2a_i\lambda_i$ and $\lim_{n\to\infty} n^{-1}\log P(n^{-1}S_n > a) = -2\sum_{i=1,\ldots,d} \lambda_i a_i^2$.

The isoperimetric inequalities in $\mathbb{R}^2$ concern the optimal inequality for the ratio of the squared perimeter and the area of a closed set $L^2 \geq 4\pi A$, with equality for the circle. In $\mathbb{R}^3$, the surface and the volume of a closed set satisfy the relationship $S^{\frac{3}{2}} \geq 6\sqrt{\pi}V$, with equality for the balls, and similar geometric inequalities are proved in higher dimensions with the minimal ratio for the hypersphere $\{x \in \mathbb{R}^n : \|x\|_n = r\}$. Talagrand's isoperimetric inequalities (1995) are probabilistic. Let A be a subset of a space $(\Omega, \mathcal{A}, P)$ and let

$$f(x, A) = \min_{y \in A} \sum_{i \leq d} 1_{\{x_i \neq y_i\}}$$

be a distance of x of Ω to A, for X having the probability distribution P

$$E_{P^*} e^{tf(X,A)} \leq P^{-1}(A) e^{\frac{t^2 d}{4}},$$

$$P^*\{x : f(x, A) \leq t\} \leq P^{-1}(A) e^{-\frac{t^2}{d}}, \ t > 0,$$

using an outerprobability P^* for non measurable functions. More generally, in Talagrand's Proposition 2.2.1 (1995), the bound is replaced by an expression with an exponent $\alpha > 0$, for $t > 0$

$$E_{P^*} e^{tf(X,A)} \geq P^{-\alpha}(A) a^d(\alpha, t),$$

$$P^*\{x : f(x, A) \geq t\} \leq P^{-\alpha}(A) e^{-\frac{2t^2}{d} \frac{\alpha}{\alpha+1}},$$

where $a(\alpha, t) = \sup_{u \in [0,1]} \{1 + u(e^t - 1)\} \{1 - u(1 - e^{\frac{-t}{\alpha}})\} \leq \exp\{\frac{t^2}{8}(1 + \alpha^{-1})\}$. The results are generalized to functions f indexed by another function such as $f_h(x, A) = \inf_{y \in A} \sum_{i \leq d} h(x_i, y_i)$, with $h(x, x) = 0$ and $h > 0$ on $\Omega^{\otimes 2d}$, and $f_h(x, A) := h(x, A) = \inf_{y \in A} h(x, y)$. The concentration measure of a Gaussian probability is obtained for $h(x, y) = K^{-1}(x - y)^2$ on $\mathbb{R}^2$, with a constant K.

Chernov's theorem extends to sequences of independent and non identically distributed random variables $(X_i)_{i=1,...,n}$ having Laplace transforms L_{X_i} are such that $n^{-1} \sum_{i=1}^n \log L_{X_i}$ converges to a limit $\log L_X$, it is written in the same form

$$\lim_{n \to \infty} n^{-1} \log P(S_n > a) = \lim_{n \to \infty} \inf_{t > 0} \{\sum_{i=1}^n \log L_{X_i}(t) - at\}$$

$$= \lim_{n \to \infty} \inf_{t > 0} \{n \log L_X(t) - at\}.$$

Bennett's inequality for independent random variables is proved as an application of Chernov's theorem under a boundedness condition. It is an exponential inequality for $P(S_n \geq t)$ under a boundedness condition for the variables X_i.

Theorem 1.5. *Let $(X_i)_{i=1,...,n}$ be a vector of independent random variables having Laplace transforms and such that $EX_i = 0$, $EX_i^2 = \sigma_i^2$ and*

$$M = \max_{p \geq 2} \sigma_i^{-1} \|X_i\|_{L_p} < \infty, \ \sigma_n^* = \max_{i=1,...,n} \sigma_i < \infty, \ n \in \mathbb{N}.$$

For every $t > 0$ and every integer n

$$P(|S_n| \geq t) \leq 2\exp\{-n\phi(\frac{t}{n\sigma_n^* M})\}$$

where $\phi(x) = (1 + x)\log(1 + x) - x$.

Proof.   First let $(X_i)_{i=1,\ldots,n}$ be a vector of independent random and identically distributed variables such that $EX_i = 0$, $EX_i^2 = \sigma^2$ and X_i have the Laplace transform L_X. A bound for L_X is obtained from an expansion of the exponential function and using the bound $|X_i| \leq b = \sigma M$, a.s.

$$L_X(\lambda) \leq 1 + \sum_{k=2}^{\infty} \frac{\lambda^k}{k!}(\sigma M)^k = 1 + \{\exp(b\lambda) - 1 - b\lambda\}$$

$$\leq \exp\{\exp(b\lambda) - 1 - b\lambda\}$$

with $1 + x \leq e^x$. From Chernov's theorem, for every $t > 0$

$$P(S_n > t) \leq \inf_{\lambda > 0} \exp\{-\lambda t + n(e^{b\lambda} - 1 - b\lambda)\},$$

where the bound is denoted $\inf_{\lambda > 0} \exp\{\psi_t(\lambda)\}$. Its minimum is reached at $\lambda_t = b^{-1}\log\{1 + (nb)^{-1}t\}$ where the bound of the inequality is written as

$$\psi_t(\lambda_t) = -n\phi(\frac{t}{nb}).$$

With non identically distributed random variables, the condition implies that their Laplace transforms L_{X_i} satisfy the condition of convergence of $n^{-1}\sum_{i=1}^{n}\log L_{X_i}$ to a limit $\log L_X$. The bound $b = M\sigma$ is replaced by $b_n = M\max_{i=1,\ldots,n}\sigma_i := M\sigma_n^*$ and the upper bound of the limit $L_X(t)$ has the same form as in the case of i.i.d. variables.   □

Bennet's inequality applies to variables X_i satisfying the same condition and such that X_i has values in a bounded interval $[a_i, b_i]$, for every i. Weighted inequalities and other inequalities for independent variables are presented by Shorack and Wellner (1986).

Varadhan's Large Deviation Principle (1984) extends Chernov's theorem in the following sense. A sequence of probabilities $(P_n)_n$ on a measurable space $(\mathbb{X}, \mathcal{X})$ follows the Large Deviation Principle with a rate left-continuous function I, with values in $\mathbb{R}_+$, if the sets $\{x : I(x) \leq \lambda\}$ are compact subsets of $\mathbb{X}$ and for every closed set C and for every open set G of $\mathbb{X}$

$$\limsup_{n\to\infty} n^{-1}\log P_n(C) \leq -\inf_{x\in C} I(x),$$

$$\liminf_{n\to\infty} n^{-1}\log P_n(G) \geq -\inf_{x\in G} I(x).$$

It follows that for every function φ of $C_b(\mathbb{X})$

$$\lim_{n \to \infty} \sup \, n^{-1} \log \int_{\mathbb{X}} \exp\{-n\varphi(x)\} \, dP_n(x) = - \inf_{x \in C}\{\varphi(x) + I(x)\}.$$

These methods has been applied to random walks, Markov or Wiener processes by several authors, in particular Deuschel and Stroock (1984), and to statistical tests and physics (den Hollander 2008).

1.9 Functional inequalities

Let $(X_i)_{i \geq 1}$ be a sequence of independent random variables defined on a probability space $(\Omega, \mathcal{F}, P)$ and with values in a separable and complete metric space $(\mathcal{X}, \mathcal{B})$ provided with the uniform norm. Let $\mathcal{F}$ be a family of measurable functions defined from $(\mathcal{X}, \mathcal{B})$ to $\mathbb{R}$ and let $S_n(f) = \sum_{i=1}^{n} f(X_i)$. For a variable X having the same distribution probability P_X as the variables X_i, $P_X(f) = Ef(X) = n^{-1}ES_n(f)$ and

$$P_X(f^2) - P_X^2(f) = varf(X) = n^{-1}\{ES_n^2(f) - E^2S_n(f)\}.$$

The weak convergence of the empirical process $\nu_n(f) = n^{-\frac{1}{2}}\{S_n(f) - P_X(f)\}$ to a Brownian bridge with sample-paths in the space $C(\mathcal{X})$ of the continuous functions on $(\mathcal{X}, \mathcal{B})$ has been expressed as a uniform convergence on $\mathcal{F}$ under integrability conditions and conditions about the dimension of $\mathcal{F}$ (Dudley, 1984; Massart, 1983; van der Vaart and Wellner, 1996, Theorem 2.5.2). For the intervals $[0, t]$ of $\mathbb{R}$, $\nu_n(t) = n^{-\frac{1}{2}}\sum_{i=1}^{n}\{1_{\{X_i \leq t\}} - P(X_i \leq t)\}$. The Kolmós-Major-Tusnady (1975a) representation theorem states the existence of a probability space where a triangular sequence of independent and identically distributed variables $(X_{in})_{i=1,\cdot,n,n \geq 1}$ and a sequence of independent Brownian bridges $(B_n)_{n \geq 1}$ are defined and such that for the empirical process ν_n of $(X_{in})_{i=1,\cdot,n}$, the process $D_n = \sup_{t \in \mathbb{R}} |\nu_n(t) - B_n(t)|$ satisfies

$$P(n^{\frac{1}{2}}|D_n| \geq x + a \log n) \leq be^{-cx}, \; x > 0$$

for positive constants a, b and c. A variant where $\log n$ is replaced by $\log d$, for an integer between 1, and n was given by Mason and van Zwet (1987). Major (1990) proved similar inequalities for the approximation near zero of the empirical process by a sequence of Poisson processes $(P_n(t))_n$ with parameter nt

$$P(n^{\frac{1}{2}} \sup_{t \in [0, n^{-\frac{2}{3}}]} |\nu_n(t) - (P_n(t) - nt)| > C) < K \exp\{-\frac{\sqrt{n}\log n}{8}\}$$

and replacing the Kolmós-Major-Tusnady representation by the tail probability for the supremum over $[t_n, \infty[$, where $nF_n(t_n) = k$ and conditionally on this equality, for all $k \geq 0$.

The class of the quadrants $C = (C_x)_{x \in \mathbb{R}^d}$, where $C_x = \{y \in \mathbb{R}^d : y \leq x\}$ for every x in $\mathbb{R}^d$, has the Vapnik-Chervonenkis index $d+1$ and exponential inequalities for $\|\nu_n\|_{C^d}$ have been considered. According to the Dvoretzky-Kiefer-Wolfowitz inequality, there exists a constant C_d such that for every x in $\mathbb{R}^d$, $P(\|\nu_n\|_{C^d} > x) \leq Ce^{-2x^2}$. Inequalities for general classes have been established by Massart (1986).

The inequalities for the maximum of sums of variables over a class $\mathcal{F}$ such that $\|S_n\|_{\mathcal{F}}$ belongs to L_p are proved in the same way as in $\mathbb{R}$

$$P(\max_{k=1,\ldots,n} \|S_k\|_{\mathcal{F}} > \lambda) \leq \frac{1}{\lambda^p} E\|S_n\|_{\mathcal{F}}^p.$$

Let $\sigma^2(\mathcal{F}) = \sup_{f \in \mathcal{F}} \{P_X(f^2) - P_X^2(f)\}$ be the maximum variance of the variables $f(X_i)$ over $\mathcal{F}$ and let $Z_n(\mathcal{F}) = \sup_{f \in \mathcal{F}} S_n(f)$, $f \in \mathcal{F}$ be a functional maximum sum variable and $v_n(\mathcal{F}) = n\sigma^2(\mathcal{F}) + 2EZ_n(\mathcal{F})$. Exponential inequality for Z_n has been also written for countable classes $\mathcal{F}$ of real functions on $\mathcal{X}$ such that $\sup_{f \in \mathcal{F}} P_X(f)$ and $\sigma^2(\mathcal{F})$ are finite.

1.10   Content of the book

The next chapters develop extensions and applications of the classical results presented in this introduction. Chapter 2 extends the Cauchy, Hölder and Hilbert inequalities for arithmetic and integral means in real spaces. Hardy's inequality is generalized, several extensions and applications are presented, in particular, new versions of weighted inequalities in real analysis. The inequalities for convex transforms of a primitive F are written with a general integration measure and the conditions for the inequalities are discussed. Similar results are established for multilinear functions. The applications in probability concern moments of the maximal variable of independent and identically distributed variables and moments of transformed time variables.

Chapter 3 presents some of the more important analytic inequalities for the arithmetic and geometric means. They include functional means for the power and the logarithm functions. Carlson's inequality (1966) provided upper and lower bounds for the logarithm mean function on $\mathbb{R}_+$, it is improved and the same approach is applied to other functions. For the expansion as a partial sum $A_n(x) = \sum_{k=0}^{n} a_k$ of differentiable functions of $C_{n+2}(\mathbb{R})$, intervals for the ratio $A_{n-1}A_{n+1}A_n^{-2}$ are considered. For the exponential function, the best interval is provided. Inequalities for the arithmetic and the geometric means extend Cauchy's results. Inequalities for the median, the mode and the mean of density or distribution functions are also established. Functional equations, Young's integral inequality and several results about the entropy and the information are proved.

Chapter 4 concerns inequalities for sums and maximum of n independent random variables. They are extended to discrete and to continuous martingales and to their maximum variables, improving Bürkholder-Davis-Gundy inequalities. The Chernov and Bennet theorems and other exponential inequalities are generalized to local martingales. They are applied to Brownian motions and Poisson processes which have deterministic squared variation processes. They are generalized in several forms to dependent and bounded variables X_i and to local martingales of $\mathcal{M}_{0,loc}^2$, using Lenglart's inequality. Under boundedness and integrability conditions, for every $T > 0$ and for every $\lambda > 0$ and $\eta > 0$

$$P(\sup_{t\in[0,T]} |M(t)| > \lambda) \leq \exp\{-\phi(\lambda\eta^{-\frac{1}{2}})\} + P(< M >_T> \eta).$$

With a normalization of the time scale by T^{-1}, this is written

$$P(\sup_{t\in[0,1]} |M(T^{-1}t)| > \lambda) \leq \exp\{-\phi(\frac{\lambda}{\sqrt{\eta T}})\} + P(< M >_1> \eta).$$

Other question related to the Brownian motion are also considered. First, solutions of diffusion equations

$$dX(t) = \alpha(t, X_t)\, dt + \beta(t)\, dB(t)$$

are explicitly established under several conditions for the function α. Then, some level crossing problems for partial sums and for the counting process of the number of crossing the levels are established.

Chapter 5 concerns inequalities in functional spaces, for sums of real functions of random variables and their supremum on the class of functions defining the transformed variables. Uniform versions of the Bürkholder-Davis-Gundy inequality and of the Chernov, Hoeffding and Bennet theorems are established, they are extended to functionals of discrete or continuous martingales. Several applications to the weak convergence of nonstandard empirical processes are detailed.

Chapter 6 deals with inequalities for processes. First, the inequalities for Gaussian processes deduced from those for their covariance functions are proved, with new results. Then, we consider the distribution of the ruin time of the Sparre Anderson ruin model and in several more optimistic stochastic models with a diffusion term. Finally, some spatial stationary measures are studied and their weak convergence is deduced from their tail behaviour using inequalities of Chapter 4.

Chapter 7 focuses on complex spaces and on the Fourier transform. The classical theory is extended to higher dimensions in order to generalize the expansions of analytic functions of several variables in series. Expansions of functions in the orthonormal basis of the Hermite polynomials and their properties are studied, with the orders of the approximations and their Fourier transforms. The isometry between $\mathbb{R}^2$ and $\mathbb{C}$ is extended to an isometry between $\mathbb{R}^3$ and a complex space where the Fourier transform is also defined. The Cauchy conditions for the differentiability of complex functions and expansions of complex functions are established in this space. The same arguments apply to higher dimensions.

Chapter 2

Inequalities for Means and Integrals

2.1 Introduction

The inequalities presented in the introduction are upper bounds for norms of real vectors and functions. Results for L_p-norms of functions are related to those for vectors as limits of step functions. Some vectorial inequalities depend on the dimension of the vector space, like the equivalence of vectorial norms, and they cannot be immediately adapted to functional spaces. In the next section, the lower bounds for norms of real vectors are also specific to finite vectors.

New functional inequalities of the same kind as those of the first chapter are developed. Most inequalities presented in this chapter rely on the convexity inequalities and they are applied to the theory of integration and to probability inequalities. They are also adapted to bilinear maps by integrating with respect to product measures. Some of them use arithmetic inequalities of the next section. In particular, generalizations of the Hardy and Minkowski inequalities provide inequalities of moments for random variables and for the maximum of n independent and identically distributed random variable. They are also applied to functions of a right-censored time variable and to Laplace transforms of dependent variables.

2.2 Inequalities for means in real vector spaces

Cauchy proved that the arithmetic mean of n real numbers is always larger than their geometric mean defined as the n-root of the product of the corresponding terms. With two terms, Cauchy's inequality is written

$$\frac{a+b}{2} \geq (ab)^{\frac{1}{2}}, \ a > 0 \ b > 0. \tag{2.1}$$

It is equivalent to $a^2 + b^2 \geq 2ab$ or $(a + b)^2 \geq 4ab$, which holds true for all real numbers a and b.

The concavity of the logarithm implies a generalization of higher powers, for every integer n

$$\left(\frac{a^n + b^n}{2}\right)^{\frac{1}{n}} \geq (ab)^{\frac{1}{2}},$$

for all $a > 0$ and $b > 0$, with equality if and only if $a = b$. For smaller exponents

$$\frac{a^{\frac{1}{n}} + b^{\frac{1}{n}}}{2} \geq (ab)^{\frac{1}{2n}}.$$

These inequalities extend to a real exponent $x > 0$, $a^x + b^x \geq (2\sqrt{ab})^x$, and to means and products of n terms.

Proposition 2.1. *For every positive real numbers* $(a_i)_{i=1,\ldots,n}$

$$n^{-1} \sum_{i=1}^{n} a_i \geq \left(\prod_{i=1}^{n} a_i\right)^{\frac{1}{n}} \tag{2.2}$$

and for every real $x > 0$

$$\left(n^{-1} \sum_{i=1}^{n} a_i^x\right)^{\frac{1}{x}} \geq \left(\prod_{i=1}^{n} a_i\right)^{\frac{1}{n}}, \tag{2.3}$$

with equality if and only if $a_1 = a_2 = \cdots = a_n$.

Proof. The first inequality is due to the concavity of the logarithm $p^{-1} \log(n^{-1} \sum_{i=1}^{n} a_i^p) \geq n^{-1} \sum_{i=1}^{n} \log a_i$, with equality if and only if all terms are equal. It is also written

$$\log\left(n^{-1} \sum_{i=1}^{n} a_i^p\right) \geq pn^{-1} \sum_{i=1}^{n} \log a_i, \ p \in \mathbb{N}$$

which yields (2.3) with an integer exponent $x = p$. Replacing p by p^{-1}, the inequality is still true and it extends to real numbers x by continuity. $\quad\square$

Proposition 2.2. *Let* μ *be a positive measure on* $I \subset \mathbb{R}$, *for every integer* p *and every positive real function* a *on* I

$$\left(\int_I a^p(t) \, d\mu(t)\right)^{\frac{1}{p}} \geq \exp\left\{\int_I \log a(t) \, d\mu(t)\right\},$$

with equality if and only if a *is constant. For every random variable* X *and for every positive real function* a *such that* $\|a(X)\|_p$ *is finite,*

$$\exp\{E \log a(X)\} \leq \|a(X)\|_p.$$

Proof. Let $(x_{i,n})_{i \leq I_n, n \geq 1}$ be partition of the interval I such that $\mu([x_{i,n}, x_{i+1,n}[) = n^{-1}$. The inequality is an application of (2.2) to the real numbers $a_{i,n}$ defining a sequence of functions $a_n(t) = \sum_{i=1}^n a_{i,n} 1_{[x_{i,n}, x_{i+1,n}[}$ converging to a as n tends to infinity. The second equation is an equivalent formulation of the same inequality with a random variable having the distribution μ. $\qquad\square$

These inequalities apply directly to bounds of moments of random variables. On a probability space $(\Omega, \mathcal{F}, P)$, let X and Y be positive real random variables, their covariance is bounded by the square root of the product of their variances from the Cauchy-Schwarz inequality. The following inequalities are deduced from Proposition 2.1, for all real variables X and Y

$$cov(X, Y) \leq \frac{1}{2}(varX + varY)$$

and for every integer $n \geq 1$

$$E(XY) \leq E\{(\frac{X^{2n} + Y^{2n}}{2})^{\frac{1}{n}}\}.$$

If variables X and Y are colinear, the first inequality reduces to the Cauchy inequality and these inequalities are strict unless $X = Y$. Let $(X_i)_{i=1,\ldots,n}$ be a random vector of dimension n, let $\bar{X}_n = n^{-1}\sum_{i=1}^n X_i$ be the empirical mean, the higher moments of the sequence of variables satisfy

$$E\{(n^{-1}\sum_{i=1}^n X_i^p)^{\frac{1}{p}}\} \geq E(\prod_{i=1}^n X_i^{\frac{1}{n}}),$$

for all integers n and $p \geq 1$. From (2.3), moments of independent and identically distributed variables satisfy

$$E(\bar{X}_n^{\frac{1}{p}}) \geq (EX^{\frac{1}{np}})^n,$$

$$E\{(n^{-1}\sum_{i=1}^n X_i^p)^{\frac{1}{p}}\} \geq (EX^{\frac{1}{n}})^n.$$

From (2.2), $E(X^p) \leq (EX^{\frac{p}{n}})^n$, for every integer n and $p \geq 1$.

Proposition 2.3. *Let $X > 0$ be a real random variable with distribution function F_X and let ϕ be a convex function of $C_1(\mathbb{R})$ with a convex derivative $\phi^{(1)}$ belonging to $L_p(F_X)$, $p > 1$. The covariance of X and $\phi(X)$ satisfies*

$$cov(X, \phi(X)) \leq \|X\|_{L_p}\|\phi^{(1)}(X)\|_{L_{p'}}$$

with the conjugate integer p'.

Proof. Let $\mu_X = EX$, there exists θ in $[0,1]$ such that $\phi(X) = \phi(\mu_X) + (X - \mu_X)\phi^{(1)}(\mu_X + \theta(X - \mu_X))$ is lower than $\theta\phi^{(1)}(X) + (1 - \theta)\phi^{(1)}(\mu_X)$. By convexity, $E\phi(X) \geq \phi(\mu_X)$,

$$E\{X\phi(X)\} \leq \mu_X\phi(\mu_X) + E\{(X - \mu_X)\phi^{(1)}(\mu_X + \theta(X - \mu_X))\},$$
$$cov\{X, \phi(X)\} \leq \|X\|_{L_p}\{\theta\|\phi^{(1)}(X)\|_{L_{p'}} + (1 - \theta)|\phi^{(1)}(\mu_X)|$$

and $|\phi^{(1)}(\mu_X)| = \|\phi^{(1)}(\mu_X)\|_{L_{p'}} \leq \|\phi^{(1)}(X)\|_{L_{p'}}$. $\square$

Let a, b and c be real numbers, Proposition 2.1 states that

$$(a^3 + b^3 + c^3) \geq 3abc \quad \text{and} \quad \frac{a + b +}{3} \geq (abc)^{\frac{1}{3}},$$

by a change of variables. Closer inequalities for sums of products rather than their powers are obtained by multiplying both sides of inequalities similar to $a^2 + b^2 \geq 2ab$ for each pair of (a, b, c) by another term and adding them, or by reparametrization of another inequality. The first inequality of the next proposition comes from Lohwater (1982).

Proposition 2.4. *For all positive real numbers a, b and c, the following inequalities are satisfied and they are all equivalent*

$$a^2 + b^2 + c^2 \geq ab + ac + bc,$$
$$a^2b^2 + a^2c^2 + b^2c^2 \geq abc(a + b + c),$$
$$c^{-1}ab + b^{-1}ac + a^{-1}bc \geq a + b + c,$$
$$a^3 + b^3 + c^3 \geq (ab)^{\frac{3}{2}} + (bc)^{\frac{3}{2}} + (ac)^{\frac{3}{2}},$$
$$a + b + c \geq (ab)^{\frac{1}{2}} + (bc)^{\frac{1}{2}} + (ac)^{\frac{1}{2}},$$

with equalities if and only if $a = b = c$.

The results of Proposition 2.4 are generalized to p terms, by the same method.

Proposition 2.5. *Let p be an integer and let $(a_i)_{i=1,...,p}$ be a vector of positive real numbers, then the following inequalities are satisfied and are all equivalent*

$$\sum_{i=1}^{p} a_i \geq \sum_{i=1}^{p} \sum_{j\neq i, j=1}^{p} (a_i a_j)^{\frac{1}{2}},$$
$$\sum_{i=1}^{p} a_i^2 \geq \sum_{i=1}^{p} \sum_{j\neq i, j=1}^{p} a_i a_j,$$

$$\sum_{i=1}^{p} a_i^3 \geq \sum_{i=1}^{p} \sum_{j \neq i, j=1}^{p} (a_i a_j)^{\frac{3}{2}},$$

$$\sum_{i=1}^{p} \sum_{j \neq i, j=1}^{p} a_i^2 a_j^2 \geq (\prod_{i=1}^{p} a_i)(\sum_{j=1}^{p} a_j),$$

$$\sum_{i=1}^{p} \sum_{j \neq i, j=1}^{p} \sum_{k \neq i, j; k=1}^{p} a_i a_j a_k^{-1} \geq \sum_{i=1}^{p} a_i,$$

and the equality holds if and only if $a_1 = a_2 = \cdots = a_p$.

Replacing the constants a_i by real functions $a_i(t)$ and integrating with respect to a positive measure μ yields functional inequalities similarly to Proposition 2.2. They are proved by convexity. For all positive real functions and for every random variable X, Proposition 2.5 implies

$$\sum_{i=1}^{p} E\{a_i(X)\} \geq \sum_{i=1}^{p} \sum_{j \neq i, j=1}^{p} E[\{a_i(X)a_j(X)\}^{\frac{1}{2}}],$$

$$\sum_{i=1}^{p} E[\{a_i(X)\}^2] \geq \sum_{i=1}^{p} \sum_{j \neq i, j=1}^{p} E\{a_i(X)a_j(X)\},$$

$$\sum_{i=1}^{p} E[\{a_i(X)\}^3] \geq \sum_{i=1}^{p} \sum_{j \neq i, j=1}^{p} E[\{a_i(X)a_j(X)\}^{\frac{3}{2}}],$$

$$\sum_{i=1}^{p} \sum_{j \neq i, j=1}^{p} E[\{a_i(X)\}^2\{a_j(X)\}^2] \geq \sum_{i=1}^{p} E\{a_i(X) \prod_{j=1}^{p} a_j(X)\},$$

$$\sum_{i=1}^{p} \sum_{j \neq i, j=1}^{p} \sum_{k \neq i, j; k=1}^{p} E\frac{a_i(X)a_j(X)}{a_k(X)} \geq \sum_{i=1}^{p} E\{a_i(X)\}$$

with equality if and only if a is constant. Let ϕ be a convex function, for every integer p and every vector $(a_i)_{i=1,\ldots,p}$ of $[0,1]^p$

$$E\{\phi(\sum_{i=1}^{p} a_i X_i)\}^k \leq E\{\sum_{i=1}^{p} a_i \phi(X_i)\}^k \leq p^k \sum_{i=1}^{p} E\{a_i^k \phi^k(X_i)\}.$$

2.3 Hölder and Hilbert inequalities

Extensions of Cauchy's inequality to positive bilinear series have been introduced by Hilbert. The next bound for the series $\sum_{n \geq 1} \sum_{m \geq 1} (m +$

$n)^{-1}x_n\,y_m$ is obtained by using the inequality $m + n \geq 2(mn)^{\frac{1}{2}}$ and the Hölder inequality.

Proposition 2.6. *Let $\alpha > 2$ and let $p > 1$ and $q > 1$ be integers, then for all positive series $(x_n)_{n\geq 1}$ and $(y_m)_{m\geq 1}$*

$$\sum_{n\geq 1}\sum_{m\geq 1}\frac{x_n y_m}{(n+m)^\alpha} \leq \frac{1}{2^\alpha}\Big(\sum_{n\geq 1}n^{-\frac{\alpha}{2}}x_n\Big)\Big(\sum_{m\geq 1}m^{-\frac{\alpha}{2}}y_m\Big)$$
$$\leq c_{p,q}\|(x_n)_n\|_p\,\|(y_n)_n\|_q,$$

with the constant

$$c_{p,q} = \Big\{\sum_{n\geq 1}n^{-\frac{\alpha p}{2(p-1)}}\Big\}^{1-\frac{1}{p}}\Big\{\sum_{n\geq 1}n^{-\frac{\alpha q}{2(q-1)}}\Big\}^{1-\frac{1}{q}}.$$

Proposition 2.7. *Let $p > 1$ and $q > 1$ be integers, let $\alpha > \min\{(p-1)p^{-1},(q-1)q^{-1}\}$ be a real number and let f and g be real functions on intervals $[a,b)$ and $[c,d)$ respectively, with $a > 0$ and $c > 0$, then there exists a constant $C_{p,q}$ such that*

$$\left|\int_a^b\int_c^d\frac{f(x)g(y)}{(x+y)^{2\alpha}}\,dxdy\right| \leq \frac{1}{2^{2\alpha}}\left|\int_a^b\frac{f(x)}{x^\alpha}\,dx\right|\left|\int_c^d\frac{g(y)}{y^\alpha}\,dy\right| \leq C_{p,q}\|f\|_p\,\|g\|_q.$$

This is a consequence of the Hölder inequality with the constant

$$C_{p,q} = \frac{1}{2^{2\alpha}}\frac{p-1}{p}\left(\frac{1}{b^{(\alpha p-1)(p-1)^{-1}}}-\frac{1}{a^{(\alpha p-1)(p-1)^{-1}}}\right)^{\frac{p-1}{p}}$$
$$\times\frac{q-1}{q}\left(\frac{1}{b^{(\alpha q-1)(q-1)^{-1}}}-\frac{1}{a^{(\alpha q-1)(q-1)^{-1}}}\right)^{\frac{q-1}{q}}.$$

The integration bounds b and d may be finite or not.

Let X and Y be independent random variables with real distribution functions F and G and densities f and g. Proposition 2.7 implies that for every $\beta > 0$

$$|E(X+Y)^{-2\alpha}| \leq 2^{-2\alpha}|EX^{-\alpha}|\,|EY^{-\alpha}|$$
$$\leq 2^{-2\alpha}\|X^{-\beta}\|_p\,\|X^{\beta-\alpha}\|_{p'}\,\|Y^{-\beta}\|_q\,\|Y^{\beta-\alpha}\|_{q'}.$$

With the distribution functions, the first inequality of Proposition 2.7 is also written as

$$\left|\int_a^b\int_c^d\frac{1}{(x+y)^{2\alpha}}\,dF(x)dG()y\right| \leq \frac{1}{2^{2\alpha}}\left|\int_a^b\frac{1}{x^\alpha}\,dF(x)\right|\left|\int_a^b\frac{1}{x^\alpha}\,dG(x)\right|.$$

Hilbert's inequality has been extended by Schur (1912) and by Hardy, Littlewood, Pólya (1952) to bound a scalar product of f and g defined with respect to a measure on $\mathbb{R}^2$. Here is another version of this inequality.

Proposition 2.8. *Let p and p' be conjugate integers and let K be a positive homogeneous weighting function on $\mathbb{R}_+^2$ such that*

$$k_p = \min\{\int_0^\infty x^{-\frac{1}{p}} K(1,x)\,dx, \int_0^\infty x^{-\frac{1}{p'}} K(x,1)\,dx\}$$

is finite. For all functions f in $L_p(\mathbb{R}_+)$ and g in $L_{p'}(\mathbb{R}_+)$

$$\left|\int_0^\infty \int_0^\infty K(x,y)f(x)g(y)\,dx\,dy\right| \leq k_p\|f\|_p\|g\|_{p'}.$$

With $K(x,y) = (x+y)^{-1}$, $k_p = \min\{p^{-1}, p'^{-1}\}$ is always smaller than the bound $\pi \sin^{-1}(p^{-1}\pi)$ of the Hilbert inequality. Let $\lambda > 0$ be real and let $K(x,y) = (x+y)^{-\lambda}$ be a kernel function, then the constant is $k_p = \lambda \min\{p^{-1}, p'^{-1}\}$.

Let (X, Y) be a random variable defined from a probability space $(\Omega, \mathcal{F}, P)$ to $\mathbb{R}^2$. Hölder's inequality can be expressed as an inequality for (X, Y) and functions of (X, Y) as X and Y are dependent.

Proposition 2.9. *Let φ and ψ be real functions on $\mathbb{R}$ and let $p > 1$, then there exists a constant k_p such that for every random variable (X, Y) with values in $\mathbb{R}^2$ and such that $\varphi(X)$ belongs to L_p and $\psi(Y)$ to $L_{p'}$*

$$|E\{\varphi(X)\psi(Y)\}| \leq k_p E\|\varphi(X)\|_p E\|\psi(Y)\|_{p'},$$

where the best constant is 1 if X and Y are independent.

More generally, a similar inequality applies to multidimensional random variables. Let (X, Y) be a random variable defined from Ω to $\mathbb{R}^{2n}$ and such that $\varphi(X)$ belongs to L_p and $\psi(Y)$ to $L_{p'}$ and let $X = (X_i)_{i=1,\ldots,n}$ and $Y = (Y_i)_{i=1,\ldots,n}$, then

$$E| < \varphi(X), \psi(Y) > | = E\left|\sum_{i=1}^n \varphi(X_i)\psi(Y_i)\right| \leq k_p E\|\varphi(X)\|_p E\|\psi(Y)\|_{p'},$$

with the $L_p(\mathbb{R}^n)$ norm $\|x\|_p = (\sum_{i=1}^n |x_i|^p)^{\frac{1}{p}}$.

The converse of the Hölder inequality for vectors can also be written in the L_p spaces. Let $p > 1$ be an integer, then a necessary condition for the existence of a constant k such that $\int_0^\infty |f(x)g(x)|\,dx \leq k\|f\|_p$, for all functions f in L_p, is the existence of a finite norm $\|g\|_{p'}$, for every g.

2.4   Generalizations of Hardy's inequality

Let $p \geq 2$ be an integer with conjugate p', such that $p^{-1} + p'^{-1} = 1$, and let f and g be positive integrable real functions on $\mathbb{R}_+$ such that f belongs to L_p and g to $L_{p'}$. Let F and G be their primitive functions satisfying

$$\lim_{x \to 0} x^{\frac{1}{p}-1}F(x) = 0, \quad \lim_{x \to \infty} x^{\frac{1}{p}-1}F(x) = 0, \tag{2.4}$$

$$\lim_{x \to 0} x^{-\frac{1}{p'}}G(x) = 0 \quad \lim_{x \to \infty} x^{-\frac{1}{p'}}G(x) = 0.$$

Hardy's inequality yields the next inequality

$$I_p = \int_0^\infty \{\frac{F(x)}{x}\}^p \, dx \leq (\frac{p}{p-1})^p \|f\|_p^p, \tag{2.5}$$

$$I_{p'} = \int_0^\infty \{\frac{G(x)}{x}\}^{\frac{p-1}{p}} \, dx \leq p^{\frac{p}{p-1}} \|g\|_{p'}^{p'}.$$

The Hardy's inequality (2.5) is proved using an integration by parts on the Hölder inequality

$$I_p = \frac{p}{p-1} \int_0^\infty \{\frac{\bar{F}(x)}{x}\}^{p-1} f(x) \, dx,$$

$$\int_0^\infty \{\frac{\bar{F}(x)}{x}\}^{p-1} f(x) \, dx \leq \bar{I}_p^{1-\frac{1}{p}} \|f\|_p.$$

By convexity, the integral

$$N_p(F) = [\int_0^\infty \{\frac{F(x)}{x}\}^p \, dx]^{\frac{1}{p}}$$

defines a norm in the space of the primitives of functions of $L_p(\mathbb{R}_+)$. It is related to a scalar product by (1.1), with the norm N_2, hence it is a weighted L_2 scalar product of the primitive functions

$$< F, G > = \int_0^\infty x^{-2}F(x)G(x)dx.$$

By the geometric equalities (1.2)

$$< F, G > = \frac{1}{4}\{N_2(F + G) - N_2(F - G)\}.$$

The integral $\int_0^\infty x^{-2}F(x)G(x)dx$ is bounded using successively Hölder's inequality and Hardy's ineqality. For all conjugate integers $p \geq 1$ and p'

$$\int_0^\infty \frac{F(x)G(x)}{x^2}dx \leq \frac{p^2}{p-1}\{\int_0^\infty f^p(x)\,dx\}^{\frac{1}{p}}\{\int_0^\infty g^{\frac{p}{p-1}}(x)\,dx\}^{1-\frac{1}{p}}. \tag{2.6}$$

For a probabilistic application, let S be a positive real random variable with distribution function H, and let $p \geq 2$ be an integer and p' be its conjugate, $p^{-1} + p'^{-1} = 1$. Let a be a function in $L_p(\mathbb{R}_+)$ and b be a function in $L_{p'}(\mathbb{R}_+)$, such that $\int a \, dH = 0$ and $\int b \, dH = 0$, the inequalities (1.14) and (2.6) are expressed in terms of mean functions of S as

$$E(\frac{\int_0^S a \, dH}{H(S)})^p \leq (\frac{p}{p-1})^p Ea^p(S), \qquad (2.7)$$

$$E\frac{\int_0^S a \, dH}{H(S)} \frac{\int_0^S b \, dH}{H(S)} \leq \frac{p^2}{p-1} \{Ea^p(S)\}^{\frac{1}{p}} \{Eb^{\frac{p}{p-1}}(S)\}^{1-\frac{1}{p}}.$$

This is a reparametrization of inequalities (2.6). The variable $X = H(S)$ has a uniform distribution in $[0, 1]$ and, with the notations $H(s) = x$, $F(s) = \int_0^s a \, dH$ and $G(s) = \int_0^s b \, dH$, we get

$$\frac{\int_0^S a \, dH}{H(S)} = \frac{F \circ H^{-1}(X)}{X},$$

$$Ea^p(S) = \int_0^\infty a^p(s) \, dH(s) = \int_0^\infty \{a \circ H^{-1}(x)\}^p \, dx,$$

moreover

$$E(\frac{\int_0^S a \, dH}{H(S)})^p = \int_0^\infty (\frac{F}{H})^p \, dH = \int_0^1 (\frac{F \circ H^{-1}(x)}{x})^p \, dx,$$

where the derivative of $F \circ H^{-1}(x)$ is $a \circ H^{-1}(x)$. The first inequality is equivalent to the Hardy inequality since $\int a \, dH = 0$. The integrals of the function b have the same reparametrization and the second inequality is deduced from the Hölder inequality. The inequalities (2.7) are also proved directly by the same arguments as (2.5), with integrals in $\mathbb{R}_+$.

There are no inequalities of the same kind as (1.13), (1.14) and (2.6) for the decreasing survival function $\bar{F}(x) = \int_x^\infty f(t) \, dt$ on $\mathbb{R}_+$ since $\bar{F}(0) = 1$. Let f be an integrable positive function on $\mathbb{R}_+$ with a survival function $\bar{F}$. Let $a > 0$, for any integer $p \geq 2$, an integration by parts implies

$$\bar{I}_p(a) := \int_a^\infty \{\frac{\bar{F}(x)}{x}\}^p \, dx = \frac{1}{p-1} [\frac{\bar{F}^p(a)}{a^{p-1}} - p \int_a^\infty \{\frac{\bar{F}(x)}{x}\}^{p-1} f(x) \, dx],$$

and by the Hölder inequality

$$\int_a^\infty \{\frac{\bar{F}(x)}{x}\}^{p-1} f(x) \, dx \leq \bar{I}_p^{1-\frac{1}{p}}(a) \, (\int_a^\infty f^p)^{\frac{1}{p}},$$

$$\bar{I}_p(a) \geq (\frac{1}{p-1})^p \{\bar{I}_p^{(\frac{1}{p}-1)}(a) \frac{\bar{F}^p(a)}{a^{p-1}} - p(\int_a^\infty f^p)^{\frac{1}{p}}\}^p.$$

This inequality is strict except if f is a constant. Let f be the density of a positive variable X, $F(x) = P(X \leq x)$ and $\bar{F}(x) = P(X \geq x)$, therefore F and $\bar{F}$ are bounded by one, $F(0) = 0$ and $\bar{F}(0) = 1$, $\lim_{x \to \infty} F(x) = 1$ and $\lim_{x \to \infty} \bar{F}(x) = 0$. The limiting condition (2.4) for F at infinity is satisfied only if $p > 1$.

Proposition 2.10. *Let $\bar{F}$ be defined in $\mathbb{R}$ and satisfy the condition $\lim_{x \to \pm\infty} x^{\frac{1}{p}-1} \bar{F}(x) = 0$. For every $p > 1$*

$$\int_{-\infty}^{\infty} \{\frac{\bar{F}(x)}{x}\}^p \, dx \geq (\frac{p}{1-p})^p \|f\|_p^p.$$

Proof. It is proved as above, integrating by parts

$$\bar{I}_p = \int_{-\infty}^{\infty} \{\frac{\bar{F}(x)}{x}\}^p \, dx = \frac{p}{1-p} \int_{-\infty}^{\infty} \{\frac{\bar{F}(x)}{x}\}^{p-1} f(x) \, dx$$

then, by the Hölder inequality

$$\int_{-\infty}^{\infty} \{\frac{\bar{F}(x)}{x}\}^{p-1} \, dx \leq \bar{I}_p^{1-\frac{1}{p}} \|f\|_p. \qquad \square$$

Since $\bar{F}$ is decreasing, the integral $\bar{I}_p$ is negative and its lower bound is necessarily negative. These results are generalized by replacing the nomalization of F and $\bar{F}$ by any distribution function and the bounds are not modified. The proof of this new result is similar to the previous ones.

Proposition 2.11. *Let F and G be distribution functions in $\mathbb{R}_+$ such that $H^{-1}F$ belongs to $L_p(H)$, F has a density f and*

$$\lim_{x \to \infty} \frac{F^p}{H^{p-1}}(x) = 0 = \lim_{x \to 0} \frac{F^p}{H^{p-1}}(x),$$

then

$$\int_0^{\infty} \{\frac{\bar{F}(x)}{H(x)}\}^p \, dH(x) \leq (\frac{p}{1-p})^p \|f\|_p^p. \qquad (2.8)$$

Let F and G be distribution functions in $\mathbb{R}$ such that $H^{-1}F$ belongs to $L_p(H)$, F has a density f and

$$\lim_{x \to \pm\infty} \frac{F^p}{H^{p-1}}(x) = 0,$$

then

$$\int_0^{\infty} \{\frac{\bar{F}(x)}{H(x)}\}^p \, dH(x) \geq (\frac{p}{1-p})^p \|f\|_p^p. \qquad (2.9)$$

The norms

$$N_p(F) = [\int_0^\infty \{\frac{F(x)}{H(x)}\}^p \, dH(x)]^{\frac{1}{p}}, \ p \geq 1,$$

and the scalar products

$$< F, G > = \int_0^\infty H^{-2}(x) F(x) G(x) dx$$

are deduced from Proposition 2.11.

Inequalities (1.13)-(1.14) and Proposition 2.11 extend to fractional and real convex power functions and to other convex functions.

Proposition 2.12. *Let $p \geq 2$ and $1 \leq q < p$ be integers and let $r = pq^{-1}$. Let f be a positive function of $L_r(\mathbb{R}_+)$ such that $\lim_{x \to 0} x^{\frac{1}{r}-1} F(x) = 0$ and $\lim_{x \to \infty} x^{\frac{1}{r}-1} F(x) = 0$, then*

$$I_r = \int_0^\infty (\frac{F(x)}{x})^r \, dx \leq (\frac{p}{p-q})^r \int_0^\infty f^r(x) \, dx, \qquad (2.10)$$

and there is equality if and only if f is a constant.

Proof. Integrating by parts implies

$$I_r = \frac{p}{q}\{I_r - \int_0^\infty (\frac{F(x)}{x})^{r-1} f(x) \, dx\},$$

$$I_r = \frac{p}{p-q} \int_0^\infty (\frac{F(x)}{x})^{r-1} f(x) \, dx.$$

Let $p = p(p-q)^{-1} > 1$, then the Hölder inequality yields

$$\int_0^\infty (\frac{F(x)}{x})^{r-1} f(x) \, dx \leq I_r^{\frac{1}{p}} (\int_0^\infty f^\beta)^{\frac{1}{\beta}}$$

with $\beta^{-1} + p^{-1} = 1$, hence $\beta = r$. It follows that

$$I_r \leq \frac{p}{p-q} I_r^{(p-q)p^{-1}} (\int_0^\infty f^r)^{\frac{1}{r}}. \qquad \square$$

The same inequality holds for the integral $I_r(y)$, as in Equation (1.13). For $\bar{I}_r$, the integration by parts leads to an opposite inequality with an additive constant, as previously. By limit of converging fractional sequences, we obtain an extension of Proposition 2.12 to real exponents.

Theorem 2.1. *Let $\lambda > 1$ be a real number and let f be a positive function in $L_\lambda(\mathbb{R}_+)$ with primitive F such that $\lim_{x \to 0} x^{\lambda^{-1}-1} F(x)$ and $\lim_{x \to \infty} x^{\lambda^{-1}-1} F(x)$ are zero, then*

$$I_\lambda = \int_0^\infty (\frac{F(x)}{x})^\lambda \, dx \leq (\frac{\lambda}{\lambda-1})^\lambda \int_0^\infty f^\lambda(x) \, dx, \qquad (2.11)$$

with equality if and only if f is constant.

Proposition 2.10 is also satisfied with a real exponent $\lambda > 1$, under the same conditions

$$\int_{-\infty}^{\infty} \{\frac{\bar{F}(x)}{x}\}^\lambda \, dx \geq (\frac{\lambda}{1-\lambda})^\lambda \|f\|_\lambda^\lambda.$$

The inequalities with a real exponent extend to weighted inequalities more general than the inequalities by Kufner and Persson (2003), with similar proofs as Theorem 2.1.

Theorem 2.2. *Let $\lambda > 1$ and $\alpha > 0$ be real numbers and let f be a positive function such that $\int_0^\infty f^\lambda(x)x^\alpha \, dx < \infty$. If $\lambda > \alpha+1$, $\lim_{x \to 0} x^{\frac{\alpha+1}{\lambda-1}} F(x) = 0$ and $\lim_{x \to 0} x^{\frac{\alpha+1}{\lambda-1}} F(x) = 0$, then*

$$I_{\lambda,\alpha} := \int_0^\infty (\frac{F(x)}{x})^\lambda x^\alpha \, dx = \frac{\lambda}{\lambda - \alpha - 1} \int_0^\infty (\frac{F(x)}{x})^{\lambda-1} f(x)x^\alpha \, dx$$

$$\leq (\frac{\lambda}{\lambda - \alpha - 1})^\lambda \int_0^\infty f^\lambda(x)x^\alpha \, dx,$$

with equality if and only if f is constant.

Proof. Integrating by parts implies and using the Hölder inequality yields

$$I_{\lambda,\alpha} = \frac{\lambda}{\alpha+1}\{I_{\lambda,\alpha} - \int_0^\infty (\frac{F(x)}{x})^{\lambda-1} x^{\frac{\alpha(\lambda-1)}{\lambda}} f(x)x^{\frac{\alpha}{\lambda}} \, dx\},$$

$$I_{\lambda,\alpha} \leq \frac{\lambda}{\lambda - \alpha - 1} I_{\lambda,\alpha}^{\frac{\lambda-1}{\lambda}} \{\int_0^\infty f^\lambda(x)x^\alpha \, dx\}^{\frac{1}{\lambda}}.$$

$\square$

Corollary 2.1. *Let $X_1, \ldots, X_n$ be a sequence of independent and identically distributed random variables with a density f on $\mathbb{R}_+$. For every $1 \leq p \leq n-1$, their maximum X_n^* satisfy*

$$E\{(X_n^*)^{-p}\} = n(\frac{n}{p})^{n-1} \int_0^\infty f^n(x)x^{n-p-1} \, dx.$$

Proof. First, let $p = n - 1$. Let F be the distribution function of the variables X_i, the distribution of their maximum X_n^* is F^n and

$$E\{(X_n^*)^{-(n-1)}\} = n \int_0^\infty (\frac{F(x)}{x})^{n-1} f(x) \, dx \leq nI_n^{1-\frac{1}{n}} \|f\|_n,$$

by Hölder's inequality. Applying Hardy's inequality to I_n leads to

$$I_n^{1-\frac{1}{n}} \leq (\frac{n}{n-1})^{n-1}(\int_0^\infty f^n(x) \, dx)^{1-\frac{1}{n}}$$

and the result for $p = n - 1$ follows. For $1 \leq p < n - 1$, the inequality is proved using Theorem 2.2, with $\lambda = n$ and $\alpha = n - p - 1$, and the inequality

$$E\{(X_n^*)^{-(p)}\} = n \int_0^\infty (\frac{F(x)}{x})^{n-1} x^{n-p-1} f(x) \, dx$$

$$\leq n I_{n,n-p-1}^{1-\frac{1}{n}} (\int_0^\infty f^n(x) x^{n-p-1} \, dx)^{\frac{1}{n}},$$

$$I_{n,n-p-1}^{1-\frac{1}{n}} \leq (\frac{n}{p})^{n-1} (\int_0^\infty f^n x^{n-p-1} \, dx)^{1-\frac{1}{n}}.$$ $\qquad \square$

This result extends to real moments $E\{(X_n^*)^{-\alpha}\}$, $\alpha > 0$. Theorem 2.2 does not apply to $E\{(X_n^*)^p\}$, with positive integer p, since the condition $\lambda > \alpha + 1$ is not fulfilled with $\lambda = n$ and $\alpha = n + p - 1$. Integrations similar to the calculus of $E\{(X_n^*)^{-p}\}$ cannot be bounded for positive moments $E\{(X_n^*)^p\}$.

The previous inequalities are specific to power functions. For a general convex function Φ, the more general inequality proved by Kufner and Persson (2003) is the next theorem, with the best constant 1. For power functions Φ on $\mathbb{R}_+$, it does not reduce to Hardy's inequality (1.13).

Theorem 2.3 (Hardy-Knopp's Inequality). *Let Φ be a convex and increasing function, for every positive function f with primitive F and such that $x^{-1} \Phi \circ f(x)$ belongs to $L_1(\mathbb{R}_+)$ and $\Phi \circ f(0) = 0$*

$$\int_0^\infty \Phi\{\frac{F(x)}{x}\} \frac{dx}{x} \leq \int_0^\infty \Phi(f(x)) \frac{dx}{x}. \qquad (2.12)$$

Proof. This is a consequence of the convex inequality

$$\Phi\{\frac{1}{x} \int_0^x f(t) \, dt\} \leq \frac{1}{x} \int_0^x \Phi \circ f(t) \, dt,$$

integrating by parts the integrals of both sides of this inequality implies

$$\int_0^\infty \Phi\{\frac{F(x)}{x}\} \frac{dx}{x} \leq \int_0^\infty \Phi(f(x)) \frac{dx}{x}. \qquad \square$$

Weighted extensions of Hardy-Knopp's inequality are straightforward.

Theorem 2.4. *Let Φ be a convex positive monotone function and let $\alpha > 0$ be a real number. For every positive function f with primitive F and such that the function $x^{-\alpha} \Phi \circ f(x)$ belongs to $L_1(\mathbb{R}_+)$*

$$\int_0^\infty \Phi\{\frac{F(x)}{x}\} \frac{dx}{x^\alpha} \leq \frac{1}{\alpha} \int_0^\infty \Phi(f(x)) \frac{dx}{x^\alpha}. \qquad (2.13)$$

Proof. For every $x > 0$, the convexity of Φ is expressed as

$$\Phi\left(\frac{F(x)}{x}\right) \leq \frac{\int_0^x \Phi(f)(t)\,dt}{x}.$$

Denoting $g = \Phi(f)$, it is still written $\Phi\{x^{-1}\int_0^x \Phi^{-1}(g)(t)\,dt\} \leq x^{-1}G(x)$, with the primitive G of g, and by Fubini's integration lemma, it implies

$$\int_0^\infty \Phi\{\frac{F(x)}{x}\}\frac{dx}{x^\alpha} = \int_0^\infty \Phi\{\frac{\int_0^x \Phi^{-1}(g)(t)\,dt}{x}\}\frac{dx}{x^\alpha}$$

$$\leq \int_0^\infty \left(\int_t^\infty \frac{dx}{x^{\alpha+1}}\right)g(t)\,dt = \frac{1}{\alpha}\int_0^\infty \Phi(f(t))\frac{dt}{t^\alpha}. \qquad \square$$

Example 2.1. With the exponential function Φ, for every function f such that $x^{-1}e^{f(x)}$ belongs to $L_1(\mathbb{R}_+)$, inequality (2.13) is written

$$\int_0^\infty e^{x^{-1}F(x)}x^{-1}\,dx \leq \int_0^\infty e^{f(x)}x^{-1}\,dx.$$

If $x^{-1}f(x)$ is integrable, it becomes

$$\int_0^\infty x^{-1}\exp\{x^{-1}\int_0^x \log\{f(t)\}\,dt\}\,dx \leq \int_0^\infty x^{-1}f(x)\,dx$$

and, for an integrable function f,

$$\int_0^\infty x^{-1}\exp\{x^{-1}\int_0^x \log\{tf(t)\}\,dt\}\,dx \leq \int_0^\infty f(x)\,dx.$$

Example 2.2. Let $f(x) = -x^{-1}$, the inequality (2.12) implies

$$\int_0^\infty x^{-(1+x^{-1})}\,dx \leq \int_0^\infty e^{-x^{-1}}x^{-1}\,dx < \infty.$$

By convexity, Theorem 2.4 extends to a Minkovski type inequality for every convex and monotone real function $\Phi > 0$

$$\int_0^\infty \Phi\{\frac{(F+G)(x)}{x}\}\frac{dx}{x^\alpha} \leq \frac{1}{\alpha}\int_0^\infty \{\Phi(f(x)) + \Phi(g(x))\}\frac{dx}{x^\alpha}.$$

The general expression of Theorem 2.4 is obtained by integrating the convex transform of F by other measures.

Theorem 2.5. *Let Φ be a convex positive monotone function and let u be a strictly positive function on $\mathbb{R}_+$ such that $U(x) = \int_x^\infty t^{-1}u(t)\,dt$ is finite for every $t > 0$. For every positive function f with primitive F*

$$\int_0^\infty \Phi\{\frac{F(x)}{x}\}u(x)\,dx \leq \int_0^\infty \Phi(f(x))\,U(x)\,dx. \qquad (2.14)$$

Let $X > 0$ be a random variable with density u and let $U(x) = P(X > x)$. Theorem 2.5 expresses that for every positive function f with primitive F, $E\Phi\{X^{-1}F(X)\} \leq \int_0^\infty \Phi(f(x)) U(x) dx$.

Let k be a function defined from $\mathbb{R}_+^2$ to $\mathbb{R}_+$ and let $K(x) = \int_0^x k(x, t) dt$. Let $u \geq 0$ be a real function defined from $\mathbb{R}_+$ and

$$v(x) = x \int_0^x \frac{k(s, x)}{K(s)} \frac{u(s)}{s} ds.$$

Kaijser, Nikolova, Perrson and Wedestig (2005) proposed further generalizations of the Hardy-Knopp inequalities, in particular for a weighted mean.

Theorem 2.6. *For every convex function φ on $\mathbb{R}_+$*

$$\int_0^\infty \phi(\frac{1}{K(x)} \int_0^x k(x, t) f(t) dt) \frac{u(x)}{x} dx \leq \int_0^\infty \phi \circ f(x) \frac{v(x)}{x} dx.$$

The constant $\{p(p-1)^{-1}\}^p$ in Hardy's inequality is larger than one and it is specific to the power function $\Phi(x) = x^p$, with a weight $u(x) = x^{-1}$. It does not appear in Theorem 2.5 where it is included in the integral $U(x)$. With the function $u = id$, the weighting function $U(x)$ is not bounded for any x, so the integral $\int_0^\infty \Phi\{x^{-1}F(x)\} dx$ cannot be bounded as a consequence of Theorem 2.5.

Let $\alpha > 1$ be a real number, let u and v be strictly positive functions on $\mathbb{R}_+$ and let

$$U(x) = \int_x^\infty t^{-1} u(t) dt, \quad V(y) = \int_y^\infty t^{-1} v(t) dt. \tag{2.15}$$

Theorem 2.5 and the Hölder inequality entail a bilinear inequality

$$\int_0^\infty \int_0^\infty \left\{ \frac{F(x)G(y)}{(x+y)^2} \right\}^\alpha u(x)v(y) dx \, dy \leq \frac{1}{2^{2\alpha}} \{ \int_0^\infty f^\alpha(x)U(x) dx \}$$
$$\times \{ \int_0^\infty g^\alpha(y)V(y) dy \}.$$

Special cases and inequalities of the same kind are obtained with other convex and monotone functions Φ.

Example 2.3. Let Φ be the exponential function

$$\int_0^\infty \int_0^\infty \exp\{ \frac{F(x) + G(y)}{x + y} \} u(x)v(y) dx \, dy$$
$$\leq \int_0^\infty \exp\{ \frac{F(x)}{x} \} u(x) dx \int_0^\infty \exp\{ \frac{G(y)}{y} \} v(y) dy$$
$$\leq \int_0^\infty \exp\{ f(x) \} U(x) dx \int_0^\infty \exp\{ g(y) \} V(y) dy.$$

Weighted probability inequalities are deduced from the above results, using the same reparametrization as inequality (2.7). Let S be a positive real random variable with distribution function H. For every function a such that $Ea(S) = 0$, Equation (2.12) is equivalent to

$$E\{H^{-1}(S)\Phi(\frac{\int_0^S a\,dH}{H(S)})\} \leq E\{H^{-1}(S)\Phi \circ a(S)\} \qquad (2.16)$$

and, for $\alpha > 0$, (2.13) is equivalent to

$$E\{H^{-\alpha}(S)\Phi(\frac{\int_0^S a\,dH}{H(S)})\} \leq E\{H^{-\alpha}(S)\Phi \circ a(S)\}.$$

The general form of the weighted inequalities is written

$$E\{\Phi(\frac{\int_0^S a\,dH}{H(S)})u \circ H(S)\} \leq E\{U \circ H(S)\,\Phi \circ a(S)\},$$

where u is a strictly positive function on $\mathbb{R}_+$ and U is defined by (2.15).

Let S and T be positive real random variables with respective distribution functions H_S and H_T, and let a and b be functions such that $\int a\,dH_S = 0$ and $\int b\,dH_T = 0$. Let u and v be strictly positive functions on $\mathbb{R}_+$ and let U and V be defined by (2.15). Example 2.3 is equivalent to

$$E[\exp\{\frac{\int_0^S a\,dH_S + \int_0^T b\,dH_T}{X_S + X_T}\}\,u(X_S)v(X_T)]$$
$$\leq E[\exp\{a(X_S) + b(X_T)\}\,U(X_S)\,V(X_T)]$$

with the uniform variables $X_S = H_S(S)$ and $X_T = H_T(T)$.

Let $\alpha > 1$ be a real number, the inequality with the power transforms becomes

$$E\left[\left\{\frac{\int_0^S a\,dH_S \int_0^T b\,dH_T}{(X_S + X_S)^2}\right\}^\alpha u(X_S)v(X_T)\right]$$
$$\leq \frac{1}{2^{2\alpha}}E[\{a(X_S)b(X_T)\}^\alpha U(X_S)\,V(X_T)].$$

2.5   Carleman's inequality and generalizations

Carleman's inequality was established for maps of series of $\ell_1(\mathbb{R}_+)$

$$\sum_{n\geq 1}\exp(\frac{\sum_{k=1}^n \log x_k}{n}) \leq \sum_{n\geq 1} x_k$$

and it has been extended to an integral form. Let f be a positive function of $L_1(\mathbb{R}_+)$, this inequality can be expressed as

$$\int_0^\infty \exp\Big(\frac{\int_0^x \log f(t)\,dt}{x}\Big)\,dx \le \int_0^\infty f(x)\,dx.$$

It is an application of Jensen's inequality for the exponential function on $\mathbb{R}_+$ in the form $\exp(x^{-1}\int_0^x g(t)\,dt) \le \exp \circ g(x)$ for every positive function g of $L_1(\mathbb{R}_+)$. It generalizes by integrating a transformed primitive with respect to a positive measure μ on $\mathbb{R}_+$.

Theorem 2.7. *Let Φ be a convex and increasing function on $\mathbb{R}_+$ and F be the primitive of a positive function f on $\mathbb{R}_+$ such that $\Phi \circ f$ belongs to $L_1(\mu)$, then*

$$\int_0^\infty \Phi\Big(\frac{F(x)}{x}\Big)\,d\mu(x) \le \int_0^\infty \Phi \circ f(x)\,d\mu(x). \qquad (2.17)$$

Example 2.4. Let $f : \mathbb{R} \mapsto \mathbb{R}_+$ be a density with derivative f', by Theorem 2.7, for every convex and increasing function Φ on $\mathbb{R}$

$$E\Phi\Big(\frac{f'}{f}(X)\Big) \ge E\Phi\Big(\frac{\log f(X)}{X}\Big).$$

In particular, the Fisher information of a parametric family of densities

$$\mathcal{F}_\Theta = \{f_\theta, \theta \in \Theta; f \in C_1(\Theta)\},$$

with derivative f'_θ with respect to the parameter, satisfies

$$I_\theta(f) = E\left\{\frac{f'_\theta}{f_\theta}(X)\right\}^2 \ge E\left\{\Big(\frac{\log f_\theta(X)}{X}\Big)^2 1_{\{X \ne 0\}}\right\}, \theta \in \Theta.$$

Carleman's inequality and Equation (2.17) are extended to an inequality involving a function Φ and the primitive of $\Phi^{-1}(f)$, in a form similar to the initial inequality with the exponential function Φ. Let Φ be a convex, positive and monotone function on $\mathbb{R}_+$ and let f be a positive function of $L_1(\mu)$, then

$$\int_0^\infty \Phi\Big(\frac{\int_0^x \Phi^{-1} \circ f(t)\,dt}{x}\Big)\,d\mu(x) \le \int_0^\infty f(x)\,d\mu(x). \qquad (2.18)$$

By the same change of variables as in the inequality (2.7), Theorem 2.7 is rewritten in the following forms.

Proposition 2.13. *Let S be a positive real random variable with distribution function H. For every function a such that $\Phi \circ a(S)$ belongs to $L_1(P)$*

$$E\Big\{\Phi\Big(\frac{\int_0^S a\,dH}{H(S)}\Big)\Big\} \le E\{\Phi \circ a(S)\},$$

if $Ea(S) = 0$

$$E\Big\{\Phi\Big(\frac{\int_0^S \Phi^{-1} \circ a\,dH}{H(S)}\Big)\Big\} \le E\{a(S)\}.$$

2.6   Minkowski's inequality and generalizations

Let $p > 1$ be an integer, Minkowski's inequality states the additivity property of the L_p norm, $\|f + g\|_p \leq \|f\|_p + \|g\|_p$. Mulholland (1950) searched to determine classes of functions $\varphi : I \subset \mathbb{R} \mapsto \mathbb{R}$ such that for all functions f and $g : J \subset \mathbb{R}^n \mapsto I$ of $L^1(\mu)$ the following inequality should be satisfied

$$\varphi^{-1}(\int \varphi(f + g)\, d\mu) \leq \varphi^{-1}\{\int \varphi(f)\, d\mu\} + \varphi^{-1}\{\int \varphi(f)\, d\mu\}, \qquad (2.19)$$

with equality if and only if f and g are proportional. On a probability space $(\Omega, \mathcal{A}, P)$, this inequality is expressed in terms of means of random variables. The question is to determine classes of functions $\varphi : I \subset \mathbb{R} \mapsto \mathbb{R}$ satisfying

$$\varphi^{-1}\{E\varphi(X + Y)\} \leq \varphi^{-1}\{E\varphi(X)\} + \varphi^{-1}\{E\varphi(Y)\}, \qquad (2.20)$$

for all random variables X and Y on $(\Omega, \mathcal{A}, P)$ and with values in I, such that $\varphi(X)$ and $\varphi(Y)$ are in $L_1(P)$, with equality if and only if X and Y are proportional.

Assume that the inequality (2.20) is satisfied. Let f and g be real functions defined on $\mathbb{R}$, and let $\widetilde{X}$ and $\widetilde{Y}$ be real random variables with joint distribution functions F and marginals $F_{\widetilde{X}}$ and $F_{\widetilde{Y}}$. With the variables $X = f(\widetilde{X})$ and $Y = g(\widetilde{Y})$, Equation (2.20) is written

$$\varphi^{-1}\{\int_{\mathbb{R}} \varphi(f + g)\, dF\} \leq \varphi^{-1}\{\int_{\mathbb{R}} \varphi(f)\, dF_{\widetilde{X}}\} + \varphi^{-1}\{\int_{\mathbb{R}} \varphi(g)\, dF_{\widetilde{Y}}\}.$$

Inequality (2.19) is obtained with variables $\widetilde{X}$ and $\widetilde{Y}$ having the same distribution μ.

The next proposition presents a weaker inequality due to monotonicity and convexity. It is extended as $E(X + Y) \leq \varphi^{-1}\{E\varphi(X)\} + \varphi^{-1}\{E\varphi(Y)\}$ for all random variables X and Y such that $\varphi(X)$ and $\varphi(Y)$ are $L_1(P)$, with equality if and only if X and Y are proportional.

Proposition 2.14. *For every convex and strictly increasing function φ and for all real sequences $(x_i)_{i \geq 1}$ and $(y_i)_{i \geq 1}$ such that $\{\varphi(x_i)\}_{i \geq 1}$ and $\{\varphi(y_i)\}_{i \geq 1}$ belong to $\ell_1(\mathbb{R})$*

$$n^{-1} \sum_{i \geq 1}(x_i + y_i) \leq \varphi^{-1}\{n^{-1} \sum_{i \geq 1} \varphi(x_i)\} + \varphi^{-1}\{n^{-1} \sum_{i \geq 1} \varphi(y_i)\},$$

with equality if and only if x_i and y_i are proportional, for every integer i.

Proof. By convexity $\varphi(n^{-1}\sum_{i\geq 1} x_i) \leq n^{-1}\sum_{i\geq 1}\varphi(x_i)$ for every sequence $(x_i)_{i\geq 1}$ satisfying the assumptions. This implies

$$n^{-1}\sum_{i\geq 1} x_i \leq \varphi^{-1}\{n^{-1}\sum_{i\geq 1}\varphi(x_i)\}, \ n^{-1}\sum_{i\geq 1} y_i \leq \varphi^{-1}\{n^{-1}\sum_{i\geq 1}\varphi(y_i)\}$$

and the result follows by their sum. $\square$

Let $\mathcal{F}_1(I)$ be the set of strictly increasing real functions φ defined on a subset I of $\mathbb{R}$ and such that the function

$$\psi_\varphi(u,v) = \varphi(\varphi^{-1}(u) + \varphi^{-1}(v))$$

is concave on $\varphi(I) \times \varphi(I)$ and let $\mathcal{F}_2(I)$ be the set of strictly increasing real functions φ defined on a subset I of $\mathbb{R}$ and such that the function ψ_φ is convex on $\varphi(I) \times \varphi(I)$.

Theorem 2.8. *Let I be a subset of $\mathbb{R}$. For every function φ of $\mathcal{F}_1(I)$ and for all random variables X and Y on $(\Omega, \mathcal{A}, P)$, with values in I and such that $\varphi(X)$ and $\varphi(Y)$ are $L_1(P)$*

$$\varphi^{-1}\{E\varphi(X+Y)\} \leq \varphi^{-1}\{E\varphi(X)\} + \varphi^{-1}\{E\varphi(Y)\}. \tag{2.21}$$

For every function φ of $\mathcal{F}_2(I)$ and for all random variables X and Y on $(\Omega, \mathcal{A}, P)$, with values in I and such that $\varphi(X)$ and $\varphi(Y)$ are $L_1(P)$

$$\varphi^{-1}\{E\varphi(X+Y)\} \geq \varphi^{-1}\{E\varphi(X)\} + \varphi^{-1}\{E\varphi(Y)\}. \tag{2.22}$$

Proof. Jensen's inequality for a concave function ψ implies that for every variable (U,V), $E\psi(U,V) \leq \psi(EU, EV)$. By the monotone change of variables $U = \varphi(X)$ and $V = \varphi(Y)$, this is equivalent to

$$E\varphi(X+Y) \leq \varphi[\varphi^{-1}\{E\varphi(X)\} + \varphi^{-1}\{E\varphi(Y)\}]$$

for an increasing function φ and the result follows by the monotonicity of the function φ^{-1}. These inequalities are reversed under the assumptions of an increasing function φ and a convex function ψ. $\square$

If φ is a decreasing function, the inequalities are inverted. In Minkowski's inequality, the function $\psi(u,v) = (u^{\frac{1}{p}} + v^{\frac{1}{p}})^p$ is increasing and concave as proved by Neveu (1970).

The first two derivatives of the function ψ are

$$\psi_u^{(1)}(u,v) = \frac{\varphi^{(1)}(\varphi^{-1}(u) + \varphi^{-1}(v))}{\varphi^{(1)} \circ \varphi^{-1}(u)},$$

$$\psi_{u,u}^{(2)}(u,v) = \frac{\varphi^{(2)}(\varphi^{-1}(u) + \varphi^{-1}(v))}{\varphi^{(1)2} \circ \varphi^{-1}(u)}$$

$$- \varphi^{(1)}(\varphi^{-1}(u) + \varphi^{-1}(v)) \frac{\varphi_u^{(2)}}{\varphi^{(1)3}}(\varphi^{-1}(u)),$$

$$\psi_{u,v}^{(2)}(u,v) = \frac{\varphi^{(2)}(\varphi^{-1}(u) + \varphi^{-1}(v))}{\varphi^{(1)} \circ \varphi^{-1}(u)\varphi^{(1)} \circ \varphi^{-1}(v)}$$

and the signs of $\psi_{u,v}^{(2)}(u,v)$ and $\psi_{u,u}^{(2)}(u,v)$ may be different. They cannot be determined from general inequalities between the derivatives of the function φ at a single point.

Example 2.5. The function $\psi(u,v) = \log(e^u + e^v)$ related to the logarithm function φ has the derivatives

$$\psi_u^{(1)}(u,v) = \frac{e^u}{e^u + e^v}, \quad \psi_{u,v}^{(2)}(u,v) = -\frac{e^{u+v}}{(e^u + e^v)^2}, \quad u \neq v,$$

it is concave for real numbers $u \neq v$. Theorem 2.8 implies

$$\exp\{E\log(X+Y)\} \leq \exp(E\log X) + \exp(E\log Y).$$

By the change of variables $X = e^U$ and $Y = e^V$, this is equivalent to

$$E\{\log(e^U + e^V)\} \leq \log(e^{EU} + e^{EV})$$

and this inequality is also a direct consequence of the concavity of the function $\psi(u,v)$.

Example 2.6. With discrete variables X and Y, Theorem 2.8 is written in the form

$$\varphi^{-1}\{\sum_i \varphi(x_i + y_i)\mu_i\} \leq \varphi^{-1}\{\sum_i \varphi(x_i)\mu_{X,i}\} + \varphi^{-1}\{\sum_i \varphi(y_i)\mu_{Y,i}\}$$

for every function φ of $\mathcal{F}_1$, where μ_i is the probabilities that $X = x_i$ and $Y = y_i$, $\mu_{X,i}$ and $\mu_{Y,i}$ are the probabilities that $X = x_i$ and, respectively, $Y = y_i$.

Let $X_1, \ldots, X_n$ be strictly positive and identically distributed random variables such that $\log X_1$ is integrable and let $\varphi(x) = \log x$. By concavity of the logarithm,

$$E\{\log(n^{-1}S_n)\} \geq n^{-1}\sum_{i=1}^{n} E(\log X_i) = E(\log X_1).$$

which entails

$$\exp\{E(\log S_n)\} \geq n \exp\{E(\log X_1)\}.$$

This inequality and the extension of the inequality of Example 2.5 to a sum of positive and identically distributed variables imply

$$\exp\{E(\log S_n)\} = n \exp\{E(\log X_1)\}.$$

Theorem 2.8 extends to sums of n variables.

Theorem 2.9. *Let* $X_1, \ldots, X_n$ *be random variables on a probability space* $(\Omega, \mathcal{A}, P)$. *For every real function* φ *of* $\mathcal{F}_1(\mathbb{R})$ *and such that the variables* $\varphi(X_i)$ *belong to* L_1 *for* $i = 1, \ldots, n$

$$\varphi^{-1}\{E\varphi(\sum_{i=1}^{n} X_i)\} \leq \sum_{i=1}^{n} \varphi^{-1}\{E\varphi(X_i)\}.$$

For every real function φ *of* $\mathcal{F}_2(\mathbb{R})$ *and such that the variables* $\varphi(X_i)$ *belong to* L_1 *for* $i = 1, \ldots, n$

$$\varphi^{-1}\{E\varphi(\sum_{i=1}^{n} X_i)\} \geq \sum_{i=1}^{n} \varphi^{-1}\{E\varphi(X_i)\}.$$

For independent and identically distributed variables $X_1, \ldots, X_n$ and a convex function φ, the inequality of Theorem 2.9 is written

$$\varphi^{-1}\{n^{-1}E\varphi(S_n)\} \leq \varphi^{-1}\{E\varphi(X_1)\}. \tag{2.23}$$

Example 2.7. Let X be a uniform variable on $[0, 1]$, it satisfies

$$E|\log X| = -\int_0^1 \log x \, dx = 2.$$

Let a be in $[0, 1]$ and let $X_1, \ldots, X_n$ be independent and identically distributed variables with the same distribution as X. Since $\log(n^{-1}S_n) < 0$

$$E|\log(n^{-1}S_n)| = -E\log(n^{-1}S_n) \leq 2$$

and inequality (1.12) is written

$$P(n^{-1}S_n > a) = P(|\log(n^{-1}S_n| < |\log a|)$$
$$\geq 1 - (|\log a|)^{-1}E|\log(n^{-1}S_n)|$$
$$\geq 1 - 2(|\log a|)^{-1},$$
$$P(n^{-1}S_n < a) = P(|\log(n^{-1}S_n)| > |\log a|) \leq 2(|\log a|)^{-1}.$$

2.7   Inequalities for the Laplace transform

The Laplace transform of a variable X is defined as $L_X(t) = Ee^{tX}$ and it is also the Laplace transform of the probability density of X. Let H be a real distribution on $\mathbb{R}$ and let U be a uniform variable on $[0, 1]$. The Laplace transform of the variable $X = H^{-1}(U)$, with distribution function H, is

$$\varphi_X(\lambda) = \int_0^1 e^{\lambda H^{-1}(u)}\, du = \int_{\mathbb{R}} e^{\lambda x}\, dH(x).$$

Let X be a symmetric variable, the odd moments EX^{2n+1} of X are zero, hence $L_X(t) = L_X(-t)$ and the odd derivatives of the Laplace transforms of X are zero. For every variable X, the derivatives of the Laplace transform L_X satisfy

$$L_X^{(k)}(t) \le \|X\|_{2k}^k L_X(2t), \; k \ge 1.$$

By convexity of the exponential function, L_X is a convex function and $L_X(t) \ge e^{tEX}$ therefore $L_X(t) \ge 1$ for every centered variable X. By the same argument, for every convex function g

$$Lg(X)(t) \ge e^{tEg(X)}.$$

The logarithm of $l_X(t)$ of $L_X(t)$ has the second derivative

$$
\begin{aligned}
l_X^{(2)}(t) &= \frac{L_X^{(2)}(t)L_X(t) - L_X^{(1)2}(t)}{L_X^2(t)} \\
&= \frac{E\{X^2 e^{tEX}\}E\{e^{tEX}\} - (E\{X e^{tEX}\})^2}{(E\{e^{tEX}\})^2} \ge 0
\end{aligned}
$$

by the Bienaymé-Chebychev inequality (1.12), therefore $\log L_X$ is a convex function. The following bounds are consequences of the convexity of the functions exponential and φ.

Proposition 2.15. *Let X be a random variable with values in a bounded interval $[a, b]$, then the Laplace transform of X satisfies*

$$\frac{L_X(t) - e^{ta}}{EX - a} \le \frac{e^{tb} - e^{ta}}{b - a} \le \frac{e^{tb} - L_X(t)}{b - EX}.$$

Proposition 2.16. *Let X be a random variable with values in a bounded interval $[a, b]$ and φ be a convex function defined on $[a, b]$, then*

$$Ee^{\varphi(X)} \le \frac{EX - a}{b - a} e^{\varphi(b)} + \frac{b - EX}{b - a} e^{\varphi(a)}$$

and the Laplace transform of X satisfies $L_X(t) \le e^{\frac{1}{8}t^2(b-a)^2}$.

Proposition 2.17. *The Laplace transform of a variable X satisfies*
$$t^{-1}L_X(t)\log L_X(t) \leq L'_X(t) \leq \{L''_X(0)L_X(2t)\}^{\frac{1}{2}}.$$

Proof. Let $Y = e^{tX}$, the first inequality is due to the convexity of the function $y\log y$, for $y > 0$, and to Jensen's inequality. □

The independence of the components of a vector $(X_1, \ldots, X_n)$ is equivalent to the factorization of their Laplace transform and their sum satisfies $L_{\sum_{i=1}^n X_i}(t) = \prod_{i=1}^n L_{X_i}(t)$, for any real t. Theorem 2.8 allows us to write a reciprocal inequality for the Laplace transform of a sum of variables.

Proposition 2.18. *Let (X, Y) be a real random variable on a probability space $(\Omega, \mathcal{A}, P)$, then their Laplace transforms satisfy*
$$L_{X+Y}(x) \geq L_X(x)L_Y(x), \quad x \geq 0,$$
with equality if and only if X and Y are independent.

Proof. The function $\varphi_\lambda(x) = e^{\lambda x}$ has the inverse $\varphi_\lambda^{-1}(x) = \lambda^{-1}\log u$, then the second derivative of the function $\psi_\alpha(u, v) = uv$ of Theorem 2.8 are $\psi_{u,v}^{(2)}(u, v) = 1$ and $\psi_{u,u}^{(2)}(u, v) = 0$ for every $0 < u \neq v$, so that ψ belongs to $\mathcal{F}_2(\mathbb{R}_+)$ and the result is the inequality (2.22). □

Example 2.8. Let X and Y be exponential variables $\mathcal{E}_\theta$, $\theta > 0$, then for every $x \geq 0$
$$\frac{\theta}{\theta - 2t} = L_{2X}(x) \geq L_X(x)L_Y(x) = \frac{\theta^2}{(\theta - t)^2}.$$
Writing the variables $X = \varphi_\alpha^{-1}(U) = \alpha^{-1}\log U$ and $Y = \alpha^{-1}\log V$, for positive variables U and V, the inequality of Proposition 2.19 is equivalent to
$$EUV \geq (EU)(EV) \tag{2.24}$$
which completes the Hölder inequality. It implies that the covariance of positive variables is always positive. For example, let $U = \cos\theta$ and $V = \sin\theta$, with a uniform variable θ on the inerval $[0, \frac{\pi}{2}]$, then $EUV = \frac{1}{2}E\sin(2\theta) = \frac{2}{\pi}$ while $EU = \frac{2}{\pi}$ and $EV = 0$.

Let X be a random variable on $[0, 1]$, Chernov's theorem implies that for every integer $n \geq 1$ and for every $a > 0$
$$\inf_{t>0}\{\log L_X(t) - at\} \geq 1 - (|\log a|)^{-1}E(|\log X|).$$

Proposition 2.19. *Let X be a random variable such that $E\log X$ is finite and strictly positive. For every $a > 1$, the Laplace transform of X satisfies*
$$\inf_{t>0}\{\log L_X(t) - at\} \leq \frac{E(\log X)}{\log a}.$$

Proof. Equation (2.23) implies that for a sum of n indepedent variables having the same distribution as X, $0 < E \log(n^{-1}S_n) \leq \log EX$ and the result is obtained from the large deviations theorem for every integer $n \geq 1$

$$P(n^{-1}S_n > a) = \inf_{t>0}\{\log L_X(t) - at\}$$

$$= P(\log(n^{-1}S_n) > \log a) \leq \frac{E\{\log(n^{-1}S_n)\}}{\log a}. \qquad \square$$

2.8 Inequalities for multivariate functions

Let F and G be the primitives of real and positive functions f and g on $\mathbb{R}_+$. The product inequality (2.6) is generalized by transforms. Let Φ and Ψ be convex functions on $\mathbb{R}_+$, let u be a positive density on $\mathbb{R}_+^2$ and let U be defined on $\mathbb{R}_+^2$ like (2.15), as

$$U(x,y) = \int_x^\infty \int_y^\infty s^{-1}t^{-1}u(s,t)\,ds\,dt,$$

then

$$\int_0^\infty \int_0^\infty \Phi(\frac{F(x)}{x})\Psi(\frac{G(y)}{y}) u(x,y)\,dx\,dy$$

$$\leq \int_0^\infty \int_0^\infty \{\int_0^x \int_0^y \Phi(f(s))\Psi(g(t))\,ds\,dt\} U(dx,dy)$$

$$\leq \int_0^\infty \int_0^\infty \Phi(f(x))\Psi(g(y))\, U(x,y)\,dx\,dy.$$

Extending Proposition 2.7, there exists a constant $k_u = \|U\|_2$ such that for every integer $p > 1$

$$|\int_0^\infty \int_0^\infty \Phi(\frac{F(x)}{x})\Psi(\frac{G(y)}{y}) u(x,y)\,dx\,dy| \leq k_u \|\Phi(f)\|_p \|\Psi(g)\|_{p'},$$

where $p^{-1} + p'^{-1} = \frac{1}{2}$. With exponential functions Φ and Ψ, this inequality becomes

$$\int_0^\infty \int_0^\infty \exp\{\frac{\int_0^x \log f}{x} + \frac{\int_0^x \log g}{y}\} u(x,y)\,dx\,dy \leq k_u \|f\|_p \|g\|_{p'}$$

for all functions f of L_p and g of $L_{p'}$ such that $p^{-1} + p'^{-1} = \frac{1}{2}$. This result is directly extended to a convex functions Φ on $\mathbb{R}_+^2$

$$\int_0^\infty \int_0^\infty \Phi(\frac{F(x)}{x}, \frac{G(y)}{y}) u(x,y)\,dx\,dy$$

$$\leq \int_0^\infty \int_0^\infty \{\int_0^x \int_0^y \Phi(f(s), g(t))\,ds\,dt\} U(dx,dy)$$

$$\leq \int_0^\infty \int_0^\infty \Phi(f(x), g(y))\, U(x,y)\, dx\, dy. \tag{2.25}$$

If the bounds in the integrals are modified, they are normalized by the surface of the domain of integration. Let f and g be defined on $[a, \infty[$ and, respectively, $[b, \infty[$ and let $U_{ab}(x,y) = \int_x^\infty \int_y^\infty (s-a)^{-1}(t-b)^{-1} u(s,t)\, ds\, dt$, for $x > a$ and $y > b$, then

$$\int_0^\infty \int_0^\infty \Phi\left(\frac{\int_a^x f}{x-a}\right)\Psi\left(\frac{\int_b^y g}{y-b}\right) u(x,y)\, dx\, dy$$

$$\leq \int_0^\infty \int_0^\infty \left\{\int_a^x \int_b^y \Phi(f(s))\Psi(g(t))\, ds\, dt\right\} U_{ab}(dx, dy)$$

$$\leq \int_0^\infty \int_0^\infty \Phi(f(x))\Psi(g(y))\, U_{ab}(x,y)\, dx\, dy.$$

In particular

$$\left| \int_0^\infty \int_0^\infty \Phi\left(\frac{\int_a^x f}{x-a}\right)\Psi\left(\frac{\int_b^y g}{y-b}\right) u(x,y)\, dx\, dy \right| \leq \|U_{ab}\|_2 \|\Phi(f)\|_p \|\Psi(g)\|_{p'},$$

for all p ans p' such that $p^{-1} + p'^{-1} = \frac{1}{2}$, and

$$\left| \int_0^\infty \int_0^\infty \Phi\left(\frac{\int_a^x f}{x-a}, \frac{\int_b^y g}{y-b}\right) u(x,y)\, dx\, dy \right| \leq \|U_{ab}\|_p \|\Phi(f,g)\|_{p'}$$

for all conjugate integers p ans p'.

As an application, let S and T be positive real random variables with respective distribution functions H_S and H_T, and let $X_S = H_S(S)$ and $X_T = H_T(T)$ be the uniform tranformed variables. Let a and b be functions such that $\int a\, dH_S = 0$ and $\int b\, dH_T = 0$. The weighted mean inequality (2.25) for the integral

$$E\left\{\Phi\left(\frac{\int_0^S a\, dH_S}{X_S}, \frac{\int_0^T b\, dH_T}{X_T}\right) u(X_S, X_T)\right\}$$

$$= E\left\{\Phi\left(\frac{\int_S^\infty a\, dH_S}{X_S}, \frac{\int_T^\infty b\, dH_T}{X_T}\right) u(X_S, X_T)\right\}$$

is also written as

$$E\left\{\Phi\left(\frac{\int_0^S a\, dH_S}{X_S}, \frac{\int_0^T b\, dH_T}{X_T}\right) u(X_S, X_T)\right\}$$

$$\leq E\{\Phi(a(X_S), b(X_T))\, U(X_S, X_T)\}.$$

Theorem 2.7 gives an inequality without weighting function, it is extended to a convex and increasing function $\Phi : \mathbb{R}_+^2 \mapsto \mathbb{R}$

$$\int_0^\infty \Phi\left(\frac{F(x)}{x}, \frac{G(x)}{x}\right) d\mu(x) \leq \int_0^\infty \Phi \circ (f,g)(x)\, d\mu(x),$$

more generally

$$\int_0^\infty \int_0^\infty \Phi(\frac{F(x)}{x}, \frac{G(y)}{y})\, \mu(dx, dy)$$

$$\leq \int_0^\infty \int_0^\infty \Phi(f(x), g(y))\, \mu(dx, dy), \qquad (2.26)$$

or

$$\int_0^\infty \int_0^\infty \Phi(\frac{\int_a^x f}{x-a}, \frac{\int_b^y g}{y-b})\, \mu(dx, dy) \leq \int_0^\infty \int_0^\infty \Phi \circ (f(x), g(y))\, \mu(dx, dy)$$

for functions f defined on $[a, \infty[$ and g defined on $[b, \infty[$. These inequalities differ from the generalized Hardy inequality (2.25) by the integration measures. They are extended to functions $f : [a, \infty[\mapsto \mathbb{R}^n$, where a belongs to $\mathbb{R}$ or $\mathbb{R}^n$, for every integer n. Let (X, Y) be a random variable with a distribution function μ on $\mathbb{R}^2$, Equation (2.26) is also written as

$$E\{\Phi(\frac{F(X)}{X}, \frac{G(Y)}{Y})\} \leq E\{\Phi(f(X), g(Y))\}$$

for every convex function Φ.

Carleman's inequality is also extended to an inequality for a multivariate function in a similar form as (2.18) but without weighting functions. Let $\Phi : \mathbb{R}^2 \mapsto \mathbb{R}^2$ be a convex and monotone function and let f and g be positive real functions defined on $\mathbb{R}$

$$\int_{\mathbb{R}^2} \Phi\{\frac{\int_a^x \int_b^y \Phi^{-1}(f(s), g(t))\, ds\, dt}{(x-a)(y-b)}\}\, dx\, dy \leq \int_a^\infty \int_b^\infty (f(x), g(y))\, dx\, dy. \tag{2.27}$$

Let $(X_i, Y_i)_{i=1,\ldots,n}$ be a sequence of independent random variables on $\mathbb{R}^2$, with respective distribution functions F_{X_i, Y_i} and denote their mean distribution function $F_{n,X,Y} = n^{-1} \sum_{i=1}^n F_{X_i, Y_i}$. Inequality (2.26) implies

$$E\Phi(\bar{X}_n, \bar{Y}_n) \leq E \int_{\mathbb{R}^2} \Phi(x, y)\, F_{n,X,Y}(dx, dy).$$

With identically distributed variables, $E\Phi(\bar{X}_n, \bar{Y}_n) \leq E\Phi(X_1, Y_1)$. As a special case

$$E\{(\bar{X}_n - E\bar{X}_n)^2 + (\bar{Y}_n - E\bar{Y}_n)^2\}^{\frac{1}{2}}$$

$$\leq n^{-1} \sum_{i=1}^n E\{(X_i - EX_i)^2 + (Y_i - EY_i)^2\}^{\frac{1}{2}}.$$

In Section 1.5, a mean integral was defined on balls of $\mathbb{R}^d$, a similar result holds on rectangles or other convex and connected domains of $\mathbb{R}^d$.

By convexity, for every x in $\mathbb{R}^d$ and for every function $f : \mathbb{R}^d \mapsto \mathbb{R}$

$$I_{p,2} = \int_{[0,\infty[} \left(\frac{|\int_{[0,x]} f(t)\, dt|}{\prod_{i=1}^d x_i} \right)^p \frac{dx}{\prod_{i=1}^d x_i}$$

$$\leq \int_{[0,\infty[} \int_{[0,x]} |f(t)|^p\, dt \frac{\prod_{i=1}^d dx_i}{\prod_{i=1}^d x_i^2}$$

$$\leq \int_{[0,\infty[} |f(t)|^p \frac{\prod_{i=1}^d dt_i}{\prod_{i=1}^d t_i}.$$

With Hardy's inequality, an integration by parts allows us to write the integral of the $(\prod_{i=1}^d x_i)^{-1} \int_{[0,x]} f(t)^p\, dt$ with respect to the Lebesgues measure on $\mathbb{R}^d$, for a positive function f, as

$$I_{p,1} = \frac{p}{p-1} \int_{[0,\infty[} \left(\frac{\int_{[0,x]} f(t)\, dt}{\prod_{i=1}^d x_i} \right)^{p-1} f(x)\, dx$$

$$I_{p,1} \leq (\frac{p}{p-1})^p \int_{[0,\infty[} f^p(x)\, dx.$$

These inequalities are generalized to integrals with respect to a positive measure μ on $\mathbb{R}^d$

$$I_{p,\mu} = \int_{[0,\infty[} \left(\frac{|\int_{[0,x]} f(t)\, d\mu(t)|}{\prod_{i=1}^d x_i} \right)^p d\mu(x)$$

$$\leq \int_{[0,\infty[} \int_{[0,x]} |f(t)|^p\, dt \frac{d\mu(x)}{\prod_{i=1}^d x_i}$$

$$\leq \int_{[0,\infty[} |f(t)|^p \int_{[t,\infty[} \frac{d\mu(x)}{\prod_{i=1}^d x_i}\, d\mu(t).$$

Let $X = (X_1, \ldots, X_d)$ be a variable with distribution μ on $\mathbb{R}_+^d$, this inequality is equivalently written

$$E\{ (\frac{F(X)}{\prod_{i=1}^d X_i})^p \} \leq E\{ |f(X)|^p \int_{[X,\infty[} \frac{d\mu(x)}{\prod_{i=1}^d x_i} \}.$$

Theorem 2.6 extends to a kernel k defined from $\mathbb{R}_+^{2d}$ to $\mathbb{R}_+$ which allows to restrict the integral on a subset of $\mathbb{R}_+^{2d}$. Let $K(x) = \int_{[0,x]} k(x,t)\, dt$ for x in $\mathbb{R}_+^d$. Let $u \geq 0$ and $v \geq 0$ be real functions defined from $\mathbb{R}_+^d$, with

$$v(x) = x \int_{[x,\infty[} \frac{k(s,x)}{K(s)} \frac{u(s)}{s}\, ds.$$

Theorem 2.10. *For every convex and real function φ on $\mathbb{R}_+^d$ and for every*
$p > 1$

$$\int_{[0,\infty[^d} \phi\left(\frac{1}{K(x)}\int_{[0,x]} k(x,t)f(t)\,dt\right)\frac{u(x)}{x}\,dx \le \int_{[0,\infty[^d} \phi \circ f(x)\frac{v(x)}{x}\,dx.$$

Replacing ϕ by its L_p norm in this inequality entails

$$\{\int_{[0,\infty[^d} \phi^p\left(\frac{1}{K(x)}\int_{[0,x]} k(x,t)f(t)\,dt\right)\frac{u(x)}{x}\,dx\}^{\frac{1}{p}}$$

$$\le \{\int_{[0,\infty[^d} \phi^p \circ f(x)\frac{v(x)}{x}\,dx\}^{\frac{1}{p}}.$$

For integrals on the ball $B_r(x)$ centered at x in $\mathbb{R}_+^d$, the Lebesgue measure
of $B_r(x)$ is a constant proportional to r^d which does not depend on x and

$$I_p(r) = \int_{[r,\infty[^d}\left(\frac{1}{\lambda_d(B_r)}\int_{B_r(x)} |f(t)|\,dt\right)^p dx \le \int_{[r,\infty[^d} |f(t)|^p\,dt,$$

since $\int_{[r,\infty[^d} 1_{B_r(x)}(t)\,dx = \lambda_d(D_r(t)) - \lambda_d(B_r)$. The same result is true for
the integration on every convex and connex set having a Lebesgue measure
independent of x.

Carleman's inequality on balls becomes

$$\int_{[r,\infty[^d} \exp\left(\frac{1}{\lambda_d(B_r)}\int_{B_r(x)} \log f(t)\,dt\right)dx \le \int_{[r,\infty[^d} f(x)\,dx$$

and on rectangular sets $[0,x]$ it is

$$\int_{[0,\infty[} \exp\left(\frac{\int_{[0,x]} \log f(t)\,dt}{\prod_{i=1}^d x_i}\right)d\mu(x) \le \int_{[0,\infty[} f(t)\,dt \int_{[t,\infty[} (\prod_{i=1}^d x_i)^{-1}\,d\mu(x)$$

and for every increasing and convex function ϕ

$$\int_{[r,\infty[^d} \phi(\frac{1}{\lambda_d(B_r)}\int_{B_r(x)} \phi^{-1}\circ f(t)\,dt)\,dx \le \int_{[r,\infty[^d} f(x)\,dx,$$

$$\int_{[0,\infty[} \phi\left(\frac{\int_{[0,x]} \phi^{-1}\circ f(t)\,dt}{\prod_{i=1}^d x_i}\right)d\mu(x) \le \int_{[0,\infty[} f(t)\,dt \int_{[t,\infty[} (\prod_{i=1}^d x_i)^{-1}\,d\mu(x).$$

Chapter 3

Analytic Inequalities

3.1 Introduction

Many inequalities rely on Taylor expansions of functions in series such as
the trigonometric series

$$\cos(x) = 1 + \sum_{k=1}^{\infty}(-1)^k \frac{x^{2k}}{(2k)!}, \quad \sin(x) = \sum_{k=0}^{\infty}(-1)^k \frac{x^{2k+1}}{(2k+1)!},$$

the expansions of the exponential and the logarithm functions, with

$$\log x = 2\arg \mathrm{th}(\frac{x-1}{x+1}) = 2\int_{-1}^{\frac{x-1}{x+1}} \frac{dy}{1-y^2}$$

$$= 2\sum_{k=0}^{\infty} \frac{1}{2k+1}\{(\frac{x-1}{x+1})^{2k+1} + 1\}, \ x > 0.$$

Numerical tables of these functions and many other functions have been
published during the 17th and 18th centuries (Hutton, 1811). Expansions in
a series provide a simple method for calculating approximations of constants
or functions by their partial sums. For example, $e = \sum_{k=0}^{\infty}(n!)^{-1}$ and it is
also expressed by de Moivre-Stirling's formula $(n^{-1}e)^n \sim \sqrt{2\pi}(n!)^{-1}$ as n
tends to infinity. The number π is expanded as

$$\pi = 4\arctan 1 = 4\sum_{k=0}^{\infty} \frac{(-1)^k}{2k+1}.$$

Approximations of the hyperbolic and trigonometric functions are easily
obtained. Inequalities for partial sums in the expansions have been considered for the evaluation of the approximation errors and they can generally
be proved by induction. Adler and Taylor (2007) presented expansions in a
series for the probability $P(\sup_{t \in A} f(t) \geq u)$ for a general parameter space

59

A and where P is a Gaussian probability distribution and f a function of $C_n(A)$.

For every integer $n \geq 1$ and for every $0 < x < y$, the ratio of $x^n - y^n$ and $x - y$ is expanded as the finite sum $\sum_{i=0}^{n-1} y^k x^{n-k-1}$ which provides the bounds

$$\frac{1}{ny^{n-1}} \leq \frac{x - y}{x^n - y^n} \leq \frac{1}{nx^{n-1}}$$

and for vectors $x = (x_1, \ldots, x_n)$ and $y = (y_1, \ldots, y_n)$ such that $0 < x_k < y_k$ for $k = 1, \ldots, n$

$$n^{-n} \prod_{k=1}^{n} y_k^{1-n} \leq \frac{\prod_{k=1}^{n}(x_k - y_k)}{\prod_{k=1}^{n}(x_k^n - y_k^n)} \leq n^{-n} \prod_{k=1}^{n} x_k^{1-n}.$$

Inequalities for the partial sum A_n of the first $n + 1$ terms of the Taylor expansion of a function have been considered. For the exponential function, Alzer (1990b) stated that

$$(n + 1)(n + 2)^{-1} A_n^2(x) < A_{n-1}(x) A_{n+1}(x) < A_n^2(x),$$

this inequality is improved in Section 3.2. The existence of constants such that

$$c_n \left(\frac{A_n(x)}{A_{n+1}(x)} \right)^2 < \frac{A_{n-1}(x)}{A_{n+1}(x)} < C_n \left(\frac{A_n(x)}{A_{n+1}(x)} \right)^2$$

is established in Section 3.2 for the partial sums of other functions, sufficient conditions for $C_n = 1$ are given in Proposition 3.1.

Cauchy's inequalities for two real sequences $(x_i)_{i=1,\ldots,n}$ and $(a_i)_{i=1,\ldots,n}$ are illustrated by several cases in Section 3.3 and they are applied to the comparison of the geometric and arithmetic means, generalizing other inequalities proved by Alzer (1990a). Section 3.4 provides inequalities for the mean, the modes and median of random variables. Section 3.5 deals with other specific points of curves related to the Mean Value Theorem and solutions of implicit functional equations are established. Carlson's inequality is an inequality of the same kind for the logarithmic mean function and it is due to a convexity argument. In Section 3.6 it is generalized to concave or convex functions. Inequalities for the functional means of power functions are then proved.

3.2 Bounds for series

Proposition 3.1. *Let $(a_n)_{n\geq 1}$ be strictly positive and strictly decreasing sequence of real functions and let $A_n = \sum_{k=0}^{n} a_k$, then for every integer n*

$$A_n^2 > A_{n-1}A_{n+1}. \qquad (3.1)$$

The same inequality holds for every strictly negative and decreasing sequence of functions $(a_n)_{n\geq 1}$.

Proof. For every integer n

$$A_n^2 - A_{n-1}A_{n+1} = a_n A_n - a_{n+1}A_{n-1}$$
$$> a_{n+1}(A_n - A_{n-1}) > 0. \qquad \square$$

Example 3.1. The power function $f_n(x) = (1+x)^n$, $x > 0$ and $n > 0$, develops as a series $A_n(x) = \sum_{k=0}^{n} a_k(x)$ with $a_k(x) = C_n^k x^k$, hence the sequence $(a_k(x))_{\frac{n}{2}\leq k\leq n}$ is decreasing on $]0,1[$. Therefore, the inequality (3.1) is fulfilled on $]0,1[$, for $A_n - A_{\frac{n}{2}}$ if n is even and for $A_n - A_{\frac{n-1}{2}}$ if n is odd.

Example 3.2. The logarithm function $\log(1-x)$ on $]0,1[$ has an expansion with functions $a_n(x) = n^{-1}x^n$ satisfying the conditions of the proposition. The logarithm function $f(x) = \log(1+x)$, $x > -1$, has an expansion $A_n(x) = \sum_{k=1}^{n}(-1)^{k+1}k^{-1}x^k = A_{1n}(x) - A_{2n}(x)$, where

$$A_{1n}(x) = \sum_{k=0}^{[\frac{n}{2}]} \frac{x^{2k+1}}{2k+1}, \quad A_{2n}(x) = \sum_{k=1}^{[\frac{n-1}{2}]} \frac{x^{2k}}{2k}.$$

For x in $]0,1[$, $A_{1n}(x)$ and $A_{2n}(x)$ are strictly decreasing and they satisfy the inequality (3.1) but not A_n.

The inequality (3.1) also holds for non decreasing sequences of functions, therefore it does not characterize a class of functions.

Example 3.3. Let $A_n(x) = \sum_{k=0}^{n} x^k = (1 - x^{n+1})(1-x)^{-1}$, for $x > 0$, $x \neq 1$, and for every integer $n \geq 1$. The sequence $(x^n)_{n\geq 1}$ is increasing for $x > 1$ and decreasing for $0 < x < 1$. Obviously, $n < A_n(x) < nx^n$ if $x > 1$ and $nx^n < A_n(x) < n$ if belongs to $]0,1[$. For every $x > 0$, $x \neq 1$, $A_n^2(x) - A_{n-1}(x)A_{n+1}(x) = x^{n-1}(x-1)^2$ is strictly positive. Reversely, $A_{n-1}A_{n+1} > c_n A_n^2$ for some function $c_n(x)$ with values in $]0,1[$ if

$$c_n(x) < 1 - x\frac{(1-x)^2}{(1-x^{n+1})}.$$

If $x > 1$, $(n+1)(x-1) < x^{n+1} - 1 < (n+1)x^{n+1}(x-1)$ and

$$1 - \frac{x}{(n+1)^2} < c_n(x) < 1 - \frac{1}{x^{2n+1}(n+1)^2},$$

if $0 < x < 1$, $(n+1)x^{n+1}(x-1) < x^{n+1} - 1 < (n+1)(x-1)$ and

$$1 - \frac{1}{x^{2n+1}(n+1)^2} < c_n(x) < 1 - \frac{x}{(n+1)^2}.$$

The exponential function develops as a series $e^x = A_n(x) + R_n(x)$ with

$$A_n(x) = \sum_{k=0}^{n} \frac{x^k}{k!} = A_{n-1}(x) + a_n(x)$$

and the sequence $(a_n(x))_{n \geq 1}$ is not decreasing for every $x > 0$.

Theorem 3.1. *The expansion of the exponential function satisfies*

$$\frac{A_n^2(x)}{2(n+1)} < A_{n-1}(x)A_{n+1}(x) < A_n^2(x)$$

for every real $x > 0$ and for every integer $n \geq 1$.

Proof. The upper bound is true for every x and for $n = 1, 2$. Let $n > 2$ be an integer, for every $x > 0$

$$(A_{n+1}^2 - A_n A_{n+2})(x) = (a_{n+1}A_{n+1} - a_{n+2}A_n)(x)$$

$$= a_{n+1}(x)\{\sum_{k=0}^{n+1} \frac{x^k}{k!} - \frac{1}{n+2}\sum_{k=1}^{n+1} \frac{x^k}{(k-1)!}\}$$

$$= a_{n+1}(x)\{1 + \sum_{k=1}^{n+1} \frac{x^k}{(k-1)!}(\frac{1}{k} - \frac{1}{n+2})\} > 0.$$

For the lower bound, let $c_n = (n+1)^{-1}k_n > 0$ with $k_n < 1$

$$(A_{n-1}A_{n+1} - c_n A_n^2)(x) = \{(1 - c_n)A_n A_{n-1} + a_{n+1}A_{n-1} - c_n a_n A_n\}(x),$$

$$(a_{n+1}A_{n-1} - c_n a_n A_n)(x) = a_n(x)\{\sum_{k=1}^{n} \frac{x^k}{(k-1)!}(\frac{1}{n+1} - \frac{c_n}{k}) - c_n\},$$

therefore $A_{n-1}A_{n+1} - c_n A_n^2 > 0$ as $c_n < (n+1)^{-1}$ if $(1 - c_n)A_n A_{n-1} - a_n c_n \geq 0$. Let $c_n = \{2(n+1)\}^{-1}$, then $A_n A_{n-1} > a_n$ and this condition is satisfied. $\square$

The bounds of Theorem 3.1 for the expansion of the exponential function have the best constant uniformly in n.

The variations between $A_{n-1}A_{n+1} - A_n^2$ have the bounds

$$\frac{A_n^2(x)}{2(n+1)} - A_n^2(y) \leq A_{n-1}(x)A_{n+1}(x) - A_{n-1}(y)A_{n+1}(y)$$

$$\leq A_n^2(x) - \frac{A_n^2(y)}{2(n+1)},$$

and the difference

$$\Delta_n(x,y) = A_{n-1}(x)A_{n+1}(x) - A_{n-1}(y)A_{n+1}(y) - A_n^2(x) - A_n^2(y)$$

satisfies

$$-A_n^2(x)\frac{2n+1}{2(n+1)} \leq \Delta_n(x,y) \leq A_n^2(y)\frac{2n+1}{2(n+1)}.$$

Let $A_n(x)$ be the partial sum of order n in the Taylor expansion of a real function f of $C^{n+1}(\mathbb{R})$ in a neighbourhood of x_0

$$A_n(x) = \sum_{k=0}^{n} \frac{x^k f^{(k)}(x)}{k!},$$

$$A_n^2(x) = f^2(x_0) + \sum_{k=1}^{\frac{n}{2}}(x-x_0)^{2k}\frac{f^{(k)2}(x_0)}{(k!)^2}$$

$$+ 2\sum_{1\leq j<m\leq n}(x-x_0)^{j+m}\frac{f^{(j)}(x_0)f^{(m)}(x_0)}{j!m!}$$

and a similar expansion can be written for $A_{n-1}A_{n+1}$, which allows to bound the difference $A_{n-1}A_{n+1} - A_n^2$.

For every sum $A_n(x) = \sum_{k=0}^{n} a_k(x)$

$$\frac{A_{n-1}A_{n+1}}{A_n^2} = 1 - \frac{a_n^2}{A_n^2},$$

$$1 - \frac{a_n^2}{n^2 \min_{0\leq k\leq n} a_k^2} \leq \frac{A_{n-1}A_{n+1}}{A_n^2} \leq 1 - \frac{a_n^2}{2\sum_{k=1}^{n} a_k^2},$$

these inequalities entails the following proposition.

Proposition 3.2. *If the series* $(a_n)_{n\geq n_0}$ *is decreasing to zero, for* $n \geq n_0$

$$1 - \frac{1}{n^2} \leq \frac{A_{n_0}^2 + (n-n_0-1)^2 a_n^2}{A_{n_0}^2 + (n-n_0)^2 a_n^2} \leq \frac{A_{n-1}A_{n+1}}{A_n^2}$$

$$\leq 1 - \frac{a_n^2}{A_{n_0}^2 + (n-n_0)^2 a_{n_0}^2}. \quad (3.2)$$

By the Stirling formula, the exponential function has a Taylor expansion with terms $a_n(x) = x^n(n!)^{-1} < 1$ for $x \leq e^{-1}n$, and a_n is decreasing if $0 < x < n+1$, under the last condition the inequality (3.1), hence (3.2) are satisfied for the exponential function.

3.3   Cauchy's inequalities and convex mappings

Cauchy (1821) established the following results for arithmetic means and transformed series.

Proposition 3.3 (Cauchy). *Let* $(x_i)_{i=1,\dots,n}$ *be a real sequence, then for every sequence of non zero real numbers* $(a_i)_{i=1,\dots,n}$

$$\min_{1 \leq k \leq n} \frac{x_k}{a_k} \leq \frac{\sum_{i=1}^n x_i}{\sum_{i=1}^n a_i} \leq \max_{1 \leq k \leq n} \frac{x_k}{a_k},$$

$$\min_{1 \leq k \leq n} \frac{x_k}{a_k} \leq \frac{(\sum_{i=1}^n x_i^2)^{\frac{1}{2}}}{(\sum_{i=1}^n a_i^2)^{\frac{1}{2}}} \leq \max_{1 \leq k \leq n} \frac{x_k}{a_k}$$

and the equality

$$\frac{x_1}{a_1} = \frac{x_2}{a_2} = \cdots = \frac{x_n}{a_n}$$

implies

$$\frac{1}{n} \sum_{k=1}^n \frac{x_k}{a_k} = \frac{\sum_{i=1}^n x_i}{\sum_{i=1}^n a_i^2} = \frac{(\sum_{i=1}^n x_i^2)^{\frac{1}{2}}}{(\sum_{i=1}^n a_i^2)^{\frac{1}{2}}}. \tag{3.3}$$

Consider two sequences of non zero real numbers $(x_i)_{i=1,\dots,n}$ and $(a_i)_{i=1,\dots,n}$, all a_i having the same sign and being different from their mean $\bar{a}_n = n^{-1} \sum_{j=1}^n a_j$. The first equality of Proposition 3.3 is also written $\bar{x}_n \bar{a}_n = n^{-1} \sum_{i=1}^n a_i x_i$ and it is obviously false without further conditions. It is equivalent to

$$\zeta_n(a, x) = \sum_{k=1}^n (\frac{\bar{a}_n}{a_k} - 1) x_k = 0$$

and generally it is not satisfied under the above conditions. It must then be an inequality.

Example 3.4. Let $a_k = k!$ and $x_k = x^k$, then $\lim_{n \to \infty} \zeta_n(a, x) > 0$ is equivalent to $x^{-1}(1-x)(e^x - 1) > \lim_{n \to \infty} n(\sum_{k=1}^n k!)^{-1} = 0$ if $0 < x < 1$, and it is equivalent to $x^{-n}(e^x - 1) > 0$ if $x > 1$.

Example 3.5. Let $x_k = x^k$ for x in $[0,1]$ and a decreasing sequence $a_k = k^{-1}$, the reverse inequality $\lim_{n \to \infty} \zeta_n(a,x) < 0$ is equivalent to $\log(1-x) < 0$.

Example 3.6. Let $x_k = x^{2k}$ and $a_k = 2^k$, the inequality $\lim_{n \to \infty} \zeta_n(a,x) > 0$ is fulfilled with equality if $x > 2$ and $(1-x^2)(2-x)^{-1} > n2^{-n}$ if $0 < x < 1$. With a faster rate of convergence to zero for the sequence $a_k = 2^{-k}$, we have $\zeta_n(a,x) > 0$ if $x \geq \frac{1}{2}$ and $\zeta_n(a,x) < 0$ if $x \leq \frac{1}{2}$.

In conclusion, the condition of a constant sequence $x_k a_k^{-1}$ is not necessary for the equality in (3.3). For sequences that do not satisfy the condition of Proposition 3.3, the equalities in Equation (3.3) may be replaced by an upper or lower inequality, according to the sequence. With a convex transformation of the series, a general inequality holds for the transformed sequence of Proposition 3.3.

Proposition 3.4. *Let* $(x_i)_{i=1,\ldots,n}$ *be a real sequence and* φ *be a convex function, then for every sequence of non zero real numbers* $(a_i)_{i=1,\ldots,n}$ *such that* $a_k^{-1}\varphi(x_k)$ *is constant*

$$\frac{1}{n} \sum_{k=1}^{n} \frac{\varphi(x_k)}{a_k} \geq \varphi\left(\frac{\sum_{i=1}^{n} x_i}{\sum_{i=1}^{n} a_i}\right).$$

However, several inequalities similar to Cauchy's Theorem 3.3 are satisfied with power, logarithm and n-roots transforms. In particular, it applies with the logarithm and in a multiplicative form such as

$$\min_{1 \leq k \leq n} x_k^{a_k^{-1}} \leq \left(\prod_{i=1}^{n} x_i\right)^{(\sum_{i=1}^{n} a_i)^{-1}} \leq \max_{1 \leq k \leq n} x_k^{a_k^{-1}}.$$

Let $(x_i)_{i=1,\ldots,n}$ and $(a_i)_{i=1,\ldots,n}$ be non zero real sequences such that $x_i > 0$ for every i. Under the condition of a constant sequence $a_k^{-1} \log x_k$

$$\prod_{i=1}^{n} x_i^{\frac{1}{a_n}} = \prod_{i=1}^{n} x_i^{\frac{1}{a_k}}.$$

Let $x = \log y$ on $]0, \infty[$. By an exponential mapping and for all non negative sequences $(y_i)_{i=1,\ldots,n}$ and $(a_i)_{i=1,\ldots,n}$

$$\exp\left(\frac{\sum_{i=1}^{n} a_i x_i}{\sum_{i=1}^{n} a_i}\right) = \prod_{i=1}^{n} y_i^{a_i(\sum_{k=1}^{n} a_k)^{-1}} \leq \frac{\sum_{i=1}^{n} a_i y_i}{\sum_{i=1}^{n} a_i},$$

with equality if and only if all x_k are equal. Assuming $a_i = 1$ for every i, this inequality becomes $\prod_{i=1}^{n} y_i^{\frac{1}{n}} \leq \bar{y}_n$, therefore the arithmetic mean of a real sequence is larger than its geometric mean, as Cauchy had already proved it. The comparison of the arithmetic mean and the geometric mean is extended as follows.

Proposition 3.5. *Let $(x_i)_{i=1,\ldots,n}$ be a real sequence in $[0,1[$*

$$\prod_{i=1}^{n}(\frac{x_i}{1-x_i})^{\frac{1}{n}} < \frac{\bar{x}_n}{1-\bar{x}_n}, \quad if\ 0 < x < \frac{1}{2},$$

$$\prod_{i=1}^{n}(\frac{x_i}{1-x_i})^{\frac{1}{n}} > \frac{\bar{x}_n}{1-\bar{x}_n}, \quad if\ \frac{1}{2} < x < 1,$$

with equality if and only if $x = \frac{1}{2}$.
For a real sequence $(x_i)_{i=1,\ldots,n}$ in $]1,\infty[$

$$\prod_{i=1}^{n}(\frac{x_i}{x_i-1})^{\frac{1}{n}} > \frac{\bar{x}_n}{\bar{x}_n-1}.$$

Proof. Let $a_i = 1 - x_i$ and let $\varphi(x) = \log\{x(1-x)^{-1}\}$ for x in $[0,1]$. The function φ is increasing on $[0,1]$. On the subinterval $[\frac{1}{2},1[$, it is convex with values in $[0,\infty[$ and

$$\prod_{i=1}^{n}(\frac{x_i}{1-x_i})^{\frac{1}{n}} = \exp\{n^{-1}\sum_{i=1}^{n}\varphi(x_i)\} \geq \exp\{\varphi(\bar{x}_n)\} = \frac{\bar{x}_n}{1-\bar{x}_n}.$$

On $[0,\frac{1}{2}]$, the function φ is concave with values in $]-\infty,0]$ and the inequality is reversed. On $]1,\infty[$, φ is replaced by the decreasing and convex function $\psi(x) = \log\{x(x-1)^{-1}\}$ and the convexity argument yields the inequality as in the first case. $\square$

Alzer (1990a) proved stronger inequalities between the arithmetic and the geometric means on $[0,\frac{1}{2}]$, $A_n(x) = \bar{x}_n$ and $G_n(x) = \prod_{i=1}^{n}x_i^{\frac{1}{n}}$, by introducing the means $A'_n(x) = 1 - \bar{x}_n$ and $G'_n(x) = \prod_{i=1}^{n}(1-x_i)^{\frac{1}{n}}$

$$\frac{G_n}{G'_n} \leq \frac{A_n}{A'_n}, \quad \frac{1}{G'_n} - \frac{1}{G_n} \leq \frac{1}{A'_n} - \frac{1}{A_n}.$$

The first inequality is proved in Proposition 3.5 and the second one is immediately deduced from the first one and from the inequality $A_n \geq G_n$, writing

$$\frac{1}{G'_n} - \frac{1}{G_n} = \frac{1}{G_n}\{\frac{G_n}{G'_n} - 1\} \leq \frac{1}{A_n}\{\frac{A_n}{A'_n} - 1\} = \frac{1}{A'_n} - \frac{1}{A_n}.$$

Proposition 3.5 extends on $\mathbb{R}_+$ by the same arguments.

Proposition 3.6. *Let* $x_n = (x_i)_{i=1,...,n}$ *be a real sequence*

$$\frac{1}{G'_n}(x_n) - \frac{1}{G_n}(x_n) < \frac{1}{A'_n}(x_n) - \frac{1}{A_n}(x_n), \quad if \ 0 < x_n < \frac{1}{2},$$

$$\frac{1}{G'_n}(x_n) - \frac{1}{G_n}(x_n) > \frac{1}{A'_n}(x_n) - \frac{1}{A_n}(x_n), \quad if \ \frac{1}{2} < x_n < 1,$$

$$\prod_{i=1}^{n}(x_i - 1)^{-\frac{1}{n}} - \prod_{i=1}^{n} x_i^{-\frac{1}{n}} > \frac{1}{\bar{x}_n - 1} - \frac{1}{\bar{x}_n}, \quad if \ x_n > 1.$$

If the sequences $\bar{x}_n$ and $n^{-1} \sum_{i=1}^{n} \log x_i$ converge respectively to finite limits μ and λ as n tends to infinity, these inequalities still hold for the limits. More generally, let X be a random variable with distribution function F on a real interval I_X and such that the means $\mu = EX$ and $\lambda = \int_{I_X} \log x \, dF(x) = E(\log X)$ are finite, then G_n converges to $e^{E(\log X)}$ and G'_n converges to $e^{E(\log(1-X))}$. By the concavity of the logarithm, $e^{E(\log X)} \leq EX$, therefore Proposition 3.6 implies the following inequalities.

$$e^{-E(\log(1-X))} - e^{-E(\log X)} < \frac{1}{1 - EX} - \frac{1}{EX}, \quad if \ I_X =]0, \frac{1}{2}[,$$

$$e^{-E(\log(1-X))} - e^{-E(\log X)} > \frac{1}{1 - EX} - \frac{1}{EX}, \quad if \ I_X =]\frac{1}{2}, 1[,$$

$$e^{-E(\log(X-1))} - e^{-E(\log X)} > \frac{1}{EX(EX - 1)}, \quad if \ I_X =]1, \infty[.$$

Similar inequalities cannot be established on other intervals.

The Cauchy distribution for a variable X is defined by its density and distribution functions on $\mathbb{R}_+$

$$f_X(x) = \frac{2}{\pi(1 + x^2)}, \ F_X(x) = \frac{2}{\pi} \arctan x. \tag{3.4}$$

Thus Cauchy's distribution is identical to the distribution of the tangent of a uniform variable U over $[0, \frac{\pi}{2}]$ and it cannot be extended by periodicity on $[0, \pi]$ or $[-\frac{\pi}{2}, \frac{\pi}{2}]$ where the tangent has negative values. The variable X has no mean since the integral $\int_0^x s f_X(s) \, ds = \frac{1}{\pi} \log(1 + x^2)$ tends to infinity with x.

It is known that the variable X^{-1} has the Cauchy distribution on $\mathbb{R}_{*+}$. The function $G(x) = \frac{1}{2}(x - x^{-1})$ is increasing from $\mathbb{R}_{*+}$ to $\mathbb{R}$, with inverse function $G^{-1}(y) = y + (1 + y^2)^{\frac{1}{2}}$, then the variable $Y = G(X)$ has also the Cauchy distribution. Let H be the function $H(x) = (1 - x)^{-1}(x + 1)$ defined on $[0, 1[\cup]1, \infty[$, it is increasing in each subinterval and it is a bijection between $[0,1[$ and $]1,\infty[$ and between $]1,\infty[$ and $]-\infty,-1[$. Let X' be a variable having the Cauchy distribution restricted to the interval $]1,\infty[$, namely, $2F_X$. The variable $Y = H(X')$ has the same restricted Cauchy distribution as X'

$$P(\frac{X'+1}{X'-1} \leq y) = P(X' \leq H^{-1}(y)) = \frac{4}{\pi}\arctan\frac{y-1}{y+1},$$

with values in $[0,1]$.

3.4 Inequalities for the mode and the median

The mode of a real function f is its maximum when it is finite, it is then reached at

$$M_f = \inf\{x \in I; f(M_f) \geq x \text{ for every } x \in I\}.$$

In the neighbourhood of its mode, a function is locally concave and it is locally convex in a neighbourhood of the mode of $-f$ where it reaches its minimum. Therefore the derivatives of f satisfy $f^{(1)}(M_f) = 0$ and $f^{(2)}(M_f) \leq 0$. For some distributions, the mode can be explicitly determined in terms of the parameters of the distribution, such as the mean μ for a Gaussian distribution $\mathcal{N}(\mu,\sigma^2)$ or the Laplace distribution with density $f_{\mu,b}(x) = (2b)^{-1}\exp\{-b^{-1}|x-\mu|\}$. Sometimes only an interval can be given, for example, the mode of the Poisson distribution with parameter λ belongs to the interval $[\lambda-1,\lambda]$. The power functions, the logarithm or the exponential have no mode. The Weibull distribution $F_{\lambda,k}(x) = 1 - \exp\{-\lambda^{-k}x^k\}$ has the mode $M_f = \lambda(1-k^{-1})^{\frac{1}{k}}$, for every real $\lambda > 0$ and for every integer k.

On a probability space $(\Omega,\mathcal{F},P)$, let X be a random variable with real values and with a right-continuous distribution function $F(x) = P(X \leq x)$. The median m_X of a variable X is defined as

$$m_X = \arg\min_{y\in\mathbb{R}} E_F|X-y|$$

with the mean $E_F|X-y| = \int_{\mathbb{R}}|x-y|\,dF(x)$. The function $g(y) = E_F|X-y| = y\{2F(y)-1\} + E_F(X1_{\{X\geq y\}}) - E_F(X1_{\{X<y\}})$ reaches its minimum at the median value, which is equivalent to $2F(m_X) - 1 = 0$ if the variable X has a density function. The median is equivalently defined by

$$m_F = \inf\{x \in \mathbb{R} : F(x) \geq \frac{1}{2}\} = \inf\{x \in \mathbb{R} : F(x) = 1 - F(x)\}.$$

At its minimum, $E_F|X - m_F| \geq 0$ and, by definition, it equals

$$E_F|X - m_F| = 2E_F(X1_{\{X \geq m_F\}}) - 1,$$

and it follows that

$$E_F(X1_{\{X \geq m_F\}}) \geq \frac{1}{2}, \; E_F(X1_{\{X < m_F\}}) \leq \frac{1}{2}.$$

Proposition 3.7. *For every random variable X of $L_1(\mathbb{R})$*

$$|EX - m_X| = E|X - m_X| \qquad (3.5)$$

and $E|X - m_X| = 0$ if and only if the mean of X equals its median.

Proof. For every real y, $E_F X - m_F = E_F|X - y| + y - m_F$ and the difference $|E_F X - m_F| \leq E_F|X - y| + |y - m_F|$ is minimum at m_F where the mean of the variable X satisfies Equation (3.5). For a variable X having a mean different from its median, $E_F|X - m_F| > 0$. $\qquad \square$

It follows for Equation (3.5) that two variables X_1 and X_2 having equal medians also have equal means if and only if $E|X_1 - m_{X_1}| = E|X_2 - m_{X_2}|$. The equality of the means of variables do not imply the equality of their medians, except if $E|X_1 - m_{X_1}| = E|X_2 - m_{X_2}|$. The distribution function of X has a left heavy tail if $EX > m_X$ and it has a right heavy tail if $EX < m_X$. For a symmetric random variable with a density function, the mean, the median and the location of the mode of its density are equal.

The median of a transformed variable by a monotone function can be explictly calculated by mapping the median of the variable. Let X be a random variable with a continuous distribution function on $\mathbb{R}$ and with median m_X, then $F(m_X) = \frac{1}{2}$ and

$$P(1 - X < m_{1-X}) = 1 - F(1 - m_{1-X}) = \frac{1}{2}$$

which entails $m_{1-X} = 1 - m_X$. This argument extends to every monotone and continuous real map φ. The median of $\varphi(X)$ is

$$m_{\varphi(X)} = \varphi(m_X).$$

In particular, $m_{\alpha X} = \alpha m_X$ for every real α. Similar to Proposition 3.6, if $m_X < EX$ we obtain

$$\frac{m_X}{m_{1-X}} < \frac{EX}{1 - EX},$$

$$\frac{1}{m_{1-X}} - \frac{1}{m_X} < \frac{1}{1 - EX} - \frac{1}{EX},$$

and if $EX < m_X$, the inequalities are reversed.

Ordered variables X_1 and X_2 have ordered distribution functions

$$X_1 < X_2 \implies F_{X_2} < F_{X_1}$$

and therefore ordered means and medians

$$EX_2 < EX_1, \ m_{X_2} \leq m_{X_1}.$$

Let F_1 and F_2 be real distribution functions with respective medians m_{F_1} and m_{F_2} and such that there exists A satisfying $F_1 \leq F_2$ on $]-\infty, A]$ and $F_2 \leq F_1$ on $[A, \infty[$, then obviously, $1 = 2F_1(m_{F_1}) \leq 2F_2(m_{F_1})$ if $m_{F_1} < A$, hence

$$m_{F_1} \leq m_{F_2} \text{ if } m_{F_2} \leq A,$$
$$m_{F_2} \leq m_{F_1} \text{ if } m_{F_1} \geq A.$$

One cannot deduce an order for the means $E_{F_1}(X)$ and $E_{F_2}(X)$ of variables under F_1 and F_2, but only for the means of variables restricted to the subintervals defined by A

$$E_{F_1}(X1_{\{X<A\}}) \leq E_{F_2}(X1_{\{X<A\}}),$$
$$E_{F_1}(X1_{\{X\geq A\}}) \geq E_{F_2}(X1_{\{X<A\}}).$$

These orders extend to subintervals where F_1 and F_2 are ordered.

Mixtures of k unimodal densities may be unimodal or multimodal according to the distance between the modes of the densities of the mixture and the number of local maxima of the mixture density is between one and k. Let us consider the mixture of two symmetric densities with proportions p and $1 - p$, and with location parameters μ and $-\mu$

$$f_{p,\mu} = pf_\mu + (1-p)f_{-\mu}$$

and let X be a random variable having the mixture density $f_{p,\mu}$. The mean of X is $EX = (2p - 1)\mu$, it is zero only if $p = .5$, when $f_{p,\mu}$ is symmetric. The first derivative of a mixture with Gaussian components equals $f_{p,\mu}^{(1)}(x) = -f_{-\mu}(x)\{(x - \mu)pe^{2x\mu} + (1 - p)(x + \mu)\}$ and it is zero as

$$e^{2x\mu} = \frac{(1-p)(x+\mu)}{p(\mu - x)} > 0$$

therefore the mode of a unimodal Gaussian mixture density $f_{p,\mu}$ satisfies

$$-\mu < M_f < \mu.$$

If p belongs to the interval $]0, .5[$, $M_f > 0$, and for p in $].5, 1[$, $M_f < 0$.

Let f_1 be the density function of a variable with distribution $\mathcal{N}(\mu_1, \sigma_1^2)$, f_2 be the density function of $\mathcal{N}(\mu_2, \sigma_2^2)$ and let $\theta = \sigma_2^2 \sigma_1^{-2}$. Replacing $x - \mu$ by $y_1 = (x - \mu)\sigma_1^{-1}$ and $x + \mu$ by $y_2 = (x + \mu)\sigma_2^{-1}$, the first derivative of $f_p = pf_1 + (1 - p)f_2$ is

$$f_{p,\mu,\sigma}^{(1)}(x) = \frac{1}{\sigma_2^2 \sqrt{2\pi}} e^{-y_2^2} \{ p\theta y_1 e^{y_2^2 - y_1^2} + (1 - p)y_2 \}$$

therefore the same inequality holds for the mode of the mixture of Gaussian densities with inequal variances.

The sub-densities pf_μ and $(1 - p)f_{-\mu}$ have an intersection at x_p where $pf_\mu(x_p) = (1 - p)f_{-\mu}(x_p)$, denoted y_p, therefore $f_{p,\mu}(x_p) = 2pf_\mu(x_p)$ and, by symmetry, $f_\mu(x_p) = f_{-\mu}(x_p)$ if and only if $p = .5$. A mixture density $f_{p,\mu}$ is bimodal if and only if it has a local minimum at x_p.

Proposition 3.8. *The Gaussian mixture density $f_{.5,\mu}$ is bimodal if $\mu > 1$ and it is unimodal if $\mu < 1$. For every $p \neq .5$ in $]0, 1[$, the Gaussian mixture density $f_{p,\mu}$ is bimodal if $\mu > (1 - x_p^2)^{\frac{1}{2}}$ and it is unimodal if $\mu < (1 - x_p^2)^{\frac{1}{2}}$. A mixture density $f_{p,\mu_1,\mu_2,\sigma^2}$, with Gaussian components f_{μ_1,σ^2} and f_{μ_2,σ^2}, is bimodal if $|\mu_1 - \mu_2| > 2(1 - x_p^2)^{\frac{1}{2}}$ and it is unimodal if $|\mu_1 - \mu_2| < 2(1 - x_p^2)^{\frac{1}{2}}$.*

Proof. The second derivative of the Gaussian mixture density $f_{p,\mu}$ at x_p is $f_{p,\mu}^{(2)}(x_p) = [\{(x_p - \mu)^2 - 1\} + \{(x_p + \mu)^2 - 1\}]y_p = 2(x_p^2 + \mu^2 - 1)y_p$ and the sign of $x_p^2 + \mu^2 - 1$ determines whether $f_{p,\mu}$ has a maximum or a minimum at x_p. □

With respective variances σ_1^2 and σ_2^2, the sign of the second derivative $f_{p,\mu,\sigma}^{(2)}$ is defined by the sign of $x_p + \mu$ with respect to a constant depending on the variances. These properties extend to mixtures of two symmetric densities with respective means μ_1 and μ_2. If their difference $|\mu_1 - \mu_2|$ is sufficiently large, a mixture density f_{p,μ_1,μ_2} is bimodal and it is unimodal if $|\mu_1 - \mu_2|$ is smaller than a threshold μ_p depending on the mixture proportion p. Proposition 3.8 can be generalized to mixtures with n symmetric components by cumulating the tails of the n components and it is necessary to increase the distance between two consecutive modes with the number of components to have a multimodal mixture density with n modes. The number of modes of mixtures with a countable number of components is necessarily smaller than the number of components.

Continuous distribution mixtures have the form

$$F_X(x) = \int_{\mathbb{R}} F_{X|W}(x, w) \, dF_W(w),$$

where F_W is the distribution function of a variable W and for every w, $F_{X|W}(\cdot, w)$ is the conditional distribution of a variable X given $W = w$. If $F_{X|W}(\cdot, w)$ is absolutely continuous with density $f_{X|W}(\cdot, w)$ for every real w, then F_X is absolutely continuous with density

$$f_X(x) = \int_{\mathbb{R}} f_{X|W}(x, w) \, dF_W(w)$$

and the Laplace transform of X satisfies

$$\varphi_X(x) = \int_{\mathbb{R}} \varphi_{X|W}(x, w) \, dF_W(w) = E\varphi_{X|W}(x, W).$$

In particular, for a scale mixture of a distribution $X = NV^{\frac{1}{2}}$, with independent random variables X and N, N being a standard normal variable and $V = X^2 Y^{-2}$ being a random variance, the Laplace transform of X is $\varphi_X(t) = \int_0^\infty e^{-\frac{x^2}{2v}} \, dF_V(v)$.

Keilson and Steutel (1974) studied several classes of continuous mixtures. They defined a scale mixture variable $X = YW$ with independent variables X and Y having absolutely continuous densities and with $W = XY^{-1}$. The mixture density of X is

$$f_X(x) = \int_0^\infty f_Y(\frac{x}{w}) w^{-1} \, dF_W(w).$$

The norms of variables X and Y of $L_p(\mathbb{R}_+)$ are such that

$$\frac{\|X\|_{L_p}}{\|X\|_{L_1}} = \frac{\|Y\|_{L_p}}{\|Y\|_{L_1}} \frac{\|W\|_{L_p}}{\|W\|_{L_1}}.$$

They proved that the variances of the variables X and Y satisfy

$$\frac{\sigma_X^2}{\|X\|_{L_1}^2} \geq \frac{\sigma_Y^2}{\|Y\|_{L_1}^2},$$

with equality if and only if the mixing distribution is degenerate, and

$$\frac{\sigma_X^2}{\|X\|_{L_1}^2} - \frac{\sigma_Y^2}{\|Y\|_{L_1}^2} = \frac{\|X\|_{L_2}^2}{\|X\|_{L_1}^2} \frac{\sigma_W^2}{\|W\|_{L_1}^2}.$$

3.5 Mean residual time

Let X be a positive real random variable with a distribution function F and a survival function $\bar{F}$ such that $\lim_{x \to \infty} x\bar{F}(x) = 0$. The mean residual time of X at t is

$$e_F(t) = E_F(X - t | X > t) = \frac{1}{\bar{F}(t)} \int_t^\infty (x - t) \, dF(x)$$

$$= \frac{1}{\bar{F}(t)} \{EX - t - \int_0^t (x - t) \, dF(x)\} = \int_t^\infty \frac{\bar{F}(x)}{\bar{F}(t)} \, dx,$$

with the convention $\frac{0}{0} = 0$, and $e_F(0) = EX$. The cumulated failure rate of F is defined as $R_F(x) = \int_0^x \bar{F}^{-1}(t) \, dF(t)$ and it defined uniquely the distribution function by $\bar{F}(x) = \exp\{-R_F(x)\}$. An exponential distribution with parameter λ has a constant failure rate $r(x) \equiv \lambda$ and a function $R_\lambda(x) = \lambda x$.

Proposition 3.9. *A distribution function F is determined from its mean residual time by*

$$\bar{F}(x) = \frac{e_F(0)}{e_F(x)} \exp\{-\int_0^x \frac{dt}{e_F(t)}\}.$$

Proof. By definition of the mean residual time, $e_F(t)\bar{F}(t) = \int_t^\infty \bar{F}(x) \, dx$. Deriving this expression yields

$$\frac{d\bar{F}(t)}{\bar{F}(t)} = -\frac{de_F(t)}{e_F(t)} - \frac{dt}{e_F(t)}$$

and the expression of $\bar{F}$ follows by integration. $\qquad\square$

An exponential distribution is therefore characterized by a constant function e_F. Let G be a distribution function of a positive real random variable satisfying the same condition as F at infinity and such that $R_F(t) \le R_G(t)$ for every t in an interval $[a, \infty[$.

Proposition 3.10. *Let t be in the interval $[a, \infty[$ and let $x > t$, the inequality $\int_t^x dR_G < \int_t^x dR_F$ for non exponential distributions F and G implies $e_F(t) < e_G(t)$.*

Proof. Let $t > a$, the difference

$$\delta_{F,G}(t, x) = \int_t^x \frac{\bar{F}(s)}{\bar{F}(t)} \, ds - \int_t^x \frac{\bar{G}(s)}{\bar{G}(t)} \, ds$$

has the derivative

$$\delta'_{F,G}(x, t) = \frac{\bar{G}(x)}{\bar{G}(t)} - \frac{\bar{F}(x)}{\bar{F}(t)} \, dx = \exp\{-\int_t^x dR_G(s)\} - \exp\{-\int_t^x dR_F(s)\},$$

for $x > t$, and $\delta'_{F,G}(x, t)$ is strictly positive if

$$\int_t^x dR_G < \int_t^x dR_F, \qquad (3.6)$$

for every $x > t$. By the mean value theorem for $\delta_{F,G}(t, x)$, the inequality (3.6) implies $e_F(t) > e_G(t)$. $\qquad\square$

If F has a density f, the function R has a derivative $r(t) = f(t)\bar{F}^{-1}(t)$.

Proposition 3.11. *The mean residual time of a variable X having a density is strictly increasing at t if and only if $r_F(t)e_F(t) > 1$.*

Proof. The derivative of $e_F(t)$ is

$$e_F'(t) = -1 + \frac{f(t)}{\bar{F}^2(t)} \int_t^\infty \bar{F}(x)\,dx = r_F(t)e_F(t) - 1.$$

□

The variance of the residual time of X at t is

$$v_F(t) = E_F\{(X - t)^2 | X > t\} - e_F^2(t) = E_F[\{X - t - e_F(t)\}^2 | X > t].$$

Let X have an exponential distribution $\mathcal{E}_\lambda$, its failure rate is $r_\lambda(t) \equiv \lambda$ and $e_\lambda(t) \equiv \lambda^{-1}$, satisfying $r_\lambda(t)e_\lambda(t) = 1$. Its variance $v_F(t) \equiv \lambda^{-2}$ equals the variance of the variable X. In the same way, for all $t > 0$ and $k \geq 1$

$$E_\lambda\{(X - t)^k | X > t\} = E_\lambda(X^k),$$

that is the lack of memory of the exponential variable. The k-th moment of its residual time after any time t is identical to the k-th moment of the exponential variable and the generating function of the residual time after t is identical to the generating function of the exponential variable. The distribution of $X - t$ conditionally on $\{X > t\}$ is therefore the exponential distribution $\mathcal{E}_\lambda$.

Proposition 3.12. *Let X be a variable X having a density, then the function v_F is strictly increasing at t if and only if $v_F(t) < r_F^{-1}(t)e_F(t) - e_F^2(t)$.*

Proof. From the expression of e_F', the derivative of $v_F(t)$ is

$$v_F'(t) = 2e_F(t) - 2r_F(t)\{v_F(t) + e_F^2(t)\}$$

and v_F is increasing at t if and only if $v_F'(t) > 0$ then the condition is satisfied.

□

3.6 Functional equations

Let $F : \mathbb{R} \to \mathbb{R}$ be a continuous monotone function and

$$\eta(x, y) = F^{-1}(F(x) + F(y)).$$

When F is increasing (respectively decreasing), η is increasing (respectively decreasing) with respect to (x, y) in $\mathbb{R}^2$. Thus, the function $F(x) = e^{-\lambda x}$, $\lambda > 0$ and x in $\mathbb{R}$, defines in $\mathbb{R}^2$ the function

$$\eta(x, y) = y - \lambda^{-1} \log\{1 + e^{-\lambda(x-y)}\} = x - \lambda^{-1} \log\{1 + e^{-\lambda(y-x)}\}.$$

Let $\eta(x, y) = x + y$, equivalently $F(x+y) = F(x) + F(y)$ hence $F(0) = 0$ and F is an odd function.

Let α and β in $\mathbb{R}$ and $\eta(x, y) = x^\alpha + y^\beta$, (x, y) in $\mathbb{R}^2$. With $y = -x$, it follows that $F(0) = 0$ and $F(x) = F(x^\alpha)$ for every x, which is true only by $\alpha = 1$. Similarly, $\beta = 1$ therefore $\eta(x, y) = x + y$ is the unique solution of the functional equation $F(x^\alpha + y^\beta) = F(x) + F(y)$.

Proposition 3.13. *For every real function ψ of $C_2(\mathbb{R}^2)$, a function f such that*

$$\frac{f(x) - 2f(\frac{x+y}{2}) + f(y)}{x - y} = \frac{\psi(x, y)}{2}, \ x \neq y$$

is defined for $x > 0$ by

$$f(x) = f(0) + xf'(0) - \frac{1}{2}\int_0^x \psi(u, 0)\, du,$$

$$f(-x) = f(0) - xf'(0) + \frac{1}{2}\int_{-x}^0 \psi(-u, 0)\, du,$$

under the necessary conditions that the partial derivatives of ψ satisfy $\psi'_x(x, x) = \psi'_y(x, x)$, and $\psi(x, 0) = \psi(0, x)$.

Proof. The functions f and ψ satisfy the following properties

$$\psi(x, -x) = \frac{f(x) + f(-x) - 2f(0)}{x} = \psi(-x, x), \tag{3.7}$$

$$0 = \lim_{|x-y|\to 0}\frac{1}{2}\{f'(x) - f'(y)\}$$

$$= \lim_{|x-y|\to 0}\frac{1}{x - y}[\{f(x) - f(\frac{x+y}{2})\} + \{f(y) - f(\frac{x+y}{2})\}]$$

$$= \psi(x, x).$$

The derivatives of f is

$$f'(x) = \lim_{|x-y|\to 0}\frac{1}{2}\{f'(x) + f'(y)\}$$

$$= \lim_{|x-y|\to 0}\frac{1}{x - y}[\{f(x) - f(\frac{x+y}{2})\} - \{f(y) - f(\frac{x+y}{2})\}],$$

$$f''(x) = \lim_{|x-y|\to 0}\frac{1}{(x - y)^2}[\{f(x) - f(\frac{x+y}{2})\} + \{f(y) - f(\frac{x+y}{2})\}]$$

$$= \lim_{|x-y|\to 0}\frac{\psi(x, y) - \psi(x, x)}{2(x - y)} = \frac{1}{2}\psi'_x(x, x) = \frac{1}{2}\psi'_y(x, x).$$

Integrating the derivatives, for every $x > 0$

$$f'(x) = f'(0) - \frac{1}{2}\psi(x,0) = f'(0) - \frac{1}{2}\psi(0,x),$$

$$f(x) = f(0) + xf'(0) - \frac{1}{2}\int_0^x \psi(u,0)\,du,$$

and in $\mathbb{R}^-$

$$f'(-x) = f'(0) + \frac{1}{2}\psi(-x,0),$$

$$f(-x) = f(0) - xf'(0) + \frac{1}{2}\int_{-x}^0 \psi(-u,0)\,du.$$

$\square$

For a function ψ defined by an odd function g as $\psi(x,y) = g(x) - g(y)$, the necessary conditions $\psi(x,-x) = \psi(-x,x)$ and $\psi(x,x) = 0$ are satisfied and there are constants a and b such that for every $x > 0$

$$f(x) = ax + b - \frac{1}{2}\int_0^x g(u)\,du,$$

$$f(-x) = -ax + b - \frac{1}{2}\int_{-x}^0 g(u)\,du.$$

As a consequence of Proposition 3.13, a function f satisfies the equality in Hadamard's inequality (Section 1.3)

$$f\left(\frac{x+y}{2}\right) = \frac{f(x) + f(y)}{2}$$

if the function $\psi = 0$, therefore f is an affine function.

Sahoo and Riedel (1998) presented several functional "open problems". The first one is to find all functions $f :]0,1[\to \mathbb{R}$ satisfying the equation

$$f(xy) + f(x(1-y)) + f((1-x)y) + f((1-x)(1-y)) = 0,\ 0 < x,y < 1. \quad (3.8)$$

Putting $y = .5$, the equation becomes

$$f\left(\frac{x}{2}\right) + f\left(\frac{1-x}{2}\right) = 0,$$

hence the function f is skew symmetric with center at $x = \frac{1}{2}$ and $f(\frac{1}{4}) = 0$. Let x tend to zero and $y = .5$ implies $f(0^+) + f(\frac{1}{2}) = 0$. Equation (3.8) also implies that f is skew symmetric around $x = \frac{1}{2}$. Letting $x = y$ tend to zero in (3.8), then

$$0 = f(x^2) + 2f(x(1-x)) + f((1-x)^2) \to 3f(0^+) + f(1^-).$$

So Equation (3.8) admits a linear solution $f(x) = a(x - \frac{1}{4})$, $a \neq 0$, hence $f(0) = -\frac{1}{4}a$ and $f(1) = \frac{3}{4}a$ and there is no other polynomial solution.

Let $f(x) = a\{g(x) - \frac{1}{4}\}$, where the straight line $g(x) = x$ is replaced by any curve that respects the symmetry of f around the line. For example, a sinusoidal function g with an even number of periods is written

$$g_k(x) = \pm \sin(\frac{2k\pi x}{\tau}), \quad \tau = \frac{(a^2 + 1)^{\frac{1}{2}}}{2\sqrt{2}},$$

with an even integer $k \geq 2$, and $f_k(x) = a\{g_k(x) - \frac{1}{4}\}$ is solution of (3.8). Obviously, the sinusoidal function can be replaced by any continuous and twice differentiable function with first two derivatives having the same signs as g_k up to a translation of its maxima and minima. Every function g that is continuous or discontinous and skew symmetric around $x = \frac{1}{4}$ and such that $g(x + \frac{1}{2}) = g(x)$ provides a solution of the equation.

Equation (3.8) on the closed interval $[0, 1]$ implies $f(x) + f(1 - x) = 0$, hence $f(\frac{1}{2}) = 0$ instead of $f(\frac{1}{2}) = \frac{3}{4}a > 0$ for the same equation on $]0, 1[$, therefore the condition on the boundary modifies the solution which reduces to zero.

The second problem is to find all functions $f :]0, 1[\to \mathbb{R}$ satisfying the equation

$$f(xy) + f((1 - x)(1 - y)) = f(x(1 - y)) + f((1 - x)y), \ 0 < x, y < 1. \ (3.9)$$

As y tends to zero, the equation implies $f((1 - x)^-) = f(x^-)$ for every x in $]0, 1[$. Let the function f be a polynomial of degree 2 $f(x) = ax^2 + bx + c$, then $a = b$ and c varies freely. With polynomials of degree ≥ 3, (3.9) reduces to the roots of higher order equation and it does not determine a function solution of the equation. Equation (3.9) is also satisfied for Neper's logarithm and for $\log_a x$ for every $a > 0$, it has therefore an infinity of solutions.

Another problem is to find functions f, g, h and $k : \mathbb{R} \to \mathbb{R}$ solutions of the equation

$$\{f(x) - f(y)\} k(x + y) = \{g(x) - g(y)\} h(x + y), \ (x, y) \in \mathbb{R}^2.$$

First, f is even if and only if g is even and, under the condition that f is odd, $f(x) = h(0)\{g(x) - g(y)\}\{2k(0)\}^{-2}$. For functions f and g of $C_1(\mathbb{R})$, dividing both terms of the equation by $x - y$ and considering the limit as $|x - y|$ tends to zero implies

$$f(x) - f(0) = \int_0^x g'(u)\frac{h(2u)}{k(2u)} \, du, \ x > 0.$$

Constraints on the functions in the Mean Value Theorem determine simpler classes of functions without the assumption of derivability.

Proposition 3.14. *The functions f and $g : \mathbb{R} \to \mathbb{R}$ satisfy the equation*

$$\frac{f(x) - f(y)}{x - y} = \frac{g(x) + g(y)}{2} \qquad (3.10)$$

if and only if g is constant c and $f(x) = f(0) + cx^2$.

Proof. Equation (3.10) with $y = 0$ implies $f(x) = ax + b + \frac{1}{2}xg(x)$ where $a = \frac{1}{2}g(0)$ and $b = f(0)$ and

$$f(x) - f(y) = a(x - y) + \frac{xg(x) - yg(y)}{2} = (x - y)\frac{g(x) + g(y)}{2}$$

which equivalent to $yg(x) - xg(y) = 0$ for all x and y, therefore $g(0) = 0$ and there exists a constant c such that $g(x) = 2cx$ for every x. □

Proposition 3.15. *Functions f, g and $h : \mathbb{R} \to \mathbb{R}$ satisfy the equation*

$$\frac{f(x) - g(y)}{x - y} = h(x + y)$$

if and only if $f = g$, h is affine and f is a second order polynomial.

Proof. Interchanging x and y in the equation implies that for every (x, y) in $\mathbb{R}^2$, $f(x) - g(y) = g(x) - f(y)$, hence $f(x) = g(x) + g(0) - f(0)$ at $y = 0$, and $f(x) + f(y) = g(x) + g(y)$ in $\mathbb{R}^2$. Combining both equalities entails $f \equiv g$. With $y = 0$, $f(x) = f(0) + xh(x)$ and

$$f(x) - f(y) = xh(x) - yh(y) = (x - y)h(x + y)$$

which is a necessary condition for the existence of a solution of the equation. Therefore

$$h(x + y) = \frac{xh(x) - yh(y)}{x - y} = h(x) + y\frac{h(x) - h(y)}{x - y},$$

$$\frac{h(x) - h(y)}{x - y} = \frac{h(x + y) - h(x)}{y},$$

letting y tend to zero implies $h(x) = xh'(x)$ for every $x \neq 0$, which is equivalent to $h(x) = ax + b$ with real constants a and b. Solutions of the equation have then the form $f(x) = f(0) + ax^2 + bx$. □

Proposition 3.16. *Functions f, g and $h : \mathbb{R} \to \mathbb{R}$ satisfy the equation*

$$f(x + y) - g(x - y) = h(x)$$

if and only if $f + g$ is constant, $h = f - g$ and the function $H = h - h(0)$ satisfies $H(rx) = rH(x)$ for every r in $\mathbb{Q}_+$.

Proof. Writing the functional equation of f, g and h with $y = x$, $y = -x$ then $y = 0$, yields

$$f(x) = h(\frac{1}{2}x) + g(0),$$

$$g(x) = -h(\frac{1}{2}x) + f(0),$$

$$f(x) + g(x) = a := f(0) + g(0),$$

$$f(x) - g(x) = h(x), \text{ with } y = 0.$$

Let $b = f(0) - g(0) = h(0)$, $F(x) = f(x) - f(0)$, $G(x) = g(x) - g(0)$ and $H(x) = h(x)_h(0)$, the above equalities imply $F(x) + G(x) = 0$, $F(x) - G(x) = H(x)$, therefore

$$F(x) = \frac{1}{2}H(x) = -G(x).$$

Introducing these functions in the functional equation yields

$$H(x) = \frac{1}{2}\{H(x+y) + H(x-y)\}, \tag{3.11}$$

for every (x, y) in $\mathbb{R}$. Moreover

$$F(x) = H(\frac{1}{2}x) = \frac{1}{2}H(x) = \frac{1}{2}F(2x)$$

and this equality extends to positive rational numbers by changing the values of x and y in the previous equality. $\square$

As a consequence of the homogeneity of the function H, the functions f, g and h solutions of Proposition 3.16 satisfy $f(rx) = rf(x) + (1-r)f(0)$, $g(rx) = rg(x) + (1-r)g(0)$ and $h(rx) = rh(x) + (1-r)h(0)$ for every r in $\mathbb{Q}_+$.

Changing g in $-g$ defines another equation and its solutions are characterized by $g(x) = \frac{1}{2}\{h(x) + f(0) - g(0)\}$ and $f(x) = \frac{1}{2}\{h(x) - f(0) + g(0)\}$, moreover $F(x) + G(x) = H(x)$, $F(x) = G(x) = H(\frac{1}{2}x) = \frac{1}{2}H(x)$ and the function H still satisfies Equation (3.11).

Corollary 3.1. *Functions f, g and $h : \mathbb{R} \to \mathbb{R}$ satisfy the equation*

$$f(x+y) + g(x-y) = h(x)$$

if and only if $f - g$ is constant, $h = f + g$ and $f(rx) = rf(x) + (1-r)f(0)$, $g(rx) = rg(x) + (1-r)g(0)$ and $h(rx) = rh(x) + (1-r)h(0)$ for every r in $\mathbb{Q}_+$.

3.7 Carlson's inequality

Carlson's inequality provides upper and lower bounds for the function

$$L(x, y) = \frac{x - y}{\log x - \log y}, \quad 0 < x \neq y \tag{3.12}$$

and it is extended by continuity to $L(x, x) = x$. It is a symmetric, homogeneous and continuous function on $\mathbb{R}_+^2$. Sahoo and Riedel (1998) gave another interesting form of the function L of Equation (3.12)

$$L(x, y) = y \int_0^1 (\frac{x}{y})^t \, dt \tag{3.13}$$

and it is extended to vectors of $\mathbb{R}_+^{*n}$

$$L(x_1, \ldots, x_n) = x_n^{n-1} \int_{[0,1]^n} \prod_{k=1}^{n-1} (\frac{x_k}{x_n})^{t_k} \prod_{j=1}^{n-1} dt_j.$$

Carlson (1972) established the lower bound $(xy)^{\frac{1}{2}} \leq L(x, y)$. Sharper upper and lower bounds are given here.

Proposition 3.17. *Let $\varphi : I \subset \mathbb{R} \mapsto \mathbb{R}$ be a function with a convex and decreasing first derivative and let*

$$L_\varphi(x, y) = \frac{x - y}{\varphi(x) - \varphi(y)}.$$

For every $x \neq y$ in $\mathbb{R}_+^2$

$$\frac{2}{\varphi'(x) + \varphi'(y)} \leq L_\varphi(x, y) \leq \frac{1}{\varphi'(x + y)}.$$

If φ has a concave and increasing first derivative, for every $x \neq y$ in $\mathbb{R}_+^2$

$$\frac{1}{\varphi'(x + y)} \leq L_\varphi(x, y) \leq \frac{2}{\varphi'(x) + \varphi'(y)}.$$

Proof. The lower bound of the first inequality is proved by convexity of the function φ'

$$L^{-1}(x, y) = \int_0^1 \varphi'(ux + (1 - u)y) \, du$$

$$\leq \int_0^1 \{u\varphi'(x) + (1 - u)\varphi'(y)\} \, du \leq \frac{1}{2}\{\varphi'(x) + \varphi'(y)\}.$$

The upper bound of $L_\varphi(x, y)$ is a consequence of the inequality $ux + (1 - u)y \leq x + y$ for every u in $[0, 1]$, hence $\varphi'(ux + (1 - u)y) \geq \varphi'(x + y)$ for a decreasing function φ'. The same arguments for an increasing and concave derivative provides the second inequality. $\square$

This general result applies to the function L defined for the logarithm and it is refined by smaller intervals.

Proposition 3.18. *For every $x \neq y$ in $\mathbb{R}_+^2$*

$$\frac{2xy}{x+y} \leq (xy)^{\frac{1}{2}} \leq \frac{(xy)^{\frac{1}{4}}}{2}(x^{\frac{1}{2}} + y^{\frac{1}{2}}) \leq L(x,y),$$

$$L(x,y) \leq (\frac{x^{\frac{1}{4}} + y^{\frac{1}{4}}}{2})^2(x^{\frac{1}{2}} + y^{\frac{1}{2}}) \leq \frac{(x^{\frac{1}{2}} + y^{\frac{1}{2}})^2}{2} \leq x + y.$$

Proof. The inequality is proved by iterative applications of the inequality

$$\frac{2xy}{x+y} \leq L(x,y) \leq x + y$$

which is proved by convexity of the inverse function $x \mapsto x^{-1}$. Applying this inequality to $(x^{\frac{1}{2}}, y^{\frac{1}{2}})$ implies

$$2\frac{(xy)^{\frac{1}{2}}}{x^{\frac{1}{2}} + y^{\frac{1}{2}}} \leq \frac{x^{\frac{1}{2}} - y^{\frac{1}{2}}}{\log x^{\frac{1}{2}} - \log y^{\frac{1}{2}}} \leq x^{\frac{1}{2}} + y^{\frac{1}{2}}$$

multiplying all terms by $\frac{1}{2}(x^{\frac{1}{2}} + y^{\frac{1}{2}})$ yields

$$(xy)^{\frac{1}{2}} \leq L(x,y) \leq \frac{(x^{\frac{1}{2}} + y^{\frac{1}{2}})^2}{2}$$

and the inequalities

$$\frac{2xy}{x+y} \leq (xy)^{\frac{1}{2}} \leq L(x,y) \leq \frac{(x^{\frac{1}{2}} + y^{\frac{1}{2}})^2}{2} \leq x + y.$$

With $(x^{\frac{1}{4}}, y^{\frac{1}{4}})$, we obtain

$$2\frac{(xy)^{\frac{1}{4}}}{x^{\frac{1}{4}} + y^{\frac{1}{4}}} \leq \frac{x^{\frac{1}{4}} - y^{\frac{1}{4}}}{\log x^{\frac{1}{4}} - \log y^{\frac{1}{4}}} \leq x^{\frac{1}{4}} + y^{\frac{1}{4}},$$

then multiplying all terms by $\frac{1}{4}(x^{\frac{1}{4}} + y^{\frac{1}{4}})(x^{\frac{1}{2}} + y^{\frac{1}{2}})$ yields

$$\frac{(xy)^{\frac{1}{4}}}{2}(x^{\frac{1}{2}} + y^{\frac{1}{2}}) \leq L(x,y) \leq (\frac{x^{\frac{1}{4}} + y^{\frac{1}{4}}}{2})^2(x^{\frac{1}{2}} + y^{\frac{1}{2}})$$

and these bounds are included in the previous ones. $\square$

More iterations like in the proof of Proposition 3.17 allow to improve the bounds of Carlson's function L. With exponential values, it follows that for all distinct $y > 0$ and $x > 0$

$$e^{-\frac{x+y}{2}} \leq e^{-\frac{x+y}{4}}\frac{e^{-\frac{x}{2}} + e^{-\frac{y}{2}}}{2} \leq \frac{e^{-x} - e^{-y}}{y - x}$$

$$\leq (\frac{e^{-\frac{x}{4}} + e^{-\frac{y}{4}}}{2})^2(e^{-\frac{x}{2}} + e^{-\frac{y}{2}}) \leq \frac{(e^{-\frac{x}{2}} + e^{-\frac{y}{2}})^2}{2} \leq (e^{-x} + e^{-y}).$$

Equivalently

$$e^{-\frac{3(x+y)}{2}} \le e^{-\frac{5(x+y)}{4}} \frac{e^{-\frac{x}{2}} + e^{-\frac{y}{2}}}{2} \le \frac{e^y - e^x}{y - x}$$

$$\le e^{-(x+y)} \left(\frac{e^{-\frac{x}{4}} + e^{-\frac{y}{4}}}{2}\right)^2 (e^{-\frac{x}{2}} + e^{-\frac{y}{2}}) \le e^{-(x+y)} \frac{(e^{-\frac{x}{2}} + e^{-\frac{y}{2}})^2}{2}$$

$$\le e^{-2(x+y)}(e^x + e^y).$$

On $[0, \frac{\pi}{2}]$, the cosine function is decreasing and concave and the sine function is increasing and concave.

Proposition 3.19. *For every $x \neq y$ in $[0, \frac{\pi}{2}]$*

$$1 < \frac{x - y}{\sin x - \sin y} < \min\left\{\frac{1}{\cos(x + y)}, \frac{2}{\cos x + \cos y}\right\},$$

$$\frac{1}{\sin(x + y)} < \frac{x - y}{\cos y - \cos x} < \frac{2}{\sin x + \sin y}.$$

The exponential on $\mathbb{R}_+$ is an increasing and convex function, and Proposition 3.17 only provides lower bounds of the function $L_{\exp}$.

Proposition 3.20. *Let $\alpha > 0$, for every $x \neq y$ in $\mathbb{R}_+^2$*

$$\frac{\alpha(x - y)}{e^{\alpha x} - e^{\alpha y}} \ge \min\left\{\frac{2}{e^{\alpha x} + e^{\alpha y}}, e^{-\alpha(x+y)}\right\}.$$

Proposition 3.21. *Let $0 < \alpha < 1$ and let $L_\alpha(x, y) = (x - y)(x^\alpha - y^\alpha)^{-1}$ be defined on $\mathbb{R}_+^2$. For every $x \neq y$*

$$\frac{(x + y)^{1-\alpha}}{\alpha} \le L_\alpha(x, y) \le \frac{2}{\alpha} \frac{(xy)^{1-\alpha}}{x^{1-\alpha} + y^{1-\alpha}}.$$

The function $\varphi(x) = x^\alpha$, $0 < \alpha < 1$, has an increasing and concave derivative and Proposition 3.17 applies, which provides the bounds. Applying this inequality to $(x^{\frac{1}{2}}, y^{\frac{1}{2}})$ and multiplying by $(x^{\frac{1}{2}} + y^{\frac{1}{2}})(x^{\frac{\alpha}{2}} + y^{\frac{\alpha}{2}})^{-1}$ implies

$$L_\alpha(x, y) \le \frac{2}{\alpha}(xy)^{\frac{1-\alpha}{2}} \frac{x^{\frac{1}{2}} + y^{\frac{1}{2}}}{\{x^{\frac{1-\alpha}{2}} + y^{\frac{1-\alpha}{2}}\}\{x^{\frac{\alpha}{2}} + y^{\frac{\alpha}{2}}\}}$$

$$\le \frac{2}{\alpha}(xy)^{\frac{1-\alpha}{2}}$$

but $(xy)^{\frac{1}{2}(1-\alpha)} \le x^{1-\alpha} + y^{1-\alpha}$ and this upper bound is larger than the upper bound of Proposition 3.21. A lower bound of L_α is

$$L_\alpha(x, y) \ge \frac{1}{\alpha} \frac{(x^{\frac{1}{2}} + y^{\frac{1}{2}})^{2-\alpha}}{x^{\frac{\alpha}{2}} + y^{\frac{\alpha}{2}}}.$$

The consecutive bounds of L_α are not decreasing intervals for every (x, y) in $\mathbb{R}^2_+$, as they are for the logarithm. The lower bounds would be increasing if $(x+y)^{1-\alpha}(x^{\frac{\alpha}{2}} + y^{\frac{\alpha}{2}}) \leq (x^{\frac{1}{2}} + y^{\frac{1}{2}})^{2-\alpha}$. By concavity, $x^{\frac{\alpha}{2}} + y^{\frac{\alpha}{2}} < 2(x+y)^{\frac{\alpha}{2}}$ and the previous inequality would be satisfied if

$$(x + y)^{1 - \frac{\alpha}{2}} < \frac{1}{2}(x^{\frac{1}{2}} + y^{\frac{1}{2}})^{2-\alpha}$$

but the right-hand term of this inequality is lower than $2^{1-\alpha}(x + y)^{1-\frac{\alpha}{2}}$ which is smaller than the left-hand term. The upper bounds would be decreasing if

$$\frac{(x^{\frac{1}{2}} + y^{\frac{1}{2}})(x^{1-\alpha} + y^{1-\alpha})}{\{x^{\frac{1}{2}(1-\alpha)} + y^{\frac{1-\alpha}{2}}\}\{x^{\frac{\alpha}{2}} + y^{\frac{\alpha}{2}}\}} \leq (xy)^{\frac{1-\alpha}{2}}$$

but it does not seem to be generally satisfied. Since $(x^{\frac{1}{2}} + y^{\frac{1}{2}})^\alpha$ is lower than $x^{\frac{\alpha}{2}} + y^{\frac{\alpha}{2}}$, the left-hand term of this inequality is smaller than

$$\frac{(x^{\frac{1}{2}} + y^{\frac{1}{2}})^{1-\alpha}(x^{1-\alpha} + y^{1-\alpha})}{x^{\frac{1-\alpha}{2}} + y^{\frac{1-\alpha}{2}}} < (x^{\frac{1}{2}} + y^{\frac{1}{2}})^{1-\alpha}(x^{1-\alpha} + y^{1-\alpha})^{\frac{1}{2}}$$

which is not always lower than $(xy)^{\frac{1}{2}(1-\alpha)}$. So no order can be established and the method of Proposition 3.18 cannot be generalized to all functions. It remains to determine a class of function for which such decreasing intervals can be defined by the same arguments as the logarithm.

3.8 Functional means

Functional means have been introduced by Stolarski (1975), Mays (1983), Sahoo and Riedel (1998). For a real function f in $\mathbb{R}$, let

$$M_f(x, y) = f'^{-1}\left(\frac{f(y) - f(x)}{y - x}\right), \quad x \neq y.$$

For $f(x) = x^{-1}$, $M_f(x, y) = (xy)^{\frac{1}{2}}$ is the geometric mean, for $f(x) = x^2$, $M_f(x, y)$ is the arithmetic mean of x and y. With other power functions $f_\gamma(x) = x^\gamma$, γ real, we obtain the following expressions called Cauchy's means

$$M_{f_\gamma}(x,y) = \left(\frac{x^\gamma - y^\gamma}{\gamma(x-y)}\right)^{\frac{1}{\gamma-1}}, \ 1 < \gamma \in \mathbb{R},$$

$$M_{f_k}(x,y) = \left(k^{-1} \sum_{j=0}^{k-1} y^j x^{k-j-1}\right)^{\frac{1}{k-1}}, \ k \in \mathbb{N},$$

$$M_{f_{\frac{1}{\gamma}}}(x,y) = \left(\frac{x^{\frac{1}{\gamma}} - y^{\frac{1}{\gamma}}}{\gamma(x-y)}\right)^{-\frac{\gamma}{\gamma-1}}, \ 1 < \gamma \in \mathbb{R},$$

$$M_{f_{\frac{1}{2}}}(x,y) = \left(\frac{x^{\frac{1}{2}} + y^{\frac{1}{2}}}{2}\right)^2,$$

and the logarithm function f yields the logarithm mean L. The functional means have been generalized to $\mathbb{R}^n$ and to functions with two parameters in Sahoo and Riedel (1998). For all real numbers x and y

$$M_{f_2}(x,y) \le M_{f_3}(x,y)$$

but no order can be established between $M_{f_{\frac{1}{2}}}(x,y)$ and $M_{f_{\frac{1}{3}}}(x,y)$ for example and, more generally, between two functional means M_{f_γ} and M_{f_α} with $\alpha \ne \gamma$. The interval established for the function $L_\alpha(x,y) = \alpha^{-1} M_{f_\alpha}^{1-\alpha}(x,y)$ in Proposition 3.21 also applies to M_{f_α}, $0 < \alpha < 1$.

Proposition 3.22. *For all $0 < \alpha < 1$ and (x,y) in $\mathbb{R}_+^{*2}$*

$$M_{f_2}(x,y) \le \frac{M_{f_\alpha}(x,y)}{2} \le \frac{xy}{(x^{1-\alpha} + y^{1-\alpha})^{\frac{1}{1-\alpha}}}.$$

Proposition 3.23. $L(x,y) \le \frac{1}{2}(x+y)$ *for every (x,y) in $\mathbb{R}_+^{*2}$, with equality if and only if $x = y$.*

Proof. Consider the difference between the logarithm mean and the arithmetic mean

$$\delta(x,y) = \frac{x-y}{\log x - \log y} - \frac{x+y}{2}, \ x, y > 0.$$

Since the respective orders of x and y do not change the sign of the logarithm mean, one can assume that $x > y$ and write $y = \theta x$ with θ in $]0,1[$. For $\theta = 1$, $\delta(x,x) = 0$ and for every $x > 0$ and θ in $]0,1[$

$$\delta(x,y) = xh(\theta),$$

$$h(\theta) = \frac{1-\theta}{\log(\theta^{-1})} - \frac{1+\theta}{2}$$

and $h(1) = 0$. The sign of $h(\theta)$ is the same as the sign of

$$g(\theta) = 2\frac{1-\theta}{1+\theta} + \log\theta,$$

increasing on $]0,1[$, therefore $g(\theta) \le g(1) = 0$. $\square$

The logarithm mean is extended to $\mathbb{R}_+^{*n}$ as

$$L_n(x) = \prod_{k=1}^{n-1} \frac{x_k - x_n}{\log x_k - \log x_n}, \quad x = (x_1, \ldots, x_n).$$

For x in $\mathbb{R}_+^{*n}$, the difference $L_n(x) - n^{-1} \sum_{k=1}^{n} x_k$ is unchanged as the sign of $x_k - x_n$ changes, therefore x_k can be supposed lower than x_n and $x_k = \theta_k x_n$, $0 < \theta_k < 1$. Let

$$\delta_n(x) = x_n^{n-1} \prod_{k=1}^{n-1} \frac{\theta_k - 1}{\log \theta_k} - x_n \frac{\sum_{k=1}^{n-1} \theta_k + 1}{n},$$

since the sign of $\delta_n(x)$ depends on whether x_n is larger or smaller than an expression depending on the $(\theta_1, \ldots, \theta_{n-1})$, there is no order between the logarithmic and the arithmetic means on $\mathbb{R}_+^{*n}$, $n \geq 3$.

For α in $]0,1[$, let $\eta_\alpha = M_{f_\alpha}$ be defined for strictly positive $x \neq y$ as the functional mean for the power function with exponent α, and let $L(x,y)$ be the limit of η_α as α tends to zero. Sahoo and Riedel (1998) mentioned Alzer's conjecture as follows. For every $\alpha \neq 0$

$$L(x,y) < \frac{\eta_\alpha(x,y) + \eta_{-\alpha}(x,y)}{2} < \frac{x+y}{2} = M_{f_2}(a,b). \tag{3.14}$$

Using the reparametrization $y = \theta x$ as in the proof of Proposition 3.23

$$h_\alpha(x,y) = \eta_\alpha(x,y) + \eta_{-\alpha}(x,y) = x[\{\frac{1-\theta^\alpha}{\alpha(1-\theta)}\}^{\frac{1}{\alpha-1}} + \{\frac{\theta^{-\alpha}-1}{\alpha(1-\theta)}\}^{\frac{-1}{\alpha+1}}].$$

For every α in $]0,1[$, the inequality

$$L \leq M_{f_2} \leq \frac{M_{f_\alpha}}{2} \leq \frac{\eta_\alpha(x,y) + \eta_{-\alpha}(x,y)}{2}$$

is a consequence of Propositions 3.22 and 3.23. However, it is not proved that the upper bound of (3.14) is true and (3.14) is still a conjecture for $\alpha > 1$.

Proposition 3.24. *For every $\alpha > 1$, $L(x,y) \leq M_{f_\alpha}(x,y)$, (x,y) in $\mathbb{R}_+^{*2}$, with equality if and only if $x = y$.*

Proof. The sign of $L - M_{f_\alpha}$ is the same as the sign of

$$g(\theta) = \frac{\alpha(1-\theta)^2}{1-\theta^\alpha} + \log \theta$$

with a negative first derivative therefore g is decreasing. Since $\lim_{\theta \to 1} = 0$, $g(\theta) > 0$ on $]0,1[$ and it is zero only if $\theta = 1$. $\qquad\square$

Proposition 3.25. *On $\mathbb{R}_+^{*2}$*

$$M_{f_{-2}}(x,y) \le L(x,y) \le M_{f_2}(x,y),$$

with equality if and only if $x = y$.

Proof. The function $L(x,y) - M_{f_{-2}}(x,y)$ is written

$$x[\frac{1-\theta}{-\log\theta} - \{\frac{\theta^{-2}-1}{2(1-\theta)}\}^{-\frac{1}{3}}]$$

with $y = \theta x$. Its sign is the same as the sign of

$$g_{-2}(\theta) = (1-\theta)(1+\theta)^{\frac{1}{3}}(2\theta^2)^{-\frac{1}{3}} + \log\theta.$$

When θ tends to zero, $(2\theta^2)^{\frac{1}{3}}g_{-2}(\theta)$ is equivalent to $1 + (2\theta^2)^{\frac{1}{3}}\log\theta$ which tends to one and $g_{-2}(\theta)$ tends to infinity. The function g_{-2} is decreasing on $]0,1[$, moreover it tends to zero as θ tends to one and the lower bound for L follows. The upper bound is due to Proposition 3.24. $\qquad\square$

Proposition 3.25 is not generalized to every $\alpha > 1$ as in Proposition 3.24. The interval for the logarithmic mean given in Proposition 3.18

$$\frac{(xy)^{\frac{1}{4}}}{2}(x^{\frac{1}{2}}+y^{\frac{1}{2}}) \le L(x,y) \le (\frac{x^{\frac{1}{4}}+y^{\frac{1}{4}}}{2})^2(x^{\frac{1}{2}}+y^{\frac{1}{2}})$$

is smaller than the interval of Proposition 3.25. Intervals for the means M_{f_2} and $M_{f_{-2}}$ are deduced form those of Proposition 3.18

$$M_{f_2}(x,y) \le \frac{(x^{\frac{1}{2}}+y^{\frac{1}{2}})^2}{2},$$

$$\frac{2xy}{x+y} \le M_{f_{-2}}(x,y) \le (xy)^{\frac{1}{2}}.$$

3.9 Young's inequalities

Young's inequality for power functions states that for all real numbers x and y

$$xy \le \frac{x^p}{p} + \frac{y^{p'}}{p'},$$

with equality if and only if $y = x^{p-1}$, where $p > 1$ and p' are conjugate integers. It has been generalized in several forms.

Theorem 3.2. *Let $f : I \subset \mathbb{R} \mapsto \mathbb{R}$ be a function of $C_1(I)$ with a strictly monotone derivative and such that $f(0) = 0$. There exists a unique function $g : f(I) \mapsto \mathbb{R}$ such that $g(0) = 0$*

$$g(y) = \sup_{x \in I}\{xy - f(x)\},$$
$$f(x) = \sup_{y \in I}\{xy - g(y)\},$$
$$xf'(x) = f(x) + g(f'(x)).$$

Proof. Let $g(y) = \sup_{x \in I}\{xy - f(x)\}$, for every y in $f(I)$ there exists a value x in $\bar{I}$ such that $y = f'(x)$ and $x = h(y)$, where h is the inverse of f'. By definition of g, $g(y) = yh(y) - f(h(y))$ and its derivative $g'(y) = h(y)$ is the inverse function of f, the definition of g is then equivalent to

$$f(x) = \sup_{y \in I}\{xy - g(y)\}.$$

Considering $y_x = f'(x)$ as a function, the derivative of xy_x equals $f'(x) + y_x' g'(y_x)$, which is equivalent to $xy = f(x) + g(y)$ under the conditions $f(0) = 0$ and $g(0) = 0$. □

Removing the condition $f(0) = 0$, the equality is replaced by an inequality for positive functions f and g

$$xy_x \leq f(x) + g(y_x).$$

By convexity of the function $\log L_X$, the Cramer transform

$$\varphi(y) = \sup_{x \in I}\{xy - \log L_X(x)\}$$

of a variable X is convex, it is a positive function and it satisfies Young's inequality

$$\log L_X(x) = \sup_{y \in I}\{xy - \varphi(y)\}.$$

For every x, there exists y_x such that

$$\varphi(y_x) = x\frac{L_X'}{L_X}(x) - \log L_X(x),$$
$$y_x = \frac{L_X'}{L_X}(x). \tag{3.15}$$

The Laplace transform satisfies $\log L_X(x) = y_x \varphi'(y_x) - \varphi(y_x)$, where y_x is defined by $\varphi'(y_x) = x$, and the inequality $\log L_X(x) \geq -\varphi(0)$ holds. At $x = 0$, $y_0 = EX$ and $\varphi(y_0) = \sup_{x \in I}\{xEX - \log L_X(x)\} = 0$ by (3.15).

Theorem 3.3. *Let $a > 0$ and let $f : [0, a] \mapsto [0, f(a)]$ be a continuous and strictly increasing function. For every b in $[0, f(a)]$, the inverse function of f satisfies*

$$ab \leq \int_0^a f(x)\, dx + \int_0^{f(a)} f^{-1}(x)\, dx,$$

$$af(a) \geq \int_0^a f(x)\, dx + \int_0^b f^{-1}(x)\, dx,$$

with equality if and only if $b = f(a)$.

Proof. The third equation of Theorem 3.2 applied to primitive functions is equivalent to

$$af(a) = \int_0^a f(x)\, dx + \int_0^{f(a)} f^{-1}(x)\, dx.$$

Since f^{-1} is a strictly increasing function

$$\int_b^{f(a)} f^{-1}(x)\, dx \geq \{f(a) - b\} f^{-1}(b)$$

which is positive if $b \leq f(a)$, which implies the inequality

$$af(a) \geq \int_0^a f(x)\, dx + \int_0^b f^{-1}(x)\, dx,$$

with equality if and only if $b = f(a)$.   □

Cunnigham and Grossman (1971) presented another inequality.

3.10   Entropy and information

Let $(\mathcal{X}, \mathcal{B})$ be a separable Banach space, with a measure μ, and let $(\mathcal{F}, \| \cdot \|)$ be a vector space of strictly positive functions of $L_1(\mathcal{X})$. The entropy is defined on $(\mathcal{F}, \| \cdot \|)$ by

$$Ent_\mu(f) = \int_\mathcal{X} (\log f) f\, d\mu - \Big(\int_\mathcal{X} f\, d\mu \Big) \log \Big(\int_\mathcal{X} f\, d\mu \Big), \quad f > 0 \in \mathcal{F}. \quad (3.16)$$

By a normalization of f as a density $g_f = (\int_\mathcal{X} f\, d\mu)^{-1} f$, the entropy of a density g is written as $Ent_\mu(g) = \int_\mathcal{X} (\log g)\, g\, d\mu = E\{\log g(X)\}$, where X is a random variable with values in $\mathcal{X}$ with density g, and

$$Ent_\mu(f) = \Big(\int_\mathcal{X} f\, d\mu \Big) Ent_\mu \Big\{ \int_\mathcal{X} f\, d\mu)^{-1} f \Big\}.$$

For every density function g in $\mathcal{F}$

$$Ent_\mu(g) = \int_{\mathcal{X}} (\log g)\, g\, d\mu.$$

Let $\mathcal{F}$ be a family of densities such that for $F = \sup\{f \in \mathcal{F}\}$, $F \log F$ is $L_1(\mathcal{X})$, then the entropy of $\mathcal{F}$ is finite and equals to

$$Ent_\mu(\mathcal{F}) = \sup_{g \in \mathcal{F}} \int_{\mathcal{X}} (\log g)\, g\, d\mu = \int_{\mathcal{X}} (\log F)\, F\, d\mu.$$

This definition is extended to probability measures P and Q on $(\mathcal{X}, \mathcal{B})$ such that P is absolutely continuous with respect to Q

$$Ent_Q(P) = \int_{\mathcal{X}} (\log \frac{dP}{dQ})\, dP.$$

For a density

$$Ent_\mu(g) = \int_{\mathcal{X}} (\log g)\, g\, d\mu \le \int_{\mathcal{X}} (\log g)\, g 1_{\{g>1\}}\, d\mu$$

and $Ent_\mu(g) \le 0$ for every density $g \le 1$, for example, for the uniform densities on intervals larger than 1, for the Gaussian densities with variances larger than $(2\pi)^{-\frac{1}{2}}$ and for the exponential densities $g(x) = ae^{-ax}$ on $\mathcal{R}_+$, with constants $0 < a \le 1$.

Another upper bound for the entropy is deduced from the concavity of the logarithm

$$Ent_\mu(f) \le \{\int_{\mathcal{X}} (\log f - \int_{\mathcal{X}} (\log f)\, d\mu)^2\, d\mu\}^{\frac{1}{2}} \{\int_{\mathcal{X}} f^2\, d\mu\}^{\frac{1}{2}}.$$

The entropy of a product of densities is

$$Ent_{\mu^{\otimes n}}(\prod_{k=1}^{n} f_k) = \sum_{k=1}^{n} Ent_\mu(f_k)$$

and more generally, for a product of functions

$$Ent_{\mu^{\otimes n}}(\prod_{k=1}^{n} f_k) = \sum_{k=1}^{n} Ent_\mu(f_k)\{\prod_{k \neq j=1}^{n} \int_{\mathcal{X}} f_j\, d\mu\}.$$

Let $\lambda > 0$, the entropy of λf is

$$Ent_\mu(\lambda f) = Ent_\mu(f) + (\int_{\mathcal{X}} f\, d\mu)\{(1-\lambda)\log(\int_{\mathcal{X}} f\, d\mu) - \lambda \log \lambda\} + \log \lambda$$

and for a density g, $Ent_\mu(\lambda g) = Ent_\mu(g) + (1-\lambda)\log \lambda$.

The Kullback-Leibler information for two densities f and g of a space of densities $(\mathcal{F}, \| \cdot \|)$ on $(\mathcal{X}, \mathcal{B}, \mu)$ is

$$I_f(g) = \int_{\mathcal{X}} (\log \frac{g}{f}) f \, d\mu. \tag{3.17}$$

For probability measures P and Q on $(\mathcal{X}, \mathcal{B})$ such that P is absolutely continuous with respect to Q, the information is

$$I_P(Q) = \int_{\mathcal{X}} (\log \frac{dQ}{dP}) \, dP.$$

The entropy of two probabilities is therefore defined as the opposite of their information $I_P(Q) = -Ent_Q(P)$.

Proposition 3.26. *The Kullback-Leibler information for two densities f and g with respect to a measure μ such that $\log f$ and $\log g$ are integrable with respect to the measure $\int_{-\infty}^{\cdot} f \, d\mu$ satisfies $I_f(g) \leq 0$, with equality if and only if $f = g$, μ a.s.*

For absolutely continuous probability measures P and Q

$$I_P(Q) = -Ent_Q(P), \quad I_P(Q) \leq 0, \quad Ent_Q(P) \geq 0,$$

with equality if and only if $P = Q$ a.s.

Proof. Since $\int_{\mathcal{X}} (g - f) \, d\mu = 0$ and $\log(1 + x) \leq x$ for every $x > 0$

$$I_f(g) = \int_{\mathcal{X}} \{(\log \frac{g}{f}) - \frac{g}{f} + 1\} f \, d\mu \leq 0. \qquad \square$$

For example, the information of two Gaussian variables with means μ_1 and μ_2, and variances σ_1^2 and σ_2^2, is

$$I_{f_1}(f_2) = (\log \frac{\sigma_1^2}{\sigma_2^2} - \frac{\sigma_1^2}{\sigma_2^2} + 1) - \frac{1}{2}(\mu_1 - \mu_2)^2 \leq -\frac{1}{2}(\mu_1 - \mu_2)^2.$$

The information of $\frac{1}{2}(f + f_0)$ with respect to f_0 is

$$I_{f_0}(\frac{1}{2}(f + f_0)) = -\log 2 + \int_{\mathbb{R}} \{\log(1 + f f_0^{-1}) - f f_0^{-1} + 1\} f_0 \, d\mu \leq 1 - \log 2.$$

By concavity

$$I_{f_0}(\sum_{k=1}^{n} \alpha_k f_k) \geq \sum_{k=1}^{n} \alpha_k \mathcal{I}_{f_0}(f_k),$$

for positive constants α_k such that $\sum_{k=1}^{n} \alpha_k = 1$, and

$$I_{f_0}(\prod_{k=1}^{n} f_k) = \sum_{k=1}^{n} I_{f_0}(f_k).$$

Chernov's theorem applied to the empirical distribution of independent variables with density f, $\widehat{F}_n(t) = \sum_{i=1}^{n} 1_{\{X_i \leq t\}}$, has a limiting function $-I(f)$, where $I(f)$ is the Fisher information of the distribution of the variables with respect to the Lebesgues measure.

Chapter 4

Inequalities for Martingales

4.1 Introduction

On a probability space $(\Omega, \mathcal{F}, P)$, let $(X_n)_{n \geq 0}$ be a sequence of real variables. Let $X_{(n)} = (X_1, \ldots, X_n)$ be the vector of its first n components and let $X_n^* = \sup_{1 \leq i \leq n} X_i$ be its maximal variable. The equivalence of the norms in vector spaces (1.3) are transposed to random vectors in $\mathbb{R}^n$. Let $\|X_{(n)}\|_{n,p} = E\{\sum_{i=1}^n X_i^p\}^{\frac{1}{p}}$, thus $n^{-1}\|X_{(n)}\|_{n,1} \leq |X_n^*| \leq \|X_{(n)}\|_{n1}$ with the equality $\|X_{(n)}\|_{n,1} = n|X_n^*|$ if and only if all components of $X_{(n)}$ are equal.

Proposition 4.1. *For all integers* $1 \leq p < q < \infty$

$$n^{\frac{1}{p}}\|X_{(n)}\|_{n,p} \leq |X_n^*| \leq \|X_{(n)}\|_{n,p} \leq n^{\frac{1}{p}}|X_n^*|,$$

$$n^{-\frac{1}{p}}\|X_{(n)}\|_{n,p} \leq \|X_{(n)}\|_{n,1} \leq n^{\frac{1}{p'}}\|X_{(n)}\|_{n,p},$$

$$n^{-\frac{1}{p'}}\|X_{(n)}\|_{n,1} \leq \|X_{(n)}\|_{n,p} \leq n^{\frac{1}{p}}\|X_{(n)}\|_{n,1},$$

$$n^{-\frac{1}{p'}-\frac{1}{q}}\|X_{(n)}\|_{n,q} \leq \|X_{(n)}\|_{n,p} \leq n^{\frac{1}{p}+\frac{1}{q'}}\|X_{(n)}\|_{n,q},$$

with conjugate integers $p^{-1} + p'^{-1} = 1$ *and* $q^{-1} + q'^{-1} = 1$.

Most bounds depend on the dimension of the vector and the inequalities are generalized as n tends to infinity with a normalization of $\|X_{(n)}\|_{n,p}$ by $n^{-\frac{1}{p}}$. For every $p \geq 1$, $|X_\infty^*| \leq \|X\|_{\infty,p}$, where $X = (X_1, X_2, \ldots)$ belongs to $\mathbb{R}^\infty$.

Several inequalities between the sum $S_n = \sum_{i=1}^n X_i$ and the sum of the squared variables $V_n = \sum_{i=1}^n X_i^2$ are first established, for a sequence of random variables $(X_n)_{n \geq 0}$. They are based on the equivalence of the norms in $\mathbb{R}^n$. The same notation is used for independent variables or for dependent variables such as the increments of a discrete martingale and the inequalities are extended to martingales indexed by $\mathbb{R}$. Other inequalities concern the distances between subsets of n-dimensional vector spaces.

4.2 Inequalities for sums of independent random variables

The Bienaymé-Chebychev inequality is an upper bound, a similar inequality is established for a lower bound of the probability $P(X > a)$, for a random variable X.

Proposition 4.2. *Let X be a random variable of $L_2(\mathbb{R})$ and let b such that $E|X| \geq b$. For every a in $]0, b[$*

$$P(|X| > a) \geq \frac{(b-a)^2}{EX^2}.$$

Proof. Splitting $|X|$ on the sets $\{|X| > a\}$ and $\{|X| \leq a\}$, we obtain

$$E|X| \leq \{EX^2 P(|X| > a)\}^{\frac{1}{2}} + E\{|X|1_{\{|X| \leq a\}},$$

moreover $E|X| - E\{|X|1_{\{|X| \leq a\}} \geq b - a$ and the result follows. □

In particular, for every a in $]0, E|X|[$

$$P(|X| > a) \geq \frac{(E|X| - a)^2}{EX^2}.$$

The next bound is lower than the bound of the Bienaymé-Chebychev inequality.

Proposition 4.3. *Let X be a random variable of $L_2(\mathbb{R})$ such that $EX = m$ and $var X > 0$. For every $a > 0$ such that $a^2 \leq var X$*

$$P(|X - m| > a) \geq \frac{a^2}{var X}.$$

Proof. By the Cauchy-Schwarz inequality

$$a < E\{|X - m|1_{\{|X-m|>a\}}\} \leq \{(var X)P(|X - m| > a)\}^{\frac{1}{2}}. □$$

The Bürkholder, Davis and Gundy inequality for integrable variables is extended under integrability to moments with real exponents with explicit and optimal bounds. Let $c_{p,n} = n^{-\frac{1}{p} - \frac{1}{2}}$ and $C_{p,n} = n^{\frac{p-1}{p} + \frac{1}{2}}$.

Proposition 4.4. *On $(\Omega, \mathcal{F}, P)$, let $(X_n)_{n \geq 0}$ be a sequence of random variables of L_α for a real $\alpha \geq 2$*

$$c_{\alpha,n} E(V_n^{\frac{\alpha}{2}}) \leq E(|S_n|^\alpha) \leq C_{\alpha,n} E(V_n^{\frac{\alpha}{2}}). \tag{4.1}$$

For positive variables $E(V_n^{\frac{\alpha}{2}}) \leq E(|S_n|^\alpha)$.

Proof. For $p \geq 2$, this is a consequence of Proposition 4.1 and the same inequalities hold for every real $\alpha \geq 2$ by limit of the inequality for rational numbers for which the inequality is obtained as for integers. The second inequality is deduced from the first one by monotonicity. $\square$

The constants in Proposition 4.4 depend on the number of variables in the sums S_n and V_n. With independent and centered random variables, $EX_iX_j = 0$, $i \neq j$, and the inequality becomes an equality with $EV_n = ES_n^2$. The inequalities are strict for $\alpha > 2$, however, for every odd integer p, $ES_n^p = 0$. Therefore Proposition 4.4 applies only to $E(|S_n|^p)$.

Proposition 4.5. *On $(\Omega, \mathcal{F}, P)$, let $(X_n)_{n \geq 0}$ be a sequence of independent and symmetric random variables of L_{2p} for every integer $p \geq 1$. There exists a constant depending only on p such that for every $n \geq 2$*

$$E(V_n^p) \leq E(S_n^{2p}) \leq C_p E(V_n^p).$$

Proof. Since the variables are centered, the odd terms in the expansion of S_n^{2p} are zero and the lower bound is an equality. The upper bound is established recursively, it is an equality for $p = 1$ and for every integer $n \geq 2$. Let us assume that it is true up to $p - 1$, for every $n \geq 2$ and for $k = 1, \ldots, p - 1$, $ES_{n-1}^{2(p-k)} \leq C_{p-1} EV_{n-1}^{p-k}$. Then $ES_n^{2p} = EX_n^{2p} + \sum_{k=1}^{p} \binom{2p}{2(p-k)} EX_n^{2k} ES_{n-1}^{2(p-k)}$ is bounded by $EX_n^{2p} + \sum_{k=1}^{p} \binom{2p}{2(p-k)} EX_n^{2k} C_{p-k} E(V_{n-1}^p) \leq C_p E(V_n^p)$, with a constant independent of n, $C_p \leq \max_{k=1,\ldots,p} \binom{2p}{2(p-k)} \binom{p}{p-k}^{-1} C_{p-k}$. $\square$

This inequality is similar to Kinchin's inequality (1.5) for sums of positive variables. In Proposition 4.6, the constants can be calculated explicitly, for $p = 2, 4, 6$ they are $C_2 = 1$, $C_4 = 3$, $C_6 = 5$, and they increase with p. The ratio $\binom{2p}{2(p-k)} \binom{p}{p-k}^{-1}$ is smaller than $2^k \binom{p}{p-k}$ and their maximum over k up to p is reached at $k = \frac{2}{3}p$.

The sequence of variables $(X_i)_{i=1,\ldots,n}$ is symmetrized by an independent sequence of mutually independent and uniform random variables $(\varepsilon_i)_{i=1,\ldots,n}$ with values in $\{1, -1\}$, as $Y_i = \varepsilon_i X_i$.

Lemma 4.1. *Let $(X_n)_{n \geq 0}$ be a sequence of independent and centered random variables of L_p for an integer $p \geq 2$. Then*

$$E\| \sum_{i=1}^{n} \varepsilon_i X_i \|_{L_{2p}} = E\| \sum_{i=1}^{n} X_i \|_{L_{2p}}.$$

Proof. By independence of the centered variables X_i, the property is true for $p = 1$, with $E(\sum_{i=1}^n \varepsilon_i X_i)^2 = E(\sum_{i=1}^n X_i^2)$, and the same property is satisfied for $E\| \sum_{i=1}^n \varepsilon_i X_i \|_{L_{2p}}$ since it develops in sums containing only products of variables X_i with an even exponent. $\qquad\square$

By symmetrization of independent and centered variables, Proposition 4.5 is extended to their sum S_n, with an even exponent.

Proposition 4.6. *Let $(X_n)_{n\geq 0}$ be a sequence of independent centered random variables of L_{2p}, $p \geq 1$. Then $E(V_n^p) \leq E(S_n^{2p}) \leq C_p E(V_n^p)$.*

Let $a = (a_i)_{i=1,\ldots,n}$ and $X = (X_i)_{i=1,\ldots,n}$ be vectors of $\mathbb{R}^n$, the sum of the weighted variables $S_{n,a} = \sum_{i=1}^n a_i X_i$ and $V_{n,a} = \sum_{i=1}^n a_i^2 X_i^2$ satisfy the inequality (4.1) with

$$E(|S_{n,a}|^\alpha) \leq \|a\|_2^\alpha E(\|S_n\|_2^\alpha), \ E(V_{n,a}^{\frac{\alpha}{2}}) \leq \|a\|_4^\alpha E(\|X\|_4^\alpha).$$

For a real $0 < \alpha < 2$, the inequalities have another form, due to the concavity of the norms indexed in $]0,1[$.

Proposition 4.7. *Let $(X_n)_{n\geq 0}$ be a sequence of independent random variables of $L_2(\mathbb{R})$ and let α be a real in $]0,2[$, $\|S_n^2 - V_n\|_{L_{\frac{\alpha}{2}}} < EV_n$. If $\alpha > 2$*

$$E|S_n^2 - V_n|^{\frac{\alpha}{2}} > E\| \sum_{i\neq j=1}^n X_i X_j |1_{\{\sum_{i\neq j=1}^n X_i X_j > 0\}}$$

$$- |\sum_{i\neq j=1}^n X_i X_j |1_{\{\sum_{i\neq j=1}^n X_i X_j < 0\}}| \geq 0.$$

Proof. Writing $(\sum_{i=1}^n X_i)^2 - \sum_{i=1}^n X_i^2 = \sum_{i\neq j=1}^n X_i X_j$, we have $E|S_n^2 - V_n|^{\frac{\alpha}{2}} < (E| \sum_{i\neq j=1}^n X_i X_j|)^{\frac{\alpha}{2}}$ and by monotonicity $\|S_n^2 - V_n\|_{L_{\frac{\alpha}{2}}}$ is lower than $\sum_{i\neq j=1}^n E|X_i X_j| < \sum_{i\neq j=1}^n (EX_i^2 EX_j^2)^{\frac{1}{2}} < \sum_{i=1}^n EX_i^2 = EV_n$. The second inequality is the consequence of the convexity of the power function and of the triangular inequality. $\qquad\square$

Proposition 4.4 is generalized as an inequality for products of sums of independent random variables, with integers $2 \leq k < m < n$. For all real numbers $\alpha \geq 2$ and $\beta \geq 2$

$$c_{\alpha,n-m}c_{\beta,m-k}E\{(V_n - V_m)^{\frac{\alpha}{2}}(V_m - V_k)^{\frac{\beta}{2}}\} \leq E(|S_n - S_m|^\alpha |S_m - S_k|^\beta)$$

$$\leq C_{\alpha,n-m}C_{\beta,m-k}E\{(V_n - V_m)^{\frac{\alpha}{2}}(V_m - V_k)^{\frac{\beta}{2}}\}.$$

Let $j_n < k_n$ be increasing indices such that $j_n k_n^{-1}$ converges to α in $]0,1[$ as j_n tends to infinity. Let S_{j_n} and S_{k_n} be the sums of independent variables with zero means and variances σ_i^2, up to j_n and k_n respectively. The normalized sums $Y_{1n} = j_n^{-\frac{1}{2}} S_{j_n}$ and $Y_{2n} = k_n^{-\frac{1}{2}} S_{k_n}$ converge weakly to a random variable $Y \sim \mathcal{N}(0, \sigma^2)$ if $\bar{\sigma}_{j_n}^2 = j_n^{-1} \sum_{i=1}^{j_n} \sigma_i^2$ and $\bar{\sigma}_{k_n}^2 = k_n^{-1} \sum_{i=1}^{k_n} \sigma_i^2$ converge to a limit σ^2 as j_n tends to infinity. Then $Y_{3n} = (k_n - j_n)^{-\frac{1}{2}} (S_{k_n} - S_{j_n})$ is independent of Y_{2n} and it converges weakly to the variable $Y_3 = (1 - \alpha)^{-\frac{1}{2}} (Y_2 - \alpha^{\frac{1}{2}} Y_1)$ where $Y_1 = \lim j_n^{-\frac{1}{2}} S_{j_n}$ and Y_3 are independent, Y_1, Y_2 and Y_3 have the same distribution. The variable

$$\frac{Y_{2n}}{Y_{1n}} = \frac{S_{k_n}}{k_n^{\frac{1}{2}}} (\frac{S_{j_n}}{j_n^{\frac{1}{2}}})^{-1} = (\frac{j_n}{k_n})^{\frac{1}{2}} + \frac{S_{k_n} - S_{j_n}}{(k_n - j_n)^{\frac{1}{2}}} (\frac{k_n - j_n}{j_n})^{\frac{1}{2}} (\frac{S_{j_n}}{j_n^{\frac{1}{2}}})^{-1}$$

converges weakly to $\alpha^{\frac{1}{2}} + (1 - \alpha)^{\frac{1}{2}} \alpha^{-\frac{1}{2}} Y_1^{-1} Y_3$ and

$$E(\frac{Y_{2n}}{Y_{1n}} - \alpha^{\frac{1}{2}})^2 = \frac{1 - \alpha}{\alpha} \frac{EY_3^2}{EY_1^2} = \frac{1 - \alpha}{\alpha}.$$

The Laplace transforms of Y_{1n}, Y_{2n} and Y_{3n} converge to the Laplace transform of X.

If the variables X_i have a mean μ_i such that $\bar{\mu}_n$ converges to a limit μ, the variables Y_{kn} are centered

$$Y_{3n} = (k_n - j_n)^{\frac{1}{2}} \{(k_n - j_n)^{-1}(S_{k_n} - S_{j_n}) - (\bar{\mu}_{k_n} - \bar{\mu}_{j_n})\}$$
$$= (k_n - j_n)^{-\frac{1}{2}}(S_{k_n} - S_{j_n}) - (k_n - j_n)^{\frac{1}{2}}(\bar{\mu}_{k_n} - \bar{\mu}_{j_n})$$

and it converges weakly to Y only if $n^{-1}\bar{\mu}_n$ converges to a finite limit, hence $\beta_n = (k_n - j_n)^{\frac{1}{2}}(\bar{\mu}_{k_n} - \bar{\mu}_{k_n})$ converges to zero as n tends to infinity.

With identically distributed variables X_i, the Laplace transforms of S_{j_n} and S_{k_n} are asymptotically equivalent to $L_X^{j_n}$ and $L_X^{k_n}$, respectively, and the Laplace transform of $S_{k_n} - S_{j_n}$ is asymptotically equivalent to $L_X^{k_n - j_n}$.

Applying the Bienaymé-Chebychev inequality (1.12) to the partial sums entails that for every $a > 0$

$$P\{(k_n - j_n)^{-\frac{1}{2}}|S_{k_n} - S_{j_n}| > a\} \leq \frac{\sum_{i=j_n+1}^{k_n} \sigma_i^2}{a^2(k_n - j_n)} \tag{4.2}$$

and it tends to zero as a tends to infinity. If the variables are not centered, Equation (4.2) applies to the variable $|S_{k_n} - S_{j_n} - \beta_n|$. With $p = 2$ and $k_n - j_n = bn$, the next bound is asymptotically independent on n

$$\sup_{n \geq 1} P\{(k_n - j_n)^{-1}|S_{k_n} - S_{j_n}| > a\} \leq \frac{n \sum_{i=j_n+1}^{k_n} \sigma_i^2}{a^2(k_n - j_n)^2} = \frac{\sum_{i=j_n+1}^{k_n} \sigma_i^2}{a^2 b(k_n - j_n)}.$$

From Proposition 4.2, if $0 < a < (k_n - j_n)^{-1} E| \sum_{i=j_n+1}^{k_n} X_i|$

$$P\{(k_n - j_n)^{-1}|S_{k_n} - S_{j_n}| \leq a\} \geq \frac{(E|S_{k_n} - S_{j_n}| - (k_n - j_n)a)^2}{E|S_{k_n} - S_{j_n}|^2}.$$

Chernov's equality for $S_{k_n} - S_{j_n}$ is also modified by the number of variables of the sum, for all $a > 0$ and $t > 0$

$$P\{S_{k_n} - S_{j_n} > a\} = E \exp\{t(S_{k_n} - S_{j_n}) - at\} = e^{-at} L_X^{k_n - j_n}(t),$$

then $\lim_{n\to\infty} \log P\{S_{k_n} - S_{j_n} > a\} = \inf_{t>0}\{(k_n - j_n) \log L_X(t) - at\}$.

Under the conditions $X_i = 0$ and $|X_i| \leq \sigma_i M$ for every i, Bennett's inequality is written

$$P((k_n - j_n)^{-1}(S_{k_n} - S_{j_n}) > t) \leq \exp\{-(k_n - j_n)\phi(\frac{t}{M \sum_{i=j_n+1}^{k_n} \sigma_i^2})\}$$

where ϕ is defined in Theorem 1.5. Removing the boundedness condition, the right member of this inequality is a sum of two terms. For all $M > 0$ and $t > 0$

$$P(\frac{S_{k_n} - S_{j_n}}{k_n - j_n} > t) \leq \exp\{-(k_n - j_n)\phi(\frac{t}{\sum_{i=j_n+1}^{k_n} \sigma_i^2 M})\}$$
$$+ \prod_{i=j_n}^{k_n} P(|X_i| > \sigma_i M).$$

By independence of $S_{k_n} - S_{j_n}$ and S_{j_n}, the inequalities are extended to product inequalities. Let $|X_i| \leq \sigma_i M$ for every i

$$\lim_{n\to\infty} P\{|S_{j_n}| > a, |S_{k_n} - S_{j_n}| > b\} \leq \frac{(\sum_{i=1}^{j_n} \sigma_i^2)(\sum_{i=j_n+1}^{k_n} \sigma_i^2)}{(ab)^2},$$

$$\lim_{n\to\infty} P(S_{j_n} > t_1, S_{k_n} - S_{j_n} > t_2) \leq \exp\{-j_n\phi(\frac{t_1}{M \sum_{i=1}^{j_n} \sigma_i^2})\}$$
$$\exp\{-(k_n - j_n)\phi(\frac{t_2}{M \sum_{i=j_n+1}^{k_n} \sigma_i^2})\}.$$

The Laplace transform of the combination of independent partial sums $Y_{n,\alpha} = \alpha S_{j_n} + (1 - \alpha)(S_{k_n} - S_{j_n})$, $0 < \alpha < 1$, is

$$L_{n,\alpha}(t) = E[\exp\{t\alpha S_{j_n}\} \exp\{t(1 - \alpha)(S_{k_n} - S_{j_n})\}]$$
$$= \prod_{i=1}^{j_n} L_{X_i}(t\alpha) \prod_{i=j_n+1}^{k_n} L_{X_i}(t(1 - \alpha)).$$

With partial sums of Gaussian variables $\mathcal{N}(0, \sigma^2)$, Chernov's inequality for $Y_{n,\alpha}$ implies

$$P\{Y_{n,\alpha} > t\} = \exp[-\frac{t^2}{\{(k_n - j_n)(1 - \alpha)^2 + j_n\alpha^2\}\sigma^2}].$$

Let S_n be the sum of independent and identically distributed vectors of $\mathbb{R}^d$ and such that the variance of the k-th component of $n^{-\frac{1}{2}}S_n$ is σ_k^2 and let $\sigma^2 = n^{-1}\sum_{k=1}^d \sigma_k^2$. From Proposition 4.4, an inequality of the same kind as (1.12) can be proved for the L_2-norm of S_n. Let $\sigma^2 = \sum_{k=1}^d \sigma_k^2$.

Theorem 4.1. *The sum of independent and identically distributed variables with values in $\mathbb{R}^d$ and mean zero satisfies*

$$P(\|S_n\|_{d,2} > a) \leq \frac{2\sigma\sqrt{d}}{a}, \ t > 0.$$

Proof. The $L_2(\mathbb{R}^d)$-norm of S_n has the variance

$$E\|n^{-1}S_n\|_{d,2}^2 = \sum_{k=1}^d E|S_{nk}|^2 = \sum_{k=1}^d EV_{nk} \leq d\sigma^2$$

and the result is obtained from (1.12). $\qquad\square$

In $\mathbb{R}^d$, Chernov's limit is replaced by an exponential inequality

$$P(n^{-1}\|S_n\|_1 > a) \leq \sum_{k=1}^d P(n^{-1}|S_{nk}| > a) \leq 2\sum_{k=1}^d \inf_{t_k > 0} L_{X_k}^n(t_k)e^{-at_k n}$$

where $t = (t_1, \ldots, t_d)$ is a vector of $\mathbb{R}^d$. When n tends to infinity, $n^{-\frac{1}{2}}S_n$ converges to a centered Gaussian variable of $\mathbb{R}^d$ with variance matrix Σ and $\inf_{t_k > 0} \exp[n\{\log L_{X_k}(t_k) - at_k\}]$ is equivalent to $\exp\{-n\frac{a^2}{2\sigma_k^2}\}$, therefore, as n tends to infinity

$$P(n^{-1}\|S_n\|_{d,2} > a) \leq 2\sum_{k=1}^d e^{-na^2(2\sigma_k^2)^{-1}}.$$

Since $En^{-\frac{1}{2}}\|S_n\|_{d,2}$ converges to $\|\Sigma\|_{d,2}$, the inequalities are also written with sharper bounds as n tends to infinity, then $P(\|n^{-\frac{1}{2}}S_n\|_{d,2} > a)$ is equivalent to $e^{-na^2(2\|\Sigma\|_2^2)^{-1}}$.

The dependence between σ-algebras $\mathcal{A}$ and $\mathcal{B}$ are measured by the co-efficients

$$\varphi(\mathcal{A}, \mathcal{B}) = \sup_{A\in\mathcal{A}, B\in\mathcal{B}: P(B) > 0} |P(A|B) - P(A)|,$$

$$\alpha(\mathcal{A}, \mathcal{B}) = \sup_{A\in\mathcal{A}, B\in\mathcal{B}: P(B) > 0} |P(A \cap B) - P(A)P(B)|.$$

The convergence of dependent variables is classically established under mixing condition between the σ-algebras they generate. Let $\mathcal{M}_{1,n}$ be the σ-algebra generated by $(X_j)_{j=1,\ldots,k}$ and $\mathcal{M}_{n+k,\infty}$ be the σ-algebra generated by $(X_j)_{j\geq n+k}$. The φ-mixing coefficients for $(S_n)_{n\geq 1}$ and its strong mixing coefficients are defined for every integer $k \geq 1$ as

$$\varphi_k = \varphi(\mathcal{M}_{1,n}, \mathcal{M}_{n+k,\infty}),$$
$$\alpha_k = \alpha(\mathcal{M}_{1,n}, \mathcal{M}_{n+k,\infty}).$$

The variables $(X_n)_{n\geq 1}$ is φ-mixing or α-mixing if the coefficients φ_k or α_k tend to zero as k tends to infinity.

Lemma 4.2 (Serfling, 1968). *Let X be a random variable and let p be an integer $1 \leq p \leq \infty$, for all Borel set $\mathcal{F}$ and integer $r > p$*

$$\|E(X|\mathcal{F}) - EX\|_p \leq 2\{\varphi(\mathcal{F}, \mathcal{A})\}^{1-\frac{1}{r}} \|X\|_r,$$
$$\|E(X|\mathcal{F}) - EX\|_p \leq 2(2^{\frac{1}{p}} + 1)\{\alpha(\mathcal{F}, \mathcal{A})\}^{\frac{1}{p}-\frac{1}{r}} \|X\|_r.$$

The moments of Proposition 4.4 is replaced by the next ones. Let X_i be in $\mathcal{M}_{1,n}$ and X_j be in $\mathcal{M}_{n+k,\infty}$ and let $p \geq 2$ be an integer. Under the condition of φ-mixing, $|E(X_iX_j) - EX_iEX_j| \leq \varphi_k^{\frac{1}{2}}\|X_i\|_p\|X_j\|_q$, where $q \geq 2$ is the conjugate integer of p, and the variance of $n^{-\frac{1}{2}}S_n$ is finite under the condition $\sum_{k\geq 0} k^2\varphi_k^{\frac{1}{2}} < \infty$. According to Billingsley (1968), the moments of S_n satisfy

$$E|S_n^p| \leq \sum_{k=0}^{\frac{p}{2}}(k+1)^{p-2}\varphi_k^{\frac{1}{2}}\{\max_{i=1,\ldots,n} E(X_i^2)\}^k, \ p \geq 2.$$

Under the α-mixing condition and

$$\sum_{n=0}^{\infty}(n+1)^{p-2}\alpha_n^{\frac{1}{2}} < \infty, \tag{4.3}$$

$|E(X_iX_j) - EX_iEX_j| \leq 12\alpha_k^{\frac{1}{2}}\|X_i\|_p\|X_j\|_q$, where $p^{-1}+q^{-1} = \frac{1}{2}$, and there exists a constant $k(\alpha, p)$ such that

$$E|S_n^p| \leq k(\alpha, p) \sum_{k=0}^{\frac{p}{2}}\{\max_{i=1,\ldots,n} E(X_i^2)\}^k.$$

Under the conditions about the convergence rate of the mixing coefficients, the normalized sum $n^{-\frac{1}{2}}(S_n - ES_n)$ converges weakly to a Gaussian variable with mean zero and a finite variance.

For dependent variables $(X_i)_{i=1,\ldots,n}$, the Laplace transform of the sum S_n is not a product of n identical terms L_X but it is sufficient that there exists a constant $\alpha > 0$ such that $n^{-\alpha} S_n$ converges a.s. to a limit to ensure a large deviations inequality.

Theorem 4.2. *Let* $(X_i)_{i=1,\ldots,n}$ *be a sequence of random variables on a probability space* $(\Omega, \mathcal{F}, P)$ *and let* $S_n = \sum_{i=1}^n X_i$. *If there exists a constant* $\alpha > 0$ *such that* $n^{-\alpha} S_n$ *converges a.s. to a limit* S_0 *having a finite Laplace transform* $L_{S_0}(t)$ *for every* $t > 0$, *then*

$$\lim_{n \to \infty} n^{-\alpha} \log P(n^{-\alpha} S_n > a) = \inf_{t>0} \{\log L_{S_0}(t) - at\}$$

and it is finite.

Proof. By concavity, $\log E e^{n^{-\alpha} S_n t} \geq n^{-\alpha} t\, E S_n$ which converges to $t E S_0$. The a.s. convergence of $n^{-\alpha} S_n$ implies

$$L_{S_n}(t) = E e^{\{S_0 + o(1)\}t} = \{L_{S_0}^{n^\alpha}(n^{-\alpha} t)\}\{1 + o(1)\}$$

with a.s. $o(1)$. The proof ends like in Chernov's theorem. $\square$

4.3 Incqualities for discrete martingales

On a filtered probability space $(\Omega, \mathcal{F}, (\mathcal{F}_n)_{n\geq 0}, P)$, let $(X_n)_{n\geq 0}$ be a real martingale and let $(V_n(X))_{n\geq 0}$ be the process of its quadratic variations. Let $U_i = X_i - X_{i-1}$, then X_n is written as the sum $X_n = X_0 + \sum_{i=1}^n U_i$, where $(U_n)_{n\geq 0}$ is a sequence of dependent random variables with respective variances $\sigma_i^2 = E(X_i - X_{i-1})^2 = E\{V_i(X) - V_{i-1}(X)\}$ and their covariance are zero by the martingale property. The conditions about the moments of the variables X_n are different from those of mixing variables. A discrete version of the Bürkholder-Davis-Gundy inequality for martingales indexed by $\mathbb{N}_*$, with the constants, is deduced for Proposition 4.4 for independent variables.

Proposition 4.8. *Let* $(X_n)_{n\geq 0}$ *be a continuous local martingale of* L_α, $\alpha \geq 2$, *on* $(\Omega, \mathcal{F}, (\mathcal{F}_n)_{n\geq 0}, P)$ *and let* $(V_n(X)))_{n\geq 0}$ *be the process of its quadratic variations. For every stopping time* N *with respect to* $(\mathcal{F}_n)_{n\geq 0}$

$$E\{c_{\alpha,N} V_N^{\frac{\alpha}{2}}\} \leq E(|X_N|^\alpha) < E(C_{\alpha,N} V_N^{\frac{\alpha}{2}}), \tag{4.4}$$

Proof. Writing X_n as the sum of the conditionally independent and centered variables $U_0 + \sum_{i=1}^n U_i$ and $V_n(X) = U_0^2 + \sum_{i=1}^n (X_i - X_{i-1})^2$, the inequality (4.4) is deduced from (4.1) in Proposition 4.4. The upper bounds are still valid when the integer n is replaced by a random stopping time N of the martingale, by Doob's theorem. □

For $p = 2$, the inequalities are equalities since $X_n^2 - V_n$ is a local martingale with zero mean and there is no coefficients in Proposition 4.8. Proposition 4.6 applies to the martingales differences $X_n - X_{n-1}$.

Proposition 4.9. *Let $(X_n)_{n\geq 1}$ be a local martingale of $\mathcal{M}_{0,loc}^{2p}$, for an integer $p \geq 1$. There exists a constant C_p such that for every $n \geq 2$*

$$E(V_n^p) \leq E(X_n^{2p}) \leq C_p E(V_n^p).$$

As a consequence, $(X_n)_{n\geq 1}$ belongs to $\mathcal{M}_{0,loc}^{2p}$ if and only if $(V_n)_{n\geq 1}$ belongs to $\mathcal{M}_{0,loc}^p$, for an integer $p \geq 1$, and for $2 \leq k \leq p$

$$P(n^{-\frac{1}{2}} X_n > x) \leq x^{-2k} n^{-k} C_k EV_n^k(X),$$
$$P(n^{-1} V_n > x) \leq (nx)^{-k} EV_n^k(X).$$

Proposition 4.10. *Let α be in $]0, 2[$ and let $(X_n)_{n\geq 0}$ be a sequence of local martingale of L_4, with a process of quadratic variations $(V_n(X)))_{n\geq 0}$. For every random stopping time N of X*

$$\|X_N^2 - V_N\|_{\frac{\alpha}{2}} \leq EV_N.$$

Let $\alpha > 2$, $\|X_N^2 - V_N\|_{\frac{\alpha}{2}} > 0$.

Proof. Let $X_N = \sum_{i=1}^n (X_i - X_{i-1})$, by concavity of the power function we have $E|X_n^2 - V_n|^{\frac{\alpha}{2}} \leq (E|\sum_{i\neq j=1}^n (X_i - X_{i-1})(X_j - X_{j-1})|)^{\frac{\alpha}{2}}$ and by the Cauchy-Schwarz inequality it is lower than

$$\sum_{i\neq j=1}^n \{E|(X_i - X_{i-1})(X_j - X_{j-1})|\}^{\frac{\alpha}{2}} \leq \{\sum_{i\neq j=1}^n E(V_i - V_{i-1})^{\frac{1}{2}}(V_j - V_{j-1})^{\frac{1}{2}}\}^{\frac{\alpha}{2}}$$

and it is bounded by $(EV_n)^{\frac{\alpha}{2}}$ by monotonocity. □

Proposition 4.8 does not extend to products, for $0 < k < m < n$ the following means of products cannot be factorized, except for martingales with independent increments. Let $\alpha \geq 2$ and $\beta \geq 2$

$$c_{\alpha,n-m} c_{\beta,m-k} E[E\{(V_n - V_m)^{\frac{\alpha}{2}}|\mathcal{F}_m\}(V_m - V_k)^{\frac{\beta}{2}}]$$
$$\leq E\{E(|S_n - S_m|^\alpha|\mathcal{F}_m)|S_m - S_k|^\beta\}$$
$$\leq C_{\alpha,n-m} C_{\beta,m-k} E[E\{(V_n - V_m)^{\frac{\alpha}{2}}|\mathcal{F}_m\}(V_m - V_k)^{\frac{\beta}{2}}].$$

A discrete version of the Birnbaum and Marshal inequality for a weighted martingale is the following. Let $X = (X_n)_n$ be a martingale and $(a_n)_n$ be a real sequence. Due to the Kolmogorov inequality, for every $\lambda > 0$, the maximum of the martingales

$$Y_n = \sum_{k=1}^{n} a_k(X_k - X_{k-1}), \qquad (4.5)$$

$$Z_n = \sum_{k=1}^{n} a_k^{-1}(X_k - X_{k-1}) \qquad (4.6)$$

have the bounds

$$P(\max_{k=1,\ldots,n} |a_k^{-1} Y_k| > \lambda) \leq \lambda^{-2} \sum_{k=1}^{n} (\frac{a_k}{a_n})^2 (V_k - V_{k-1})(X), \qquad (4.7)$$

$$P(\max_{k=1,\ldots,n} |a_k Z_k| > \lambda) \leq \lambda^{-2} \sum_{k=1}^{n} (\frac{a_n}{a_k})^2 (V_k - V_{k-1})(X)$$

therefore $P(\max_{k=1,\ldots,n} |a_k^{-1} Y_k| > \lambda) \leq \lambda^{-2} V_n(X)$ if $(a_n)_n$ is increasing and $P(\max_{k=1,\ldots,n} |a_k Z_k| > \lambda) \leq \lambda^{-2} V_n(X)$ if $(a_n)_n$ is decreasing.

Proposition 4.11. *Let* $X = (X_n)_n$ *be in* $\mathcal{M}^2_{0,loc}$, *let* $(A_n)_{n \geq 1}$ *be a predictable process. Let* $Y_n = \sum_{k=1}^{n} A_k(X_k - X_{k-1})$, *then for every* $\lambda > 0$ *and for every stopping time* N *of* X

$$P(Y_N > \lambda) \leq \lambda^{-p} C_p E(V_N^{\frac{p}{2}}(Y)),$$

$$P(Y_N^* > \lambda) \leq \lambda^{-p} C_p E\{\sum_{i=1}^{N} V_n^{\frac{p}{2}}(Y)\}.$$

The first inequality is an immediate consequence of Equations (1.16) and (4.4). The inequality $P(Y_N^* > \lambda) \leq E\{\sum_{i=1}^{N} P(Y_n > \lambda)\}$ implies the second bound of Proposition 4.11.

Proposition 4.12. *Let* φ *be a real convex function on* $\mathbb{R}$, *let* $X = (X_n)_n$ *be in* $\mathcal{M}^2_{0,loc}$ *and let* $Y_{n+1} = Y_n + \varphi(X_{n+1} - X_n)$, *then* Y *and its quadratic variations are submartingales. For every* $\lambda > 0$

$$P(\sup_{n \geq 1} |Y_n| > \lambda) \leq 2\lambda^{-2} \sum_{n=1}^{\infty} E\varphi^2(X_n - X_{n-1}).$$

Proof. For every $m > n \geq 0$

$$E\{Y_m | \mathcal{F}_n\} = Y_n + E[E \sum_{i=n}^{m-1} E\{\varphi(X_{i+1} - X_i) | \mathcal{F}_i\} | \mathcal{F}_n]$$

$$\geq \varphi(Y_n) + E[E \sum_{i=n}^{m-1} \varphi(E\{X_{i+1} | \mathcal{F}_i\}) - \varphi(X_i) | \mathcal{F}_n] = Y_n.$$

The quadratic variations $V_n(Y) = \sum_{1 \leq i \leq n} (Y_i - Y_{i-1})^2$ have conditional means

$$E\{V_{n+1}(Y)|\mathcal{F}_n\} = V_n(Y) + E(Y_{n+1}^2|\mathcal{F}_n) - Y_n^2$$
$$\geq V_n(Y) + \varphi^2(E(X_{n+1}|\mathcal{F}_n)) - \varphi^2(X_n) = V_n(Y). \quad \square$$

By Jensen's inequality

$$E\{e^{t(X_{i+1}-X_i)} e^{t(X_i - X_{i-1})}|\mathcal{F}_i\} = e^{t(X_i - X_{i-1})} E\{e^{t(X_{i+1}-X_i)}|\mathcal{F}_i\}$$
$$\geq e^{t(X_i - X_{i-1})},$$

it follows that the Laplace transform of X_n is

$$L_{X_n}(t) = E \prod_{i=1}^{n} e^{t(X_i - X_{i-1})} = E \prod_{i=1}^{n} E\{e^{t(X_i - X_{i-1})}|\mathcal{F}_{i-1}\} \geq 1$$

and $\log L_{X_n}(t) \geq E \sum_{i=1}^{n} \log E\{e^{t(X_i - X_{i-1})}|\mathcal{F}_{i-1}\}$, which is the mean of a sum of the random variables $\log L_i(t) = \log E\{e^{t(X_i - X_{i-1})}|\mathcal{F}_{i-1}\}$.

Theorem 4.3. *Let $(X_n)_{n \geq 0}$ be a real martingale with a Laplace transform such that $n^{-1} \log L_{X_n}(t)$ converges to a limit $\log L_X(t)$ as n tends to infinity. Then*

$$\lim_{n \to \infty} n^{-1} \log P(X_n > a) = \inf_{t > 0} \{n \log L_X(t) - at\}.$$

Bennett's inequality for independent random variables can be adapted to martingales. Let $\sigma_i^2 = E\{V_i(X) - V_{i-1}(X)\}$, the variance of X_n is $n\bar{\sigma}_n^2$, where the mean variance $\bar{\sigma}_n^2 = n^{-1} \sum_{i=1}^{n} \sigma_i^2$ is supposed to converge to a limit σ^2 as n tends to infinity.

Theorem 4.4. *Let $(X_n)_{n \geq 0}$ be a real martingale with mean zero and such that there exists a constant M for which the variations of the martingale satisfy $\sigma_n^{-1}|X_n - X_{n-1}| \leq M$ a.s., for every integer n. For every $t > 0$*

$$P(X_n > t) \leq \exp\{-\phi(\frac{t}{n\bar{\sigma}_n M})\},$$

where $\phi(x) = (1 + x)\log(1 + x) - x$.

Proof. The Laplace transform L_{X_n} has the same expansion as L_X in Bennet's Theorem 1.5, under the boundedness condition for $|X_n - X_{n-1}|$. This condition implies $\sum_{i=1}^{n} EX_n^k \leq M^k \sum_{i=1}^{n} \sigma_i^k \leq (Mn\bar{\sigma}_n)^k$, where the bound is denoted b_n^k, and the expansion of L_{X_n} differs from the expansion of the Laplace transform of n independent variables

$$E\{e^{\lambda X_n}\} \leq 1 + \sum_{k=2}^{\infty} \frac{\lambda^k}{k!} b_n^k = 1 + \{\exp(b_n\lambda) - 1 - b_n\lambda\}$$
$$\leq \exp\{\exp(b_n\lambda) - 1 - b_n\lambda\}.$$

From Chernov's theorem, for every $t > 0$

$$\log P(X_n > t) = \inf_{\lambda > 0} \{\exp(b_n \lambda) - 1 - b_n \lambda - \lambda t\}.$$

The first derivative with respect to λ of $h_t(\lambda) = \exp(b_n \lambda) - 1 - b_n \lambda - \lambda t$ is $h'_t(\lambda) = b_n \exp(b_n \lambda) - b_n - t$, hence the function h_t is minimum at the value $\lambda_{n,t} = b_n^{-1} \log\{1 + b_n^{-1} t\}$ where $\exp(b_n \lambda_{n,t}) = 1 + b_n^{-1} t$ and

$$\exp\{h_t(\lambda_{n,t})\} = \exp\{-\phi(b_n^{-1} t)\} = \exp\{-\phi((Mn\bar{\sigma}_n)^{-1} t)\}. \qquad \square$$

Under a weaker condition, another version of Theorem 4.4 is deduced from its application to the restriction of the process X to the set $\{< X >_T \leq \eta\}$.

Theorem 4.5. *Let* $(X_n)_{n \geq 0}$ *be a real martingale with mean zero and such that there exists a constant M for which the variations of the martingale satisfy* $|X_n - X_{n-1}| \leq MV_n$ *a.s., for every integer n. For every $t > 0$*

$$P(X_n > t) \leq \exp\{-\phi(\frac{t}{\sqrt{\eta}M})\} + P(V_n > \eta).$$

For every monotone function H, Theorem 4.4 implies

$$P(H(X_n) > t) \leq \exp\{-\phi(\frac{H^{-1}(t)}{n\bar{\sigma}_n M})\}, \; t > 0.$$

Let $X = (X_n)_n$ be a martingale, let $(A_n)_n$ be a sequence of predictable random variables, and let Y_n be the transformed martingale

$$Y_n = \sum_{k=1}^n A_k (X_k - X_{k-1}). \qquad (4.8)$$

The variance of Y_n is

$$\bar{\sigma}_n^2(Y) = \sum_{i=1}^n E\{A_i^2(V_i - V_{i-1})(X)\} \leq \|(A_i)_{1 \leq i \leq n}\|_4^2 \|(\sigma_i)_{1 \leq i \leq n}\|_4^2$$

and $\bar{\sigma}_n^2(Y)$ converges to a finite limit σ_Y^2 if $(A_n)_n$ belongs to $L_4(\mathbb{R}^n)$ and if $\bar{\sigma}_n^2(X)$ converges to a limit σ^2. Under the boundedness condition $|Y_i - Y_{i-1}| \leq M\{A_i^2(V_i - V_{i-1})(X)\}^{\frac{1}{2}}$, we have $EY_n^k = \sum_{i=1}^n E(Y_i - Y_{i-1})^k$ where each term is bounded by

$$E|Y_i - Y_{i-1}|^k \leq M^k E[\{A_i^2(V_i - V_{i-1})(X)\}^{\frac{k}{2}}] \leq \{Mn\bar{\sigma}_n(Y)\}^k.$$

Theorem 4.4 applies to the martingale Y_n using the bound $b_n = n\sigma_n(Y)M$ in the expansion of the Laplace transform of Y_n. If $(X_n)_{n \geq 0}$ is a super-martingale and satisfies the boundedness condition of Theorem 4.4, the same bound holds.

Proposition 4.13 (Neveu, 1972). *Let* $(X_n)_{n\geq0}$ *be a positive super-martingale and let* $(A_n)_{n\geq0}$ *be a predictable process such that* $|A_n| \leq 1$ *a.s. For every* $c > 0$

$$P(\sup_{n\geq0} |Y_n| > c) \leq 9c^{-1}EX_0.$$

Let φ be a real concave function on $\mathbb{R}$ let $X = (X_n)_n$ be in $\mathcal{M}^2_{0,loc}$ and let $Y_{n+1} = Y_n + \varphi(X_{n+1} - X_n)$. The mean quadratic variations of the supermartingale Y satisfy

$$E\{V_{n+1}(Y)|\mathcal{F}_n\} = V_n(Y) + E(Y^2_{n+1}|\mathcal{F}_n) - Y^2_n$$
$$\leq V_n(Y) + \varphi^2(E(X_{n+1}|\mathcal{F}_n)) - \varphi^2(X_n)_n = V_n(Y),$$

hence $varY_n \leq \sum_{i=1}^n EV_i(Y) \leq nE(X_1-X_0)^2$ which is denoted $n\sigma^2$. Under the condition $|Y_i - Y_{i-1}| \leq M\{E(Y_i - Y_{i-1})^2\}^{\frac{1}{2}}$, Theorem 4.4 applies to the supermartingale Y_n with the bound $b_n = \bar{\sigma}_n M = M \sum_{i=1}^n \{E(Y_i-Y_{i-1})^2\}^{\frac{1}{2}}$. Assuming that φ is a real Lipschitz function on $\mathbb{R}$ implies that $(Y_n)_n$ is a supermartingle sequence and there exists a constant $c_\varphi > 0$ such that $E\varphi^2(X_{n+1} - X_n) \leq c_\varphi^2 E\{(X_{n+1} - X_n)^2|\mathcal{F}_n\}$ and the same inequality holds for the variances, $E\{V_{n+1}(Y) - V_n(Y)|\mathcal{F}_n\} \leq c_\varphi^2 E\{V_{n+1}(X) - V_n(X)|\mathcal{F}_n\}$. Under the conditions of Theorem 4.4, for every Lipschitz function φ

$$P(X_n > t) \leq \exp\{-\phi(\frac{t}{c_\varphi n\bar{\sigma}_n M})\}, \ t > 0.$$

4.4　Inequalities for martingales indexed by $\mathbb{R}_+$

Let $(M_t)_{t\geq0}$ be in $\mathcal{M}^2_{loc}$, it is written as the sum of a continuous process M^c and a jump process $M^d = \sum_{0<s\leq t} \Delta M_s$, with real jumps $\Delta M_s = M_s - M_{s-}$. The increasing predictable process of its quadratic variations is the sum of $M^{c2}_t - <M^c>_t$ and $\sum_{0<s\leq t}(\Delta M_s)^2 - <M^d>_t$, both belonging to $\mathcal{M}_{0,loc}$. The stochastic integral of a predictable process A with respect to a martingale or a local martingale M

$$Y_t = \int_0^t A_s \, dM_s$$

is defined as the process satisfying

$$<Y>_t = \int_0^t A^2_s \, d<M>_s,$$

where the integral of A^2 with respect to the increasing process of the quadratic variations of M is the Stieltjes integral. If the uniform norm

$\|A\|_{[0,t]} = \sup_{s \in [0,t]} |A_s|$ is a.s. finite and M belongs to $\mathcal{M}^k_{0,loc}$, then Y belongs to $\mathcal{M}^k_{0,loc}$.

The constant in the Bürkholder-Davis-Gundy inequality for martingales indexed by $\mathbb{R}$ is obtained from the following lemma.

Lemma 4.3. *Let $p \geq 2$ and $q \geq 2$ be integers and let f be a real function in $L_p(\mathbb{R}) \cap L_q(\mathbb{R})$. For every $x > 0$, there exist $C_{1,x}$ and $C_{2,x}$ depending on p and q such that $C_{1,x}\|f1_{[0,x]}\|_p \leq \|f1_{[0,x]}\|_q \leq C_{2,x}\|f1_{[0,x]}\|_p$.*

Proof. Let $t > 0$, $f(t) = \lim_{n \to \infty} \sum_{i \leq n} a_i r_i(t)$ with indicators functions r_i on disjoint intervals, the result is a consequence of the Hölder inequality with $\|\sum_{i \leq n} r_i 1_{[0,x]}\|_p \leq x^{\frac{1}{p}}$. □

Proposition 4.14. *On a filtered probability space $(\Omega, \mathcal{F}, (\mathcal{F}_t)_{t \geq 0}, P)$, let $(M_t)_{t \geq 0}$ be a continuous process in $\mathcal{M}^\alpha_{0,loc}$, $\alpha \geq 2$, having a process of quadratic variations $(< M >_t)_{t \geq 0}$. There exist functions $C_{1,x} > 0$ and $C_{2,x}$ such that for every random stopping time T of M*

$$E(C_{1,T} < M >_T^{\frac{\alpha}{2}}) \leq E(|M_T|^\alpha) \leq E(C_{2,T} < M >_T^{\frac{\alpha}{2}}).$$

Proof. The difference $m_t = M_t - < M >_t$ belongs to $\mathcal{M}_{0,loc}$ on $(\Omega, \mathcal{F}, (\mathcal{F}_t)_{t \geq 0}, P)$ hence $Em_t = 0$ and the equality $E(|M_T|^2) = E < M >_T$ is satisfied for $p = 2$. For $\alpha > 2$, let $\pi_n = (t_i)_{i \leq m_n}$ be a partition of $[0, t]$ in subintervals $I_{n,i}$ with path $t_{n,i+1} - t_{n,i} = h_n(t)$ and let $m_n = (h_n)^{-1}$ tend to infinity. The martingale is approximated by a sum of conditionally independent variables $X_{n,i} = M_{t_{n,i}} - M_{t_{n,i-1}}$

$$M_t = \sum_{k=1}^{m_n(t)} X_{n,i} 1_{I_{n,i} \cap [0,t]}, \quad V_t = \sum_{k=1}^{m_n(t)} E(X_{n,i}^2 | \mathcal{F}_{i-1}) 1_{I_{n,i} \cap [0,t]}.$$

From Lemma 4.3, the constant in the upper bound of inequality (4.4) with an integer exponent does not depend on the number of variables in the sums but only on the exponent α and on t, therefore the inequality is also true as n tends to infinity and by replacing the index t by a stopping time of the martingale M, by Doob's theorem. The inequality with a real exponent is obtained by limit from the inequality for rational numbers which is a consequence of the inequality for integers. □

Proposition 4.15. *Let $(X_t)_{t \geq 0}$ be a local martingale of $\mathcal{M}^{2p}_{0,loc}$, for an integer $p \geq 1$. For every stopping time T and for every $p \geq 2$, there exists a constant C_p such that*

$$E(V_T^p) \leq E(X_T^{2p}) \leq C_p E(V_T^p).$$

This is a consequence of Proposition 4.9 applied to the martingale sequence $X_{t_n \wedge T} = \sum_{k=1}^{m_n} (X_{t_{m_k} \wedge T} - X_{t_{m_k-1} \wedge T})$, based on a partition of $[0, T]$. For a real $0 < \alpha < 2$, there is no inequality similar to those of the previous proposition. Arguing as in the proof of Proposition 4.14 and applying Proposition 4.10, entails the next bound.

Theorem 4.6. *Let $(M_t)_{t \geq 0}$ belong to $\mathcal{M}_{0,loc}^2$, for every real $0 < \alpha < 2$ and for every stopping time T of $(M_t)_{t \geq 0}$*

$$\| M_T^2 - <M>_T \|_{L_{\frac{\alpha}{2}}} < E <M>_T .$$

Let $\alpha > 2$, for every $t > 0$, $\| M_t^2 - <M>_t \|_{\frac{\alpha}{2}} > 0$.

In $(\Omega, \mathcal{F}, (\mathcal{F}_t)_{t \geq 0}, P)$, let $(M_t)_{t \geq 0}$ be in $\mathcal{M}_{0,loc}^k$, for every integer k. For $0 < s < t$, the variations of the Laplace transform of the martingale are

$$E\{e^{\lambda M_t} - e^{\lambda M_s} | \mathcal{F}_s\} = L_{M_s}(\lambda) E\{\exp^{\lambda(M_t - M_s)} - 1 | \mathcal{F}_s\}$$
$$= L_{M_s}(\lambda) \sum_{k=2}^{\infty} \lambda^k E\{(M_t - M_s)^k | \mathcal{F}_s\},$$

from Proposition 4.15, they are bounded by the sum

$$L_{M_s}(\lambda) \sum_{k=2}^{\infty} \lambda^k C_k E\{(<M>_t - <M>_s)^{\frac{k}{2}} | \mathcal{F}_s\}.$$

A continuous version of Chernov's theorem for local martingales is deduced from the result for discrete martingales. On $(\Omega, \mathcal{F}, (\mathcal{F}_t)_{t \geq 0}, P)$, let $(M_t)_{t \geq 0}$ be in $\mathcal{M}_{0,loc}^k$, for every integer k. For every $a > 0$

$$P(M_t > a) = \inf_{\lambda \geq 0} \{L_{M_t}(\lambda) - e^{-\lambda t}\}.$$

Proposition 4.15 is not sufficient to establish an inequality of the same kind as Bennet's inequality and it is proved using the same argument as Lenglart's inequality, under a moments condition.

Theorem 4.7. *On a probability space $(\Omega, \mathcal{F}, (\mathcal{F}_t)_{t \geq 0}, P)$, let $(M_t)_{t \geq 0}$ be in $\mathcal{M}_{0,loc}^k$, for every integer k, and such that there exists a constant c for which*

$$\{E(M_t^k)\}^{\frac{1}{k}} < c\{E(<M>_T^{\frac{k}{2}})\}^{\frac{1}{k}}, \tag{4.9}$$

for every $k \geq 2$ and for every t in $[0, T]$. For all $a > 0$ and $\eta > 0$

$$P(\sup_{t \in [0,T]} t^{-\frac{1}{2}} M_t > a) \leq \exp\{-\phi(\frac{a\sqrt{T}}{c\sqrt{\eta}})\} + P(t^{-\frac{1}{2}} <M>_T > \eta).$$

Proof. For all $t > 0$ and $\lambda > 0$, the Laplace transform $L_{X_t}(\lambda)$ is finite and, under the conditions, it satisfies

$$L_{M_t}(\lambda) = E\{\exp^{\lambda M_t}\} = 1 + \sum_{k=2}^{\infty} \frac{\lambda^k}{k!} E(M_t^k)$$

$$\leq 1 + \sum_{k=2}^{\infty} \frac{\lambda^k}{k!} c^k E(< M >_t^{\frac{k}{2}}),$$

the bound is increasing with t since $< M >$ is an increasing process. For the restriction of the process X to the set $\{< X >_T \leq \eta\}$,

$$L_{X_t}(\lambda) \leq e^{\lambda c \eta^{\frac{1}{2}} T^{-\frac{1}{2}}} - \lambda c \eta^{\frac{1}{2}} T^{-\frac{1}{2}} \leq \exp(e^{\lambda c \eta^{\frac{1}{2}} T^{-\frac{1}{2}}} - 1 - \lambda c \eta^{\frac{1}{2}} T^{-\frac{1}{2}}).$$

The function $I_T(\lambda) = e^{\lambda c \eta^{\frac{1}{2}} T^{-\frac{1}{2}}} - 1 - \lambda c \eta^{\frac{1}{2}} T^{-\frac{1}{2}} - a\lambda$ has the minimum value $\exp\{-\phi((c\sqrt{\eta})^{-1} a\sqrt{T})\}$ under the condition (4.9). □

Proposition 4.16. *Let $M = (M_t)_{t \geq 0}$ be in $\mathcal{M}_{0,loc}^k$, for every integer k and let $T > 0$ be a stopping time $T > 0$ such that $\max_{k \geq 2}(ET^{-1} \int_0^T d|M|^k)^{\frac{1}{k}}$ is bounded by a constant C. For every $x > 0$*

$$P(\sup_{t \in [0,T]} |M(t)| > x) \leq \exp\{-\phi(\frac{x}{CT})\}.$$

Proof. The Laplace transform of M_t is

$$L_{M_t}(\lambda) = Ee^{\lambda M_t} \leq 1 + \sum_{k \geq 2} \frac{\lambda^k}{k!} \int_0^T d|M|^k$$

$$\leq \exp\{e^{\lambda CT} - 1 - \lambda CT\}$$

and the proof ends using the same argument as for Theorem 4.7. □

Proposition 4.16 applies to the stochastic integral Y of a predictable process $A = (A_t)_{t \geq 0}$ with respect to a local martingale $M = (M_t)_{t \geq 0}$, under the required integrability conditions for Y. The Bienaymé-Chebychev inequality for $Y_t = \int_0^t A_s \, dM_s$ in $\mathcal{M}_{0,loc}^2$ is written for the supremum of the martingale over an random interval $[0, T]$ determined by an integrable stopping time of the process Y

$$P(\sup_{0 \leq t \leq T} t^{-\frac{1}{2}} Y_t > \lambda) \leq E \frac{1}{\lambda^2 T} \int_0^T A_u^2 \, d < M >_u \tag{4.10}$$

$$\leq \frac{1}{\lambda^2} E(\sup_{0 \leq t \leq T} A_t^2 \frac{< M >_T}{T}).$$

If $< M >_t$ has a derivative, it is written as $< M >_t = \int_0^t B_s \, ds$ and the bound in inequality (4.10) can be precised by the Mean Value Theorem. For every integer $p \geq 1$, there exists θ in $]0, 1[$ such that

$$P(\sup_{0 \leq t \leq T} t^{-\frac{1}{2}} Y_t > \lambda) \leq \frac{1}{\lambda^{2p}} E\{(\frac{1}{T} \int_0^T A_u^2 \, d < M >_u)^p\}$$

$$\leq \frac{1}{\lambda^{2p}} E\{(A_{\theta T}^2 B_{\theta T})^p\}.$$

For stopping times $0 < T_1 < T_2$ of the process Y and for all $\lambda_1 > 0$ and $\lambda_2 > 0$, there exist θ_1 and θ_2 in $]0, 1[$ such that

$$P(\sup_{0 \leq t \leq T_1} t^{-\frac{1}{2}} \int_0^t A_u \, dM_u > \lambda_1, \sup_{T_1 \leq t \leq T_2} (t - T_1)^{-\frac{1}{2}} \int_{T_1}^t A_u \, dM_u > \lambda_2)$$

$$\leq \frac{1}{(\lambda_1 \lambda_2)^2} E\{\frac{1}{T_1} \int_0^{T_1} A_u^2 \, d < M >_u\} E\{\frac{1}{T_2 - T_1} \int_{T_1}^{T_2} A_u^2 \, d < M >_u\}$$

$$= \frac{1}{(\lambda_1 \lambda_2)^2} E\{A_{\theta_1 T_1}^2 B_{\theta_1 T_1}\} E\{A_{\theta_2 (T_2 - T_1)}^2 B_{\theta_2 (T_2 - T_1)}\}.$$

Proposition 4.17. *Let $M = (M_t)_{t \geq 0}$ and $Y_t = \int_0^t A_s \, dM_s$ be in $\mathcal{M}_{0,loc}^k$, for every integer k, where $A = (A_t)_{t \geq 0}$ is a predictable process. Let $T > 0$ be a stopping time $T > 0$ such that $\sup_{t \in [0,T]} |A|$ is a.s. bounded by a constant B and there exists a constant C such that $\max_{k \geq 2} (ET^{-1} \int_0^T d|M|^k)^{\frac{1}{k}} < C$. For every $x > 0$*

$$P(\sup_{t \in [0,T]} |Y(t)| > x) \leq \exp\{-\phi(\frac{x}{BCT})\}.$$

Proof. Under the condition $|A(t)| \leq B$ for every t in $[0, T]$ and $E|\int_0^t A \, dM|^k \leq B^k E \int_0^t d|M|^k$ is finite for every $t > 0$ and for every integer k, the Laplace transform of Y_t is written

$$L_{Y_t}(\lambda) = E e^{\lambda \int_0^t A_s \, dM_s} \leq 1 + \sum_{k \geq 2} \frac{\lambda^k B^k}{k!} \int_0^T d|M|^k$$

$$\leq \exp\{e^{\lambda BCT} - 1 - \lambda BCT\}.$$

The proof ends by minimizing $\log L_{Y_t}(\lambda) - \lambda a$ for $\lambda > 0$ as above. $\qquad \square$

4.5   Poisson processes

On a probability space $(\Omega, \mathcal{F}, P)$, the Poisson process $(N_t, t \geq 0)$ is a right-continuous process with left-hand limits defined as $N(t) = \sum_{i \geq 1} 1\{T_i \leq t\}$,

$t > 0$, for a sequence of random variables $0 = T_0 < T_1 < \cdots < T_i < \cdots$, and $N(0) = 0$. Its natural filtration $(\mathcal{F}_t)_{t \geq 0}$ is defined by the σ-algebras $\mathcal{F}_t = \sigma(T_i \leq t, i \geq 1)$. A homogeneous Poisson process has a constant intensity $\lambda > 0$ and independent increments and the time variables T_k are sums of k independent exponential variables with parameter λ^{-1}. Its probability distribution satisfying

$$P(N(t+h) - N(t) = 1) = h\lambda + o(h) = p(h),$$

$$P(N(t+h) - N(t) > 1) = o(h),$$

for every $t > 0$, when h tends to zero. These properties imply that for all $k \geq 1$ and $t > 0$

$$P_k(t) := P(N_t = k) = e^{-\lambda t}(\lambda t)^k (k!)^{-1}$$

and $P(N_t = 0) = e^{-\lambda t}$. The probabilities P_k satisfy the differential equation $P_k'(t) = -\lambda P_k(t) + \lambda P_{k-1}(t)$, with $P_k(0) = 0$, $k \leq 1$.

The Poisson process is also represented in terms of the ordered statistics of the sample of a variable having the uniform distribution. For a Poisson process with parameter 1, the variable $\xi_{n:i} = T_i T_{n+1}^{-1}$ is the i-th order statistics of a vector $(\xi_i)_{i=1,\ldots,n}$ of independent and uniform variables on $[0, 1]$ and $(T_1 T_{n+1}^{-1}, \ldots T_n T_{n+1}^{-1})$ is independent of T_{n+1} (Breiman, 1968). The same result holds for a Poisson process with parameter λ.

A renewal process has independent increments and identically distributed inter-arrival times, it has the properties of a heterogeneous Poisson process with an increasing functional intensity $\Lambda(t) = EN(t) > 0$. Its distribution function is defined by $P(N_t = k) = e^{-\Lambda(t)}\Lambda^k(t)(k!)^{-1}$, $t > 0$, for every integer $k \geq 1$ and $P(N_t = 0) = e^{-\Lambda_t}$. It is a process with independent increments and the joint density at $(x_1, \ldots, x_n)$ of the independent waiting times $X_k = T_k - T_{k-1}$, $k = 1, \ldots, n$, is the product of the exponential densities $\prod_{k=1,\ldots,n} e^{-\{\Lambda_{x_k} - \Lambda_{x_{k-1}}\}} = e^{-\Lambda_{x_n}}$, with $x_0 = 0$. The transformed times $\Lambda(T_k)$ of the process are the time variables of a Poisson process with intensity one which satisfy Breiman's properties. If the function Λ has a derivative λ

$$P(N(t+h) - N(t) = 0) = 1 - e^{-\Lambda(t+h) - \Lambda(t)} = h\lambda(t) + o(h),$$

for every $t > 0$, when h tends to zero. This entails

$$P_k'(t) = -\lambda(t)P_k(t) + \lambda(t)P_{k-1}(t), \ k \leq 1.$$

Its generating function is

$$G_{N_t}(u) = e^{-\Lambda_t} \sum_{k \geq 0} \frac{u^k \Lambda_t^k}{k!} = \exp\{\Lambda_t(u - 1)\}.$$

Its Laplace transform

$$L_{N_t}(u) = \exp\{\Lambda_t(e^u - 1)\},$$

is an increasing function of u and its minimum for $u > 0$ is reached as u tends to zero.

Theorem 4.8. *On the probability space* $(\Omega, \mathcal{F}, (\mathcal{F}_t)_{t \geq 0}, P)$, *let N be a Poisson process with mean* $EN(t) = \Lambda(t)$, *then for every* $x > 0$

$$t^{-1} \log P(t^{-1} N_t > a) = \inf_{u > 0} \{t^{-1} \log L_{N_t}(u) - au\}$$

$$= a\{1 - \log a + \log(t^{-1}\Lambda_t)\} - t^{-1}\Lambda_t$$

and its limit as t tends to infinity is finite if $\lim_{t \to \infty} t^{-1}\Lambda_t$ *is finite.*

Proof. The minimum of $h_a(u) = t^{-1} \log L_{N_t}(u) - au = t^{-1}\Lambda_t(e^u - 1) - au$ is reached at $u_a = \log(t\Lambda_t^{-1} a)$ where $h_a'(u_a) = t^{-1}\Lambda_t e^{u_a} - a = 0$ and

$$h_a(u_a) = a - t^{-1}\Lambda_t + a \log\{(at)^{-1}\Lambda_t\}. \qquad \square$$

The difference $M_t = N_t - \Lambda(t)$ belongs to $\mathcal{M}_{0,loc}^2$ and the function of quadratic variations of M_t is $\Lambda(t)$. By Proposition 4.14, for every random stopping time T of N and for every $\alpha \geq 2$, there exist constants $0 < c_\alpha < C_\alpha$ such that the moments of M_t satisfy

$$c_\alpha E(\Lambda_T^{\frac{\alpha}{2}}) \leq E(|M_T|^\alpha) \leq C_\alpha E(\Lambda_T^{\frac{\alpha}{2}}).$$

The exponential submartingale related to the Poisson process is

$$Y_t(u) = \exp\{uN_t - L_{N_t}(u)\} = \exp\{uN_t - \exp\{-\Lambda_t(1 - e^u)\}\}.$$

Let $(N_t, t \geq 0)$ be a Poisson process with a functional cumulative intensity $\Lambda(t) > 0$ and let T be a stopping time of N. For every $x > 0$

$$\sup_{T \geq 0} P(\sup_{0 \leq t \leq T} |N_t - \Lambda(t)| > x) \leq 2x^{-2} E\{\sup_{T \geq 0} \Lambda(T)\}.$$

By the martingale property, $E\{(M_t - M_s)^2 | \mathcal{F}_s\} = \Lambda_t - \Lambda_s$. For stopping times $0 < T_1 < T_2$ of the process N and for all $x_1 > 0$ and $x_2 > 0$, the independence of the increments of the Poisson process implies that for all $x_1 > 0$ and $x_2 > 0$

$$P(\sup_{0 \leq s \leq T_1} |N_s - \Lambda_s| > x_1, \sup_{T_1 \leq t \leq T_2} |N_t - \Lambda_t| > x_2)$$

$$\leq \frac{2}{x_1^2 x_2^2} E\{\Lambda_{T_2} - \Lambda_{T_1}\} E\{\Lambda_{T_1}\}.$$

Let $(N_{ij})_{j=1,\ldots,J_{in};i=1,\ldots,I_n}$ be an array of point processes on $\mathbb{R}$ such that $I_n = O(n^{1-\alpha})$ and $J_{in} = O(n^{\alpha})$ for all $i = 1,\ldots,I_n$, with a total number $n = \sum_{i=1,\ldots,I_n} J_{in}$. The predictable compensator of the processes N_{ij} are supposed to have the same form $\widetilde{N}_{ij}(t) = \int_0^t h(Y_{ij}(s))\,d\Lambda(s)$, where $(Y_{ij})_{j=1,\ldots,J_{in},i=1,\ldots,I_n}$ is an array of predictable processes and h is a left-continuous positive function with right-hand limits. Moreover $E\{N_{ij}(t)N_{ik}(t)\} = E < N_{ij}(t), N_{ik}(t) >$ develops as

$$\frac{1}{2}(< N_{ij} + N_{ik} >_t - < N_{ij} >_t - < N_{ik} >_t\}$$

$$= \frac{1}{2}\int_0^t \{h_i(Y_{ij}(s) + Y_{ik}(s)) - h_i(Y_{ij}(s)) - h_i(Y_{ik}(s))\}\,d\Lambda(s),$$

denoted $\int_0^t k_i(Y_{ij}, Y_{ik})\,d\Lambda$, and $E\{N_{ij}(t)N_{i'k}(t)\} = \int_0^t k_{ii'}(Y_{ij}, Y_{i'k})\,d\Lambda$. The processes

$$X_{ij}(t) = \int_0^t \{h(Y_{ij})\}^{-1}1_{\{h(Y_{ij})>0\}}\,dN_{ij},$$

$$\widetilde{X}_{ij}(t) = \int_0^t \{h(Y_{ij})\}^{-1}1_{\{h(Y_{ij})>0\}}\,d\Lambda,$$

define a centered local martingale of L_2, $M_{ij} = X_{ij} - \widetilde{X}_{ij}$ with variance $EM_{ij}^2(t) = E\int_0^t \{h_i(Y_{ij})\}^{-1}1_{\{h_i(Y_{ij})>0\}}\,d\Lambda$ and covariances

$$EM_{ij}(t)M_{ik}(s) = E\int_0^{s\wedge t} \frac{k_i(Y_{ij}, Y_{ik})}{h_i^2(Y_{ij})h_i^2(Y_{ik})}1_{\{h_i(Y_{ij})h_i(Y_{ik})>0\}}\,d\Lambda,$$

$$EM_{ij}(t)M_{i'k}(s) = E\int_0^{s\wedge t} \frac{k_{ii'}(Y_{ij}, Y_{i'k})}{h_i^2(Y_{ij})h_{i'}^2(Y_{i'k})}1_{\{h_i(Y_{ij})h_{i'}(Y_{i'k})>0\}}\,d\Lambda.$$

The martingale $S_n = (S_{in})_{i=1,\ldots,I_n}$ with components $S_{in} = \sum_{j=1,\ldots,J_{in}} M_{ij}$ has a variance matrix Σ_n with components $\Sigma_{ii'} = \sum_{j,j'=1}^{J_{in}} E(M_{ij}M_{i'j'})$, for i and i' in $\{1,\ldots,I_n\}$. For every $t = (t_1,\ldots,t_{I_n})$

$$P(\sup_{t\in[0,T]} S_n(t) > x) \leq \frac{E\|\Sigma_n(T)\|_2}{\|x\|_2^2}.$$

4.6 Brownian motion

Proposition 4.15 applies to the Brownian motion $B - (B_t)_{t\geq 0}$ defined on the filtered probability space $(\Omega, \mathcal{F}, (\mathcal{F}_t)_{t\geq 0}, P)$ with its natural filtration. Let $p \geq 2$, for every stopping time T of B, we have

$$E(|B_T|^p) \leq C_p E(T^{\frac{p}{2}}). \tag{4.11}$$

For products of the increments of the process on disjoint intervals, the inequalities are factorized and for every increasing sequence of stopping times $(T_j)_{j \leq k}$, for integers $0 < m_k < n_k$ and $p_k \geq 2$, $k > 1$

$$\prod_{j=1}^{k} E(|B_{T_j} - B_{T_{j-1}}|^{p_j}) \leq C_p E\{\prod_{j=1}^{k}(T_j - T_{j-1})^{\frac{p}{2}})\} \qquad (4.12)$$

and the same inequality for the maximum value of the Brownian motion on the intervals $]T_{j-1}, T_j]$. From Theorem 4.6, for every real number α in $]0, 2[$ and for every $t > 0$, $\|B_t^2 - t\|_{L_{\frac{\alpha}{2}}} \leq t$. Moreover, for $x > 0$

$$P(|B_t^2 - t| > x) = 2e^{-\frac{(t+x)^2}{t}}$$

and for $x = 0$ this probability is strictly positive. The Laplace transform of the Brownian motion is $L_{B_t}(u) = \exp\{\frac{1}{2}\theta^2 t\}$, $t \geq 0$, $u \geq 0$.

Proposition 4.18. *On a filtered probability space $(\Omega, \mathcal{A}, (\mathcal{F}_t)_{t \geq 0}, P)$, let $(X_t)_{t \geq 0}$ be a martingale with independent increments such $EX_0 = 0$. For all u and t such that the Laplace transform $L_{X_t}(u)$ of X_t is finite at u, the process $Y_t(u) = \exp\{uX_t - \log L_{X_t}(u)\}$ is a $(\mathcal{F}_t)_{t \geq 0}$-submartingale if $L_{X_s}L_{X_{t-s}} \geq L_{X_t}$ for every $0 < s < t$ and it is a supermartingale if $L_{X_s}L_{X_{t-s}} \leq L_{X_t}$ for every $0 < s < t$.*

Proof. Let $(\mathcal{F}_t)_{t \geq 0}$ be the filtration generated by $(X_t)_{t \geq 0}$, for all $u > 0$ and $t > s > 0$, $L_{X_t}(u) - L_{X_s}(u) = E(e^{uX_t} - e^{uX_s})$ and

$$E\{(Y_t - Y_s)(u)|\mathcal{F}_s\} = Y_s E\{\frac{Y_t}{Y_s}(u) - 1|\mathcal{F}_s\}$$

$$= Y_s[\frac{L_{X_s}}{L_{X_t}}(u)\exp\{uE(X_t - X_s)|\mathcal{F}_s\} - 1]$$

$$= Y_s\{\frac{L_{X_s}L_{X_{t-s}}}{L_{X_t}}(u) - 1\},$$

therefore $E\{Y_t(u)|\mathcal{F}_s\} = Y_s L_{X_s}(u)L_{X_{t-s}}(u)L_{X_t}^{-1}(u)$, $u > 0$. $\qquad \square$

Corollary 4.1. *Let $(X_t)_{t \geq 0}$ be a martingale with independent increments and $EX_0 = 0$ and let L_{X_t} be the Laplace transform of X_t. If for every u the process $(Y_t(u))_{t \geq 0} = (\exp\{uX_t - \log L_{X_t}(u)\})_{t \geq 0}$ is a martingale with respect to the filtration generated by the process $(X_t)_{t \geq 0}$, then $(X_t)_{t \geq 0}$ is the Brownian motion.*

Example 4.1. Let $(X_n)_n$ be a random walk where the variables X_n have the Laplace transform L and let $\mathcal{F}_n$ be the σ-algebra generated by $(X_i)_{1 \leq i \leq n}$.

For every t such that $L(t)$ is finite, consider the processes $Y_n(t) = \exp\{tX_n - n\log L(t)\}$, it satisfies

$$E\{Y_{n+1}(t) - Y_n(t)|\mathcal{F}_n\} = Y_n(t)L^{-1}(t)E[\exp\{t(X_{n+1} - X_n)\}|\mathcal{F}_n]$$
$$> L^{-1}(t)Y_n(t),$$

therefore $E\{Y_{n+1}(t)|\mathcal{F}_n\} \geq \{1 + L^{-1}(t)\}Y_n(t)$ and the process $(Y_n)_{n\geq 0}$ is a submartingale.

Let β be a right-continuous function of $L_2(\mathbb{R})$ with left-hand limits, the process

$$Y(t) = \int_0^t \beta(s)\,dB_s$$

is a transformed Brownian motion and the increasing function of its quadratic variations is

$$V_Y(t) = \int_0^t \beta^2(s)\,ds \leq t\sup_{[0,t]}|\beta|^2.$$

Theorem 4.6 implies

$$\|(\int_0^T \beta_s\,dB_s)^2 - \int_0^T \beta_s^2\,ds\|_{L_{\frac{\alpha}{2}}} \leq E\int_0^T \beta_s^2\,ds$$

for every real $0 < \alpha < 2$ and for every stopping time T. Propositions 4.14 and 4.15 applies to the process Y and the next bounds are analogous to Equation (4.11).

Proposition 4.19. *Let $\alpha \geq 2$ and T be a stopping time of B, there exist functions $C_{1,t} > 0$ and $C_{2,t}$ such that*

$$E\{C_{1,T}\int_0^T \beta^2\,ds\}^{\frac{\alpha}{2}}\} \leq E(|\int_0^T \beta\,dB|^\alpha) \leq E\{C_{2,T}\beta^{\frac{\alpha}{2}}(T)\}. \qquad (4.13)$$

Morevover, for every stopping time T and for every $p \geq 2$, there exists a constant C_p such that

$$E(V_T^p) \leq E(X_T^{2p}) \leq C_p E(V_T^p).$$

The Laplace transform of B_t is $L_{Y_t}(\lambda) = \exp\{\frac{1}{2}\lambda^2 V_Y(t)\}$ and, for every $\lambda > 0$, the process

$$Y_\lambda(t) = \exp\{\lambda\int_0^t \beta(s)\,dB_s - \frac{\lambda^2}{2}\int_0^t \beta^2(s)\,ds\}$$

is a local martingale with respect to $(\mathcal{B}_t)_{t\geq 0}$ with mean 1.

Proposition 4.20. *Let $a > 0$, the variable $T_{Y,a} = \inf\{s : Y(s) = a\}$ is a stopping time for the process Y and*

$$Ee^{\lambda T_{Y,a}} \geq \exp\{\frac{a\sqrt{2}}{\sqrt{\lambda}\|\beta\|_\infty}\}.$$

Proof. The martingale property of the process Y_λ and a convexity argument provide a lower bound for the Laplace transform of $T_{Y,a}$

$$e^{a\lambda} = E\exp\{\frac{1}{2}\lambda^2 V_Y(T_{Y,a})\} \le E[\frac{1}{T_{Y,a}}\int_0^{T_{Y,a}} \exp\{\frac{T_{Y,a}}{2}\lambda^2\beta^2(s)\}\,ds]$$

$$\le E[\exp\{\frac{T_{Y,a}}{2}\lambda^2\|\beta\|_\infty^2\}].$$

$\square$

Let $T > 0$ be a random variable and let $x > 0$, several inequalities can be written for the tail probability of $\sup_{0\le t\le T} t^{-\frac{1}{2}}Y_t$. By the Bienaymé-Chebychev inequality

$$P(\sup_{t\in[0,T]} t^{-\frac{1}{2}}Y_t > x) \le x^{-2}E\sup_{t\in[0,T]} |\beta(t)|^2.$$

From Lenglart's theorem, for every $\eta > 0$

$$P(\sup_{t\in[0,T]} Y_t > x) \le \frac{\eta}{x^2} + P(\int_0^T \beta^2(s)\,ds > \eta)$$

and the previous inequality is a special case with $\eta = E\sup_{0\le t\le T}|\beta(t)|^2$. The Gaussian distribution of the process Y allows us to write Chernov's theorem in the form of

$$P(t^{-\frac{1}{2}}Y_t > x) = E\inf_{\lambda>0}\exp\{\frac{\lambda^2}{2}\int_0^t \beta^2(s)\,ds - \lambda x\}$$

$$= E\exp\{-\frac{x^2 t}{2\int_0^t \beta^2(s)\,ds}\}$$

$$P(\sup_{0\le t\le T} Y_t > x) \le E\exp\{-\frac{x^2}{2\sup_{0\le t\le T}|\beta(t)|^2}\}.$$

For the Brownian motion, it is simply written as

$$P(\sup_{t\in[0,T]} t^{-\frac{1}{2}}B_t > x) = \exp(-\frac{x^2}{2}),$$

for every stopping time $T > 0$, where the upper bounds tend to zero as x tends to infinity. With a varying boundary $f(t)$, function on $\mathbb{R}$, Chernov's theorem implies

$$P\{|B_t| > f(t)\} = 2\exp\{-\frac{f^2(t)}{2t}\},$$

it tends to 1 as t tends to infinity if $t^{-\frac{1}{2}}f(t)$ tends to zero as t tends to infinity, and it tends to zero as t tends to infinity if $|t^{-\frac{1}{2}}f(t)|$ tends to infinity with t.

Proposition 4.21. *Let* $Y(t) = \int_0^t \beta(s)\,dB_s$ *and* $T = \arg\max_{t \in [0,1]} Y_t$

$$P(T \leq t) = \frac{2}{\pi}\arcsin(\frac{V_t}{V_1})^{\frac{1}{2}},$$

where $V_t = \int_0^t \beta^2(s)\,ds$.

Proof. The variable T is a stopping time of the process Y with independent increments, and its distribution is a generalization of (1.21) for the Brownian motion. The variance of $Y_{t-u} - Y_t$ is V_u and the variance of $Y_{t+v} - Y_t$ is $V_{t+v} - V_t$. The normalized processes $X_0 = V_u^{-\frac{1}{2}}(B_{t-u} - B_t)$ and $X_1 = V_v^{-\frac{1}{2}}(B_{t+v} - B_t)$ are independent and identically distributed, with a standard distribution which does not depend on t, u and v, therefore

$$P(T \leq t) = P(\sup_{u \in [0,t]} Y_{t-u} - Y_t \geq \sup_{v \in [0,1-t]} Y_{t+v} - Y_t)$$

$$= P\{V_t^{\frac{1}{2}}X_1 \geq (V_1 - V_t)^{\frac{1}{2}}X_2\}$$

and

$$P(T \leq t) = P\{\frac{X_2}{(X_1^2 + X_2^2)^{\frac{1}{2}}} \leq (\frac{V_t}{V_1})^{\frac{1}{2}}\} = \frac{2}{\pi}\arcsin(\frac{V_t}{V_1})^{\frac{1}{2}}. \qquad \square$$

Proposition 4.22. *Let* $Y_t = \int_0^t (\int_0^s \beta_y^2\,dy)^{-1}\beta_s\,dB_s$, *where* β *is a positive function of* $\mathcal{C}_b(I)$, *with a bounded interval* I *of* $\mathbb{R}_+$. *For very interval* $[S,T]$ *of* $\mathbb{R}_+$

$$\sup_{S \leq s \leq t \leq T} P(Y_t - Y_s > x) = \exp\{-\frac{x^2}{2(T-S)}\}.$$

Proof. The martingale $(Y_t)_{t \in [0,T]}$ has an increasing process of quadratic variations V such that $V_t - V_s = E(Y_t^2 - Y_s^2|\mathcal{F}_s) = t - s$ for every positive $s \leq t$, the process Y has therefore the distribution of a Brownian motion. The variance $var(Y_t - Y_s) = t - s$ is bounded by $T - S$ on the interval $[S,T]$. Let $x > 0$, for every $t \geq s \geq 0$

$$P(Y_t - Y_s \geq x) \leq \exp\{-\frac{x^2}{2(T-S)}\}$$

and the bound is reached at (S,T). $\qquad \square$

Let $(X_i)_{i \geq 1}$ be a sequence of independent random variables defined on a probability space $(\Omega, \mathcal{F}, P)$ and with values in a separable and complete metric space $(\mathcal{X}, \mathcal{B})$ provided with the topology of the space $D(\mathcal{X})$. When the variables have the same distribution function F, the empirical process

$$\nu_n(t) = n^{-\frac{1}{2}}\{S_n(t) - \mu_n(t)\}$$

t in $\mathbb{R}$, defined by

$$S_n(t) = \sum_{i=1}^{n} 1_{\{X_i \leq t\}}, \quad \mu_n(t) = ES_n(t) = nF(t),$$

converges weakly to the process $W_F = W \circ F$ with covariance function $C(s,t) = F(s \wedge t) - F(s)F(t)$, where W is the standard Brownian bridge on $[0,1]$. For every real number $x > 0$, $P(\sup_{0 \leq t \leq T} \nu_n(t) > x) \leq x^{-2}F(T)$. Let $V_n(t) = varS_n(t) = nV(t)$, with $V(t) = F(t)\{1 - F(t)\}$. The normalized process $W_n(t) = \{V_n(t)\}^{-\frac{1}{2}}\{S_n(t) - \mu_n(t)\}$, t in $\mathbb{R}$, converges weakly to $V^{-\frac{1}{2}}W_F$ with covariance function $\rho(s,t) = \{V(s)V(t)\}^{-\frac{1}{2}}C(s,t)$. The Laplace transform of the standard Brownian bridge W and of the transformed Brownian bridge W_F at t are

$$L_{W_t} = e^{ut}e^{-\frac{u^2 t}{2}}, \qquad L_{W_F(t)} = e^{uF(t)}e^{\frac{u^2 V(t)}{2}}$$

and by Chernov's theorem

$$\lim_{n \to \infty} P(|\nu_n(t)| > x) = 2\exp[-\frac{\{F(t) - x\}^2}{2V(t)}].$$

4.7 Diffusion processes

On a probability space $(\Omega, \mathcal{F}, P)$, let $X = (X_t)_{t \leq 0}$ be a continuous diffusion process with sample paths in $\mathbb{R}_+$ and defined by a random initial value X_0, $\mathcal{F}_0$-measurable and belonging to $L_2(P)$, and by the differential equation

$$dX_t = \alpha_t \, dt + \beta_t \, dB_t \tag{4.14}$$

where B is the Brownian motion, α and β are real functions on $\mathbb{R}_+$ such that α belongs to $L_1(\mathbb{R}_+)$ and β to $L_2(\mathbb{R}_+)$. The solution of this diffusion equation is $X_t = X_0 + \int_0^t \alpha_s \, ds + \int_0^t \beta_s \, dB_s$, then the process X has a Gaussian distribution with mean $EX_t = EX_0 + \int_0^t \alpha_s \, ds$ and variance $varX_t = varX_0 + \int_0^t \beta_s^2 \, ds$.

For every real function F belonging to $C_2(\mathbb{R})$, the process is transformed into a diffusion

$$F(X_t) - F(X_0) = \int_0^t F'(X_s)\alpha_s \, ds + \int_0^t F'(X_s)\beta_s \, dB_s + \frac{1}{2}\int_0^t F'(X_s)\beta_s^2 \, ds, \tag{4.15}$$

it is the solution of the stochastic differential equation

$$dF(X_t) = F'(X_t)(\alpha_t + \frac{1}{2}\beta_t^2) \, dt + F'(X_t)\beta_t \, dB_t.$$

More generally, the sample paths of a Gaussian process solution of a SDE belong to functional sets determined by the expression of the drift of the equation. With a linear drift

$$dX(t) = \alpha(t)X(t)\,dt + \beta(t)\,dB(t), \tag{4.16}$$

the stochastic integral

$$X_t = X_0 \exp\{\int_0^t \alpha(s)\,ds\} + \int_0^t \beta(s)\,dB(s) \tag{4.17}$$

is the solution of the diffusion Equation (4.16). Let $\mathcal{F}_t$ be the σ-algebra generated by $\{B_s, 0 \leq s \leq t\}$, the Brownian motion B is a $(\mathcal{F}_t)_{t\geq 0}$-martingale with independent increments and with mean zero and the variance $EB_t^2 = t$. The initial value X_0 is supposed to be independent of $B(t) - B(s)$ for every $t > s > 0$, then $E(X_t) = E(X_0) \exp\{\int_0^t \alpha(s)\,ds\}$ and the centered process

$$M_t = X_t - X_0 \exp\{\int_0^t \alpha(s)\,ds\} = \int_0^t \beta(s)\,dB(s)$$

is a transformed Brownian motion on $(\Omega, \mathcal{F}, (\mathcal{F}_t)_{t\geq 0}, P)$. It satisfies the properties of Section 4.6, in particular $E(M_t^2) = \int_0^t \beta^2(s)\,ds$, and for every $0 < s < t < u$, $E\{(M_u - M_t)(M_t - M_s)\} = 0$ and

$$E(M_t^2|\mathcal{F}_s) = M_s^2 + \int_s^t \beta^2(s)\,ds.$$

The function of its quadratic variations is $< M >_t = \int_0^t \beta^2(s)\,ds$ and the exponential process related to M, $\mathcal{E}_{M_t}(\lambda) = \exp\{\lambda M_t - L_{M_t}(\lambda)\}$, is a martingale, where L_{M_t} is the Laplace transform of M_t.

Let $A(t) = \int_0^t \alpha(s)\,ds$, the process $X = M + X_0 e^A$ has the variance

$$E(X_t^2) = var(X_0)\,e^{2A(t)} + \int_0^t \beta^2(s)\,ds$$

and the quadratic variations

$$E\{(X_t - X_s)^2|\mathcal{F}_s\} = EX_0^2\,(e^{A(t)} - e^{A(s)})^2 + \int_s^t \beta^2(s)\,ds,$$

the predictable process of its quadratic variations is

$$< X >_t = X_0^2\,e^{2A(t)} + < M >_t .$$

The Laplace transform of X_t is

$$L_{X_t}(\lambda) = L_{X_0}(\lambda e^{A(t)})\,L_{M_t}(\lambda).$$

If $E\|\beta\|_\infty$ is finite, the process $< M >$ has the order $O(t)$. Proposition 4.11 applies to M and X, for every $T > 0$

$$P(\sup_{t\in[0,T]} t^{-\frac{1}{2}}|M_t| > \lambda) \le \lambda^{-2} \sup_{T>0} ET^{-1} \int_0^T \beta^2 \le \lambda^{-2}E \sup_{t\in[0,T]} \beta^2(t),$$

$$P(\sup_{t\in[0,T]} t^{-\frac{1}{2}}|X_t - EX_t| > 2\lambda) \le P(\sup_{t\in[0,T]} t^{-\frac{1}{2}}|M_t| > \lambda) \tag{4.18}$$

$$+ P(\sup_{t\in[0,T]} t^{-\frac{1}{2}}e^{A(t)}|X_0 - EX_0| > \lambda)$$

$$\le \lambda^{-2} \sup_{t\in[0,T]} \{\beta^2(t) + t^{-1}e^{-2A(t)}varX_0\}.$$

These inequalities still hold as T tends to infinity.

Consider a diffusion process with a drift function α depending on the time index and on the sample path of the process and with a random diffusion term, determined by the differential equation

$$dX(t) = \alpha(t, X_t)\, dt + \beta(t)\, dB(t). \tag{4.19}$$

Under the condition that $E|\int_0^T |X_s^{-1}\alpha(s, X_s)|\, ds|$ is finite, the stochastic integral

$$X_t = X_0 e^{A(t,X_t)} + \int_0^t \beta(s)\, dB(s), \tag{4.20}$$

$$A(t, X_t) = \int_0^t X_s^{-1}\alpha(s, X_s)\, ds, \quad A(t, X_t) = 0$$

is an implicit solution of the diffusion Equation (4.19). On $(\Omega, \mathcal{F}, (\mathcal{F}_t)_{t\ge0}, P)$, X is a process with mean $E(X_t) = E\{X_0 e^{A(t,X_t)}\}$ and variance $var(X_t) = var\{X_0 e^{A(t,X)}\} + \int_0^t \beta^2(s)\, ds$, where

$$var\{X_0 e^{A(t,X)}\} = E[X_0^2 var\{e^{A(t,X)}|\mathcal{F}_0\}] + var[X_0 E\{e^{A(t,X)}|\mathcal{F}_0\}].$$

The process X is not a martingale since

$$|E\{(X_t - X_s)|\mathcal{F}_s\}| = |X_0|\, E[\exp\{\int_s^t X_u^{-1}\alpha(u, X_u)\, du\}|\mathcal{F}_s]$$

$$\ge |X_0|\, \exp[\int_s^t E\{X_u^{-1}\alpha(u, X_u)|\mathcal{F}_s\}\, du] > 0.$$

The quadratic variations process of X is then

$$< X >_t = X_0^2\, e^{2A(t)} + \int_0^t \beta^2(s)\, ds$$

and an inequality similar to (4.18) is satisfied.

Diffusion equations with polynomial functions $\alpha(x)$ defined on the sample-paths space of the process are easily solved and they are not exponential. Let γ be a function of $L_1(\mathbb{R}_+)$ with primitive Γ and let β be a real function of $L_2(\mathbb{R}_+)$. Equation (4.19) defined by

$$\alpha(t, X_t) = \gamma(t)(X_t - x)^p, \ p \geq 2$$

has a solution sum of a $\mathcal{F}_0$-measurable drift

$$X_{p,t} = x + \{(X_0 - x)^{-(p-1)} - (p-1)\Gamma(t)\}^{-\frac{1}{p-1}} \tag{4.21}$$

and the Gaussian martingale $M_t = \int_0^t \beta_s \, dB_s$, $X_t = X_{p,t} + M_t$. This result is extended to diffusions with polynomial drifts.

Theorem 4.9. *The diffusion Equation* (4.19) *defined for $p \geq 2$ by a drift*

$$\alpha(t, X_t) = \sum_{k=0}^{p} \gamma_k(t)(X_t - x)^k$$

and a volatility $\sum_{k=0}^{p} \beta_{k,t} \, dB_{k,t}$, with p independent Brownian motions, has the solution

$$X_t = \sum_{k=0}^{p} X_{k,t} + \sum_{k=0}^{p} \int_0^t \beta_{k,s} \, dB_{k,s},$$

where $X_{k,t}$ is defined by (4.21) *for $k \geq 2$, $X_{1,t}$ is defined by the exponential process* (4.17) *and $X_{0,t} = X_0 + \int_0^t \gamma_0(s) \, ds$.*

The mean of X_t is the mean of the drift and its variance conditionally on $\mathcal{F}_0$ is $\sum_{k=0}^{p} \int_0^t \beta_{k,s}^2 \, ds$. The inequalities proved for the Brownian apply to the noise process of the diffusion and thus to the centered process X_t, like in Equation (4.18).

Stochastic differential equations with jumps are defined on an interval $[0, \tau]$ and a metric space $(\mathbb{X}, \| \cdot \|)$, by real functions α in L_1, β and γ in L_2, from a Brownian motion B and a point process N independent of B. The process X_t solution of the stochastic differential equation

$$dX_t = \alpha(t, X_t)dt + \beta(t)dB_t + \gamma(t, X_t)dN_t \tag{4.22}$$

has a continuous part defined in Theorem 4.9 and a discrete part. The solution of (4.22) is the sum of the solution of the continuous diffusion Equation (4.21) and of a jump process. If γ is a deterministic function, the discontinuous part of the solution is $\int_0^t \gamma_s \, dN_s$ and it is independent of its continuous part. With a process γ such that $E \int_0^t X_s^{-1} \gamma_s \, dN_s$ is finite for every t in the interval $[0, \tau]$, the jump part of the solution is

$$Y_t = X_0 \prod_{0 < T_i \leq t} \{1 - X_{T_i}^{-1} \gamma(T_i, X_{T_i})\}$$

where the jumps of the point process N are $(T_i)_{i \geq 1}$.

4.8 Level crossing probabilities

The previous sections provide bounds for the tail probabilities of sums of variables and martingales. The exact tail probabilities of distributions of empirical laws have been computed by many authors who established tables used in goodness-of-fit testing for samples of variables, such as tables for the Kolmogorov-Smirnov statistics. They give exact or asymptotic expansions for the tail probabilities of empirical distributions, calculated from the order statistics of the variable sample. Other results concern the limiting probabilities of the number of level crossings. The first result in this field concerned sums of independent and identically distributed variables with mean zero. Let

$$N_n = \min\{k : S_k \geq S_m, 1 \leq k, m \leq n\}$$

be the first index where the maximum of the partial sums $S_k = X_1 + \cdots + X_k$ is achieved, for $k = 1, \ldots, n$. Considering the binomial distribution with values -1 and 1 with probabilities $\frac{1}{2}$. Lévy (1939) established that for every α in $[0, 1]$

$$\lim_{n \to \infty} P(n^{-\frac{1}{2}} N_n < \alpha) = \frac{2}{\pi} \arcsin(\alpha^{\frac{1}{2}}).$$

Erdös and Kac (1947) extended this result to other distributions. These results are easily proved from the weak convergence of the process $n^{-\frac{1}{2}} N_n$ to the standard Brownian motion and from the arcsine law (1.21) for the time when the Brownian motion reaches its maximum.

Consider the sum S_n of n exponential variables with parameter 1, $\lim_{n \to \infty} n^{-1} S_n = 1$ a.s. and the variables S_n are the increasing times of a Poisson process. For a threshold $t > 0$, the number of sums such that $S_n < t$ is N_{t^-}, where $N_t = \sum_{k \geq 1} 1_{\{S_k \leq t\}}$, therefore $N_t = n$ when t belongs to the random interval $[S_n, S_{n+1}[$. The process N_t has a homogeneous Poisson distribution with intensity t and mean $EN_t = t$, and the process $t^{-1} N_t$ converges a.s. to one. Applying Proposition 1.1 to the Poisson process implies that for every x in $]0, 1[$

$$\lim_{t \to \infty} P\{t^{-1}(N_t - t) > x - 1\} = \lim_{t \to \infty} P\{t^{-1} N_t > x\} \leq \frac{1}{x}$$

and the limit is given by Chernov's theorem.

These results are generalized to Markov processes. Lamperti (1958) studied the limit of the probability $G_n(t) = P(n^{-1} N_n \leq t)$ for the number of transitions in a recurrent state s of a Markov process having occurrence probabilities p_n. Under the necessary and sufficient conditions that the limits $\alpha = \lim_{n \to \infty} En^{-1}N_n$ and $\delta = \lim_{x \to 1}(1 - x)F'(x)\{1 - F(x)\}^{-1}$ are finite, where F is the generating function of the occurrence probabilities, the probability function $G_n(t)$ converges to a distribution function $G(t)$ with a density generalizing the arcsine law and depending on the parameters α and δ. Other limits of interest are those of the current time $\gamma(t) = t - S_{N_t}$ and the residual time $\delta(t) = S_{N_t+1} - t$ of sums of independent and identically distributed variables X_i with distribution function F. As $t \to \infty$, the distribution function of the current time $\gamma(t)$ converges to the distribution function

$$F_{\gamma_\infty}(y) = \lim_{t \to \infty} P(\gamma(t) \leq y) = \frac{1}{EX} \int_y^\infty \{1 - F(t^-)\}\, dt$$

(Pons, 2008) and the distribution function of the residual time δ_t converges to the distribution function F_{δ_∞}

$$\bar{F}_{\delta_\infty}(y) = \lim_{t \to \infty} P(\delta(t) \leq y) = \frac{1}{EX} \int_0^y \{1 - F(t^-)\}\, dt$$

(Feller, 1966), therefore

$$\lim_{t \to \infty} P(S_{N_t} \leq t + y) = F_{\gamma_\infty}(y), \qquad \lim_{t \to \infty} P(S_{N_t} \geq t - y) = F_{\delta_\infty}(y).$$

Karlin and McGregor (1965) provided expressions for the limits $\lim_{t \to \infty} P(t^{-1}\gamma(t) \leq x)$ and $\lim_{t \to \infty} P(t^{-1}\delta(t) \leq x)$, under the condition that $1 - F(t) \sim t^{-\alpha}L^{-1}(t)$ as t tends to infinity, where L is a slowly varying function and $0 < \alpha < 1$.

By the weak convergence of the normalized process $(t^{-\frac{1}{2}}(N_t - t))_{t \geq 0}$ to a standard Brownian motion $(B_t)_{t \geq 0}$, the same limit applies to B and to $t^{-\frac{1}{2}}(N_t - t)$. For every x in $[0, 1]$

$$\lim_{t \to \infty} P\{t^{-\frac{1}{2}}B_t > x\} = \lim_{t \to \infty} P\{t^{-\frac{1}{2}}B_t < -x\}$$

$$= \lim_{t \to \infty} P\{t^{-1}N_t < 1 - x\} = \exp\{-\frac{x^2}{2}\}$$

from the probability of large deviations for the Brownian motion. Replacing x by a function h defined from $\mathbb{R}$ to $\mathbb{R}_+$ implies

$$P\{B_t > h(t)\} = \exp\{-\frac{h^2(t)}{2t}\}.$$

Assuming $\lim_{t \to \infty} t^{-\frac{1}{2}}h(t) = 0$, it follows that $\lim_{t \to \infty} P\{B_t > h(t)\} = 1$ and the Brownian motion crosses every function $h(t) = o(\sqrt{t})$, as t tends to infinity. Their almost sure limits are usually expressed by the law of iterated logarithm (Kiefer, 1961)

$$\limsup_{t \to \infty} \frac{B_t}{\sqrt{2t \log \log t}} = 1, \qquad \liminf_{t \to \infty} \frac{B_t}{\sqrt{2t \log \log t}} = -1, \text{ a.s.}$$

By the same argument, the probability of the event $\{B_n > \sqrt{2n \log \log n}\}$ is $(\log n)^{-1}$ which tends to zero as n tends to infinity

$$\sum_{n \geq 2} P\{B_n > \sqrt{2n \log \log n}\} = \sum_{n \geq 2} \exp\{-\log \log n\}$$

$$= \sum_{n \geq 2} \frac{1}{\log n} > \sum_{n \geq 1} \frac{1}{n} = \infty.$$

By Borell-Cantelli's lemma, we obtain

$$\limsup_{t \to \infty} \frac{|B_t|}{\sqrt{2t \log \log t}} \geq 1, \text{ a.s.} \qquad (4.23)$$

This inequality proved for the discrete process extends to the Brownian motion indexed by $\mathbb{R}$ by limit as n tends to infinity. It is also true for other functions than the iterated logarithm. Section A.2 presents some examples from Robbins and Siegmund (1970). The next boundary is more general.

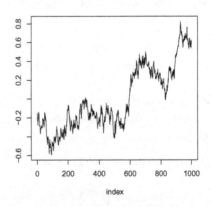

Fig. 4.1   A sample path of the Brownian motion.

Proposition 4.23. *For every real function h such that $\int_0^\infty h_t^{-1} \, dt$ is finite, the Brownian motion satisfies*

$$\limsup_{t \to \infty} \frac{|B_t|}{\sqrt{2t \log h(t)}} \leq 1, \text{ a.s.}$$

The proof follows the same arguments as the proof of (4.23). For example, a function such that $h(t) \leq \exp\{t^{-\alpha}\}$, $\alpha > 1$, fulfills the summability condition of Proposition 4.23. Conversely, for every real function $h > 0$ such that $\int_0^x h_t^{-1} dt$ tends to infinity as x tends to infinity, the Brownian motion crosses a.s. the frontiers $\pm\sqrt{2t \log h(t)}$

$$\limsup_{t \to \infty} \frac{|B_t|}{\sqrt{2t \log h(t)}} \geq 1, \text{ a.s.}$$

This behaviour extends to transformed Brownian motions.

Proposition 4.24. *Let* $Y_t = A_t^{-1} \int_0^t A_s \, dB_s$, *where* A *is a.s. decreasing and positive process. For very real function* h *on* $\mathbb{R}$ *such that* $\int_0^x h_t^{-1} dt$ *tends to infinity as* x *tends to infinity, the process* Y *satisfies*

$$\limsup_{t \to \infty} \frac{|Y_t|}{\sqrt{2t \log h(t)}} \geq 1, \text{ a.s.}$$

Indeed the variance of Y_t is $E \int_0^t (A_t^{-1} A_s)^2 \, ds \geq t$ and the proof follows the same arguments as in the proof for the Brownian motion, using Chernov's theorem

$$\int_0^x P\{Y_t > \sqrt{2 \log h_t}\} \, dt > \int_0^x \frac{1}{h_t} \, dt$$

which tends to infinity with x. With an a.s. increasing positive process A and a function h such that $\int_0^\infty h_t^{-1} dt$ is finite

$$\limsup_{t \to \infty} \frac{|Y_t|}{\sqrt{2t \log h(t)}} \leq 1, \text{ a.s.}$$

This result applies to sequences of martingales indexed by $\mathbb{N}$. Let $S_n = \sum_{i=1}^n X_i$ be a martingale with conditional means zero and conditional variances 1, the law of iterated logarithm for S_n is written as

$$\limsup_{n \to \infty} (2n \log \log n)^{-\frac{1}{2}} |S_n| \geq 1, \text{ a.s.}$$

Freedman (1975) extended the results to a martingale $S_n = \sum_{i=1}^n X_i$, with the quadratic variations $V_n = \sum_{i=1}^n var(X_n | \mathcal{F}_{n-1})$ and such that $|X_n| \leq 1$

$$\limsup_{n \to \infty} (2V_n \log \log V_n)^{-\frac{1}{2}} |S_n| \leq 1 \text{ a.s. on } \{V_n = \infty\}.$$

Under similar conditions, a supermartingale S_n such that $|X_n| \leq 1$ satisfies $\limsup_{n \to \infty} (2V_n \log \log V_n)^{-\frac{1}{2}} |S_n| \geq 1$, a.s.

Proposition 4.23 also applies to the martingale S_n with increasing threshold functions φ. By the a.s. asymptotic equivalence of V_n and EV_n, for every function φ such that $\sum_{n \geq 1} \exp\{-\frac{1}{2}\varphi_n^2\}$ or $\int_1^\infty \exp\{-\frac{1}{2}\varphi_t^2\} \, dt$ are finite

$$\limsup_{n \to \infty} V_n^{-\frac{1}{2}} \varphi_n^{-1} S_n \leq 1, \qquad \liminf_{n \to \infty} V_n^{-\frac{1}{2}} \varphi_n^{-1} S_n \geq -1, \quad \text{a.s.}$$

Replacing S_n by the sum of the independent transformed increments of the martingale $Y_n = \sum_{i=1}^n \varphi(X_i)$, with an $L_2(\mathbb{R})$ function φ such that the empirical mean $\bar{\mu}_n = n^{-1} \sum_{i=1}^n E\varphi(X_i)$ converges a.s. to a finite limit, the law of iterated logarithm for the process $(Y_n)_{n \geq 0}$ is written in terms of the quadratic variations $T_n = \sum_{i=1}^n var\{\varphi(X_i)|\mathcal{F}_{i-1}\}$

$$\lim_{n \to \infty} \sup (2T_n \log\log\varphi_n)^{-\frac{1}{2}} (Y_n - \bar{\mu}_n) \leq 1, \quad \text{a.s.}$$

$$\lim_{n \to \infty} \inf (2T_n \log\log\varphi_n)^{-\frac{1}{2}} (Y_n - \bar{\mu}_n) \geq -1, \quad \text{a.s.}$$

It is generalized to varying threshold functions like S_n.

4.9   Martingales in the plane

Walsh (1974) and Cairoli and Walsh (1975) presented inequalities for martingales and stochastic integrals with respect to (local) martingales in the plane. A partial order is defined as the order with respect to both coordinates and the notion of martingale is defined for this order, it implies that the martingale property holds with respect to both marginal filtrations. For every $z = (s,t)$ in $\mathbb{R}_+^2$, let $(\mathcal{F}_z)_{z>0}$ be a family of increasing σ-algebra and let $(\mathcal{F}_z^k)_{z>0}$ be the marginal defined σ-algebras defined by $\mathcal{F}_z^1 = \mathcal{F}_{s,\infty}$ and $\mathcal{F}_z^2 = \mathcal{F}_{\infty,t}$. A martingale M with respect to the filtration on $(\Omega, \mathcal{F}, P)$ is an integrable real process, adapted to $(\mathcal{F}_z)_{z>0}$ such that $E(M_z|\mathcal{F}_{z'}) = M_z$ for every $z' \leq z$ in the total order of $\mathbb{R}_+^2$. The increment of the martingale on the rectangle $R_{z',z} = [z', z]$ is

$$\Delta M_{[z',z]} = M_z + M_{z'} - M_{(s',t)} - M_{(s,t')} \qquad (4.24)$$

and a weak martingale is defined by the property $E(M_{[z',z]}|\mathcal{F}_{z'}) = 0$. The process M is a strong martingale if $E(M_{[z',z]}|\mathcal{F}_{z'}^1 \vee \mathcal{F}_{z'}^2) = 0$.

Let $R_z = [0, z]$ and $R_{z'} = [0, z']$ be rectangles from 0 to z and z' respectively, then $R_z \cap R_{z'} = [0, z \wedge z']$ with the minimum in both coordinates and $E(M_{[z \wedge z', z \vee z']}|\mathcal{F}_{z \wedge z'}) = M_{R_{z \wedge z'}}$. If a weak martingale satisfies the condition $M(0) = M(0,t) = M(s,0) = 0$ for every $z = (s,t)$, where the values of M are considered as its increments on empty rectangles, $E(M_z|\mathcal{F}_{(0,t)}) = E(M_z|\mathcal{F}_{(s,0)}) = 0$. Under the condition of conditionally independent marginal σ-algebras, a local square integrable strong martingale M has a predictable compensator $< M >$ such that $M^2 - < M >$ is a martingale. Lenglart's theorem applies to a strong martingale in the plane.

On $(\Omega, \mathcal{F}, P)$, a counting process N in $\mathbb{R}_+^d$ is defined by the cardinal $N(A)$ of N in the subsets A of $\mathbb{R}_+^d$. The notions of martingales (respectively weak and strong martingales) are generalized to $\mathbb{R}^n$ with similar properties. Point processes with independent increments are defined as d-dimensional stochastic discrete measures. The cumulative intensity function Λ of a Poisson process N is a positive measure in $\mathbb{R}_+^d$ such that $N - \Lambda$ is a weak martingale with respect to the filtration generated by N on the rectangles of $\mathbb{R}_+^d$ and the function $\Lambda(A) = \int_A \lambda(s)\,ds$ is a positive measure in $\mathbb{R}_+^d$. The process $N - \Lambda$ has independent rectangular increments distributed like Poisson variables having as parameters the measure of the sets, with respect to Λ, then the property of conditional independence of the marginal σ-algebras is satisfied and the process $N - \Lambda$ is a martingale. Moreover, $\Lambda(A)^{-\frac{1}{2}} \{N(A) - \Lambda(A)\}_{A \in \mathbb{R}^d}$ converges weakly to a Gaussian process as the volume of A tends to infinity, when $\sup_{A \in \mathbb{R}^d} \|A\|^{-1} \Lambda(A)$ is bounded. From the martingale property, the variance of $N(A)$ is $\Lambda(A)$ and the covariance of $N(A)$ and $N(B)$ is

$$E\{N(A)N(B)\} - \Lambda(A)\Lambda(B) = E(N - \Lambda)^2(A \cap B) = \Lambda(A \cap B),$$

the normalized process has therefore a covariance equal to the correlation of N on the subsets of $\mathbb{R}_+^d$

$$R(A, B) = \frac{\Lambda(A \cap B)}{\{\Lambda(A)\Lambda(B)\}^{\frac{1}{2}}}.$$

The independence of the increments is a property of the Poisson process, and every point process with independent increments has a similar behaviour with a stochastic or deterministic compensator instead of the function Λ. More generally, every point process N has a predictable compensator $\widetilde{N}$ such that $M = N - \widetilde{N}$ is a weak martingale, it is not a martingale if the increments of the process N are dependent.

Because of the independence of the increments of N, as $\|A\|$ tends to infinity, the variable $N(A)$ is the sum of the number of points of N in an increasing number of nonintersecting subsets A_i of A with bounded volume, and the variables $N(A_i)$ are independent. It follows that $\|A\|^{-1} N(A)$ converges a.s. to $\Lambda(A)$ and $\|A\|^{-1} \widetilde{N}(A)$ converges a.s. to the same limit. The normalized variable $X(A) = \|A\|^{-\frac{1}{2}} \{N(A) - \widetilde{N}(A)\}$ has the mean zero and it converges weakly to a Gaussian variable as $\|A\|$ tends to infinity. Its covariance is

$$E\{X(A)X(B)\} = \frac{E\{N(A) - \widetilde{N}(A)\}\{N(B) - \widetilde{N}(B)\}}{\{\|A\|\,\|B\|\}^{\frac{1}{2}}}.$$

Under the condition of independent increments, the covariances and the variances of the counting process N and of the entered process X satisfy $E\{N(A)N(B)\} = EN(A)\,EN(B) + var\,N(A \cap B)$, therefore

$$E\{X(A)X(B)\} = \frac{1}{\{\|A\|\,\|B\|\}^{\frac{1}{2}}} var\{N(A \cap B) - \widetilde{N}(A \cap B)\}.$$

Chapter 5

Functional Inequalities

5.1 Introduction

Analytic inequalities for operators in functional spaces are often deduced from their derivatives, such as bounds for the entropy or the information of classes (Section 5.1). The derivative of an operator T on a vector space $(\mathcal{F}, \| \cdot \|)$ at f is

$$T'(f)(h) = \lim_{t \to \infty} \frac{T(f + th) - T(f)}{t},$$

where h belongs to $\mathcal{F}$ and t to $\mathbb{R}$ where the limit for the topology of the norm is the convergence to zero of $\|t^{-1}\{T(f + th) - T(f)\} - T'(f)(h)\|$.

Let $(X_1, \ldots, X_n)$ be a sequence of random and identically distributed variables defined on a probability space $(\Omega, \mathcal{A}, P)$ and with values in a separable Banach space $(\mathcal{X}, \mathcal{B})$ provided with the Borel σ-algebra $\mathcal{B}$. Let us denote $\mathcal{P}_\mathcal{X}$ the space of probability distributions on $\mathcal{X}$, in $\mathcal{P}_\mathcal{X}$, P is the probability distribution of the variables X_i, $\widehat{P}_n(t) = \sum_{i=1}^n 1_{\{X_i \leq t\}}$ is their empirical distribution, $\nu_{n,P} = \sqrt{n}(\widehat{P}_n - P)$ is the normalized empirical process and G_P is the Brownian bridge associated to P.

Let $N(\mathcal{X}, \varepsilon)$ be the minimum number of sets covering the space $\mathcal{X}$ with ball of radius less than ε for the metric of $\mathcal{X}$, the ε-entropy of $\mathcal{X}$ is $H(\mathcal{X}, \varepsilon) = \log N(\mathcal{X}, \varepsilon)$. Under the condition that $\int_0^1 H(\mathcal{X}, \varepsilon) \, d\varepsilon$ is finite and under regularity conditions for the variables, the normalized sum $n^{-\frac{1}{2}} S_n$ converges weakly to a centered Gaussian variable (Dudley, 1974; Giné, 1974). This result was applied to the sums of continuous processes, with the uniform topology.

Processes are also defined on functional families of $\mathcal{L}_P^2 = \{f : \mathcal{X} \mapsto \mathbb{R};$ $\int_\mathcal{X} f^2 \, dP < \infty\}$. Pollard's conditions and notations for the functional empirical process are adopted. A pseudo-distance $\rho_P(f, g) = \sigma_P(f - g)$ is used on $\mathcal{F}$, with $\sigma_P^2(f) = P(f^2) - P^2(f)$. The sample-path space of the Brownian bridge is $\ell^\infty(\mathcal{F}) = \{h : \mathcal{F} \mapsto \mathbb{R}; h \in C_b\}$, and $\ell^\infty(\mathcal{F})$ is a measurable space provided with the uniform norm on $\mathcal{F}$ and the Borel σ-algebra for this norm. Then the space $\mathcal{U}_b(\mathcal{F}, P)$ of the uniformly continuous and bounded real functions on $(\mathcal{F}, \rho_P)$ is a separable Banach space in $\ell^\infty(\mathcal{F})$ (Dudley, 1984). We suppose that $\mathcal{F}$ has a measurable envelope function $F = \sup\{f \in \mathcal{F}\}$ in L_2. Let

$$N_F(\varepsilon, \mathcal{F}) = \sup_{Q \in \mathcal{P}_\mathcal{X}} \max \ \{m : \exists \text{ distinct } f_1, \ldots, f_m \in \mathcal{F} \cap C_b(\mathcal{X});$$

$$Q(|f_i - f_j|^2) > \varepsilon^2 Q(F^2), \forall i \neq j\}$$

be the Pollard entropy function of $(\mathcal{F}, F)$, a set of functions $(f_j)_j$ bracketing all functions of a class $\mathcal{F}$ for a constant ε is called an ε-net of $\mathcal{F}$. The entropy dimension of $\mathcal{F}$ is

$$d_F^{(2)}(\mathcal{F}) = \inf \left\{ \delta > 0; \limsup_{\varepsilon \to 0} \varepsilon^\delta N_F(\varepsilon, \mathcal{F}) < \infty \right\}$$

and its exponent entropy is

$$e_F^{(2)}(\mathcal{F}) = \inf \left\{ \delta > 0; \limsup_{\varepsilon \to 0} \varepsilon^\delta \log N_F(\varepsilon, \mathcal{F}) < \infty \right\}.$$

The supremum of the Brownian bridge W on a class $(\mathcal{F}, F)$ such that $e_F^{(2)}(\mathcal{F}) < 2$ and $\sup_{f \in \mathcal{F}} \sigma^2(f) < \sigma^2$ satisfies

$$P(\|W\|_\mathcal{F} > t) \leq C_\sigma \exp\{-\frac{t^2}{2\sigma^2}\}.$$

Similar inequalities are proved in the next section for $\sup_{f \in \mathcal{F}} |S_n(f)|$ under entropy and integrability conditions, generalizing Bennet's inequality. For $s \geq 2$, the entropy function of order s for $(\mathcal{F}, F)$ is defined as

$$N_F(\varepsilon, \mathcal{F}, s) = \sup_{Q \in \mathcal{P}_\mathcal{X}} \max \ \{m : \exists \text{ distinct } f_1, \ldots, f_m \in \mathcal{F} \cap C_b(\mathcal{X});$$

$$Q(|f_i - f_j|^s) > \varepsilon^s Q(F^s), \forall i \neq j\}.$$

5.2　Exponential inequalities for functional empirical processes

A subset $\mathcal{C}$ of $\mathcal{X}$ is a Vapnik-Cervonenkis class if $\|\widehat{P}_n - P\|_\mathcal{C}$ converges a.s. to zero as n tends to infinity. It is a P-Donsker class if there exists a sequence of uniformly continuous Brownian bridges $G_P^{(n)}$ such that for every $\varepsilon > 0$ there exists an integer n_0 such that $P(\|\nu_n - G_P^{(n)}\|_\mathcal{C} > \varepsilon) \leq \varepsilon$, for every $n \geq n_0$.

Let $\mathcal{F}$ be a class of functions on a metric space $(\mathbb{X}, \mathcal{X}, \|\cdot\|)$, measurable with respect to the σ-algebra generated by the closed balls of bounded functions for the uniform topology, centered at uniformly continuous functions in $L_2(\mathbb{X})$. Pollard (1981) proved that sufficient conditions implying that $\mathcal{F}$ is a P-Donsker class are the existence of an envelope function F of $\mathcal{F}$ belonging to $L_2(\mathbb{X}, P)$ and

$$\int_0^1 (\log N_F(\varepsilon, \mathcal{F}))^{\frac{1}{2}} \, d\varepsilon < \infty.$$

From Pollard (1982), this integral is finite if the envelope F belongs to $L_2(P)$ and $N_F(\varepsilon, \mathcal{F}) \le A\varepsilon^{-2k}$ on $[0,1]$, with constants A and k. Assouad (1981), Dudley (1984), Massart (1983), van der Vaart and Wellner (1996) present examples of Vapnik-Cervonenkis and Donsker classes.

The moments of $\sup_{f \in \mathcal{F}} |S_n(f)|$ satisfy inequalities similar to those established for the sum of variables, with $V_n(f) = \sum_{i=1}^n f^2(X_i)$. The bounds depends on the same constants $c_{\alpha,n}$ and $C_{\alpha,n}$ as in Proposition 4.4.

Proposition 5.1. *Let* $(X_i)_{i=1,\dots,n}$ *be a vector of independent random and identically distributed variables on a probability space* $(\Omega, \mathcal{A}, P)$, *with values in a metric space* $(\mathbb{X}, \mathcal{X}, \|\cdot\|)$. *Let* $\alpha \ge 2$ *and let* $\mathcal{F}$ *be a class of measurable* $L_\alpha(P)$ *functions on* $(\mathbb{X}, \mathcal{X}, \|\cdot\|)$ *such that* $P(f) = 0$ *for every* f *in* $\mathcal{F}$ *and* $E_P\|f(X_i)\|_{2,\mathcal{F}} = \sigma_P^2(F)$, *then*

$$c_{\alpha,n} E\{\sup_{f \in \mathcal{F}} V_n^{\frac{\alpha}{2}}(f)\} \le E(\sup_{f \in \mathcal{F}} |S_n(f)|^\alpha) \le C_{\alpha,n} E\{\sup_{f \in \mathcal{F}} V_n^{\frac{\alpha}{2}}(f)\}.$$

Proof. Writing $E \sup_{f \in \mathcal{F}} |S_n(f)| \le \{E \sup_{f \in \mathcal{F}} |S_n(f)|^2\}^{\frac{1}{2}}$ and using the inequality $|S_n(f)|^2 \le C_{2,n} V_n(f)$ for every f of $\mathcal{F}$ and we obtain

$$E \sup_{f \in \mathcal{F}} |S_n(f)| \le \{C_{2,n} E \sup_{f \in \mathcal{F}} V_n(f)\}^{\frac{1}{2}} = \{C_{2,n} E V_n(F)\}^{\frac{1}{2}}.$$

For $\alpha \ge 2$, the inequalities are proved by the same arguments, as in the proof of Proposition 4.4. $\square$

The constant $C_{\mathcal{F}}$ that appears in the next functional inequalities is an integral function of the dimension of the class $\mathcal{F}$. Here the main constant of dimension is

$$C_{\mathcal{F}} = \{\int_0^1 N_F(\varepsilon, \mathcal{F}) \, d\varepsilon\}^{\frac{1}{2}}. \tag{5.1}$$

Proposition 5.2. *Let* $(X_i)_{i=1,\dots,n}$ *be a sequence of independent and identically distributed random variables on* $(\Omega, \mathcal{A}, P)$, *with values in a metric* space $(\mathbb{X}, \mathcal{X}, \|\cdot\|)$. *Let $\mathcal{F}$ be a class of measurable functions on $\mathbb{X}$ such that* $\|f(X_i)\|_{\mathcal{F}}$ *belongs to L_2, and $C_{\mathcal{F}}$ is finite. For every $t > 0$*

$$E \exp\{t \sup_{f \in \mathcal{F}} S_n(f)\} \leq C_{\mathcal{F}} L_{F(X)}^{\frac{n}{2}}(2t).$$

Proof. From the equivalence of the norms L_p of random variables, $\|f(X_i)\|_{\mathcal{F}}$ belongs to L_p for every $p \geq 2$. Let $t > 0$, by independence of the variables, $E \exp\{t \sup_{f \in \mathcal{F}} S_n(f)\} \leq [E \exp\{t \sup_{f \in \mathcal{F}} f(X_i)\}]^n$. Let ε be in $]0, \delta[$, with $\delta < 1$. Let $\mathcal{F}_\varepsilon$ be an ε-net of $\mathcal{F}$ adapted to P and let π_ε be the projection from $\mathcal{F}$ to $\mathcal{F}_\varepsilon$, then for every f in $\mathcal{F}$ and for $p \geq 2$

$$\|(f - \pi_\varepsilon(f))(X)\|_{L_2} \leq \varepsilon \|F(X)\|_{L_2},$$
$$E \exp\{t \sup_{f \in \mathcal{F}} f(X)\} \leq [E \exp\{2t \sup_{f_\varepsilon \in \mathcal{F}_\varepsilon} f_\varepsilon(X)\}]^{\frac{1}{2}}$$
$$[E \exp\{2t \sup_{f \in \mathcal{F}} (f - \pi_\varepsilon(f))(X)\}]^{\frac{1}{2}}, \quad (5.2)$$

by the the Cauchy-Schwarz inequality. Moreover

$$E[\exp\{2t \sup_{f_\varepsilon \in \mathcal{F}_\varepsilon} f_\varepsilon(X)\}] \leq E \exp\{2tF(X)\} \int_0^\delta N_F(\varepsilon, \mathcal{F}) \, d\varepsilon$$

and the exponential function in $E \exp\{2t \sup_{f \in \mathcal{F}} (f - \pi_\varepsilon)(X)\}$ is expanded as a sum bounded, using Proposition 5.1, as $1 + \frac{1}{2}\varepsilon^2 \|F(X)\|_{L_2}^2 + o(\varepsilon^2)$ when δ tends to zero.

Since the bounds are valid for every $\delta < 1$, the integral over $[0, 1]$ gives an upper bound for the product of (5.2). $\qquad \square$

Functional Bienaymé-Chebychev inequalities for the sum $S_n(f)$ are consequences of Proposition 5.1, they are proved like Proposition 5.2, by projections on each function of $\mathcal{F}$ in an ε-net of the class, a bound for the supremum is obtained by summing a uniform bound of the projections over all functions of the ε-net

$$P(\sup_{f \in \mathcal{F}} |n^{-\frac{1}{2}} S_n(f)| \geq t) \leq 2C_{\mathcal{F}} \frac{\sigma_P^2(F)}{t^2},$$

$$P(\sup_{f \in \mathcal{F}} |S_n(f) S_m(f)| \geq (n \wedge m)t) \leq 2C_{\mathcal{F}} \frac{\sigma_P^2(F)}{t},$$

and for every integer $p \geq 2$

$$P(\sup_{f \in \mathcal{F}} |n^{-\frac{1}{2}} S_n(f)| \geq t) \leq 2C_{\mathcal{F}} \frac{E\{V_n^p(F)\}}{t^{2p}}. \quad (5.3)$$

Similar results hold uniformly in n, for every stopping time N of S_n

$$P(\sup_{1\leq n\leq N}\sup_{f\in\mathcal{F}}|n^{-\frac{1}{2}}S_n(f)|\geq t)\leq 2C_{\mathcal{F}}E(N)\frac{\sigma_P^2(F)}{t^2},$$

$$P(\sup_{1\leq n\leq N}\sup_{f\in\mathcal{F}}|n^{-\frac{1}{2}}S_n(f)|\geq t)\leq\frac{2C_{\mathcal{F}}}{t^{2p}}E\{\sum_{n=1}^{N}n^{-p}V_n^p(F)\}.$$

Let $\|F_2-F_1\|_p=[\frac{1}{n}\sum_{i=1}^{n}E\{(F_2-F_1)(X_i)\}^p]^{\frac{1}{p}}$.

Proposition 5.3. *Let X be a random variable on a probability space $(\Omega,\mathcal{A},P)$, with values in a metric space $(\mathbb{X},\mathcal{X},\|\cdot\|)$. Let $\mathcal{F}$ be a class of measurable functions on $(\mathbb{X},\mathcal{X},\|\cdot\|)$ such that there exist functions F_1 and F_2 in $\mathcal{F}$, belonging to $L_p(\mathbb{X})$ and satisfying $F_1\leq f\leq F_2$ for every f of $\mathcal{F}$. For every $p\geq 2$*

$$P(\sup_{f\in\mathcal{F}}|f(X)-\frac{1}{2}\{F_1(X)+F_2(X)\}|\geq t)\leq 2C_{\mathcal{F}}\frac{\|F_2-F_1\|_p^p}{(2t)^p},$$

$$P(\sup_{f\in\mathcal{F}}|f(X)-Ef(X)|\geq t)\leq 2C_{\mathcal{F}}\frac{\|F_2-F_1\|_p^p}{t^p}.$$

Proof. The variable $Y(f)=|f(X)-\frac{1}{2}\{F_1(X)+F_2(X)\}|$ is uniformly bounded on $\mathcal{F}$ by $\frac{1}{2}(F_2-F_1)(X)$ and $\sup_{f\in\mathcal{F}}|f(X)-Ef(X)|$ is bounded by $(F_2-F_1)(X)$ which belongs to $L_p(\mathbb{X})$, the functional Bienaymé-Chebychev inequalities yield the result. □

Under the conditions of Proposition 5.3, the empirical process of a non centered sample $(X_i)_{i=1,\ldots,n}$ of measurable random variables with values in $\mathbb{X}$ satisfies similar inequalities, due to (5.3)

$$P(\sup_{f\in\mathcal{F}}|\nu_n(f)-\frac{1}{2}\{F_1(X)+F_2(X)\}|>t)\leq 2C_{\mathcal{F}}\frac{\|F_2-F_1\|_2^2}{4t^2},$$

$$P(\sup_{f\in\mathcal{F}}|\nu_n(f)|>t)\leq 2C_{\mathcal{F}}\frac{\|F_2-F_1\|_2^2}{t^2}.$$

The first inequality is due to the bound of the empirical process by the empirical mean of the envelope of the variable $Y(f)$

$$E\{\sup_{f\in\mathcal{F}}|\nu_n(f)|\}^2\leq 2C_{\mathcal{F}}\frac{1}{2n}\sum_{i=1}^{n}E\{(F_2-F_1)(X_i)\}^2.$$

The second inequality is a consequence of the uniform bounds on $\mathcal{F}$ $X(F_1)\leq X(f)\leq X(F_2)$, which implies

$$P(X(F_2)>x)\leq P(X(f)>x)\leq P(X(F_1)>x)$$

and $E|\nu_n(f)|^2 \leq \|F_2 - F_1\|_2^2$, from Proposition 5.1.

Bennett's inequality for independent random variables is extended to the sum $S_n(f)$, with a uniform norm over a class of functions $\mathcal{F}$.

Theorem 5.1. *Let $(X_i)_{i=1,...,n}$ be a sequence of independent and identically distributed random variables on a probability space $(\Omega, \mathcal{A}, P)$, with values in a metric space $(\mathbb{X}, \mathcal{X}, \|\cdot\|)$. Let $\mathcal{F}$ be a class of measurable functions on $(\mathbb{X}, \mathcal{X}, \|\cdot\|)$ with envelope F in L_p for every $p \geq 1$, and such that $P(f) = 0$ in $\mathcal{F}$, $C_{\mathcal{F}}$ is finite and there exists a constant M for which $f(X_i)| \leq \sigma_P(F)M$, a.s. in $\mathcal{F}$. For every $t > 0$*

$$P(\sup_{f \in \mathcal{F}} |S_n(f)| \geq t) \leq 2C_{\mathcal{F}} \exp\{-n\phi(\frac{t}{n\sigma_P(F)M})\}$$

where $\phi(x) = (1+x)\log(1+x) - x$.

Proof. From Chernov's theorem

$$P(\sup_{f \in \mathcal{F}} S_n(f) \geq t) = E \exp\{\lambda \sup_{f \in \mathcal{F}} S_n(f) - \lambda t\}.$$

For every $t > 0$ and ε in $]0, \delta[$, with $\delta < 1$, let $\mathcal{F}_\varepsilon$ be an ε-net of $\mathcal{F}$ adapted to P. A bound for the Laplace transform of $F(X)$ is obtained from an expansion of the exponential function, $L_{F(X)} \leq \exp\{\exp(b\lambda) - 1 - b\lambda\}$, with the bound $b = \sigma_P(F)M$ for the uniform L_p-moments of the variables $f(X_i)$. By Proposition 5.2

$$P(\sup_{f \in \mathcal{F}} S_n(f) \geq t) \leq C_{\mathcal{F}} \inf_{\lambda > 0} \exp\{\psi_t(2\lambda)\},$$

where the function $\psi_t(\lambda) = n(\exp(b\lambda) - 1 - b\lambda) - \lambda t$ satisfies

$$\inf_{\lambda > 0} \psi_t(\lambda) = -n\phi(\frac{t}{nb})$$

therefore

$$\inf_{\lambda > 0} \exp\{\frac{1}{2}\psi_t(2\lambda)\} = \exp\{-\frac{n}{2}\phi(\frac{t}{nb})\}. \qquad \square$$

Corollary 5.1. *Under the conditions of Theorem 5.1, for a sequence of independent, identically distributed and non centered random variables $(X_i)_{i=1,...,n}$, and for a class $\mathcal{F}$ of functions, for every $x > 0$*

$$P(\sup_{f \in \mathcal{F}} |S_n(f)| \geq x + n\sup_{f \in \mathcal{F}} |P(f)|) \leq 2C_{\mathcal{F}} \exp\{-n\phi(\frac{x}{n\sigma_P(F)M})\}.$$

This result is a consequence of Theorem 5.1 and of the inequality $|\sup_{\mathcal{F}}|S_n(f)| - n\sup_{\mathcal{F}}|P(f)|| \leq \sup_{\mathcal{F}}|S_n(f) - nP(f)|$, hence $P(\sup_{\mathcal{F}}|S_n(f)| \geq x + n\sup_{\mathcal{F}}|P(f)|) \leq P(\sup_{\mathcal{F}}|S_n(f) - nP(f)| \geq x)$.

Theorem 5.1 is also written without boundedness condition for $\sup_{f \in \mathcal{F}}|f(X_i)|$ and with a condition for the upper bound of $f^2(X_i)$. For every $x > 0$

$$P(\sup_{f \in \mathcal{F}} |n^{-\frac{1}{2}}S_n(f)| \geq x) \leq 2C_{\mathcal{F}}\exp\{-n\phi(\frac{x}{\sqrt{n\eta}})\} + P(F^2(X) > \eta), \ x > 0.$$

With non identically distributed random variables, the boundedness condition implies that the Laplace transforms $L_{|f(X_i)|}$ satisfy the condition of convergence of $n^{-1}\sum_{i=1}^n \log L_{|f(X_i)|}$ to a limit, $\log L_{|f(X)|}$. The bound in the expression of the Laplace transforms is $b_n = M\max_{i=1,\ldots,n}\sigma_{P,i}$ with $\sigma_{P,i}^2 = E_P\|X_i\|_{\mathcal{F}}$, and the upper bound of the limit $L_{|f(X)|}(t)$ is similar to the bound for the Laplace transform of X with i.i.d. variables, replacing σ_P by $\sigma_{P,n}^* = \max_{i=1,\ldots,n}\sigma_{P,i}$. The constant depending on the entropy function is unchanged by considering a maximum over the probability distributions $(P_{X_i})_{i=1,\ldots,n}$.

Let $(X_i)_{i=1,\ldots,n}$ be a sequence of independent variables on $(\Omega, \mathcal{A}, P)$ and let $\mathcal{F}$ be a class of measurable functions on $(\mathbb{X}, \mathcal{X}, \|\cdot\|)$ such that the variables $f(X_i)$ have the respective means $E_P f(X_i) = P_i(f)$ and variances $E_P f^2(X_i) - E_P f(X_i) = \sigma_{P,i}^2(f)$, and such that $\|f(X_i)\|_{\mathcal{F}}$ belongs to L_p for every integer p. For every $x > 0$

$$P(\sup_{f \in \mathcal{F}} |\nu_n(f)| \geq x) \leq 2C_{\mathcal{F}}\exp\{-n\phi(\frac{x}{\sqrt{n\eta}})\}$$
$$+ P(\max_{i=1,\ldots,n} |F(X_i)| > \eta).$$

The next version of Corollary 5.1 applies to functional classes having lower and upper envelopes.

Corollary 5.2. *Let $(X_i)_{i=1,\ldots,n}$ be independent variables with values in a metric space $(\mathbb{X}, \mathcal{X}, \|\cdot\|)$ and let $\mathcal{F}$ be a class of measurable functions on $\mathbb{X}$ such that $C_{\mathcal{F}}$ is finite and there exist functions F_1 and F_2 in $\mathcal{F}$ having a finite Laplace transform and satisfying $F_{1i}(X_i) \leq f(X_i) \leq F_{2i}(X_i)$ for every f of $\mathcal{F}$. For every $t > 0$*

$$P(\sup_{f \in \mathcal{F}} |\nu_n(f)| \geq t) \leq 2C_{\mathcal{F}}\exp\{-n\psi(\frac{t\sqrt{n}}{\{\sum_{i=1}^n E(F_{2i} - F_{1i})(X_i)\}^{\frac{1}{2}}})\}.$$

Proof. Under the conditions of Corollary 5.2, $E\{f(X_i) - Ef(X_i)\}^2$ is bounded by

$$EF_{2i}^2(X_i) - \{EF_{1i}(X_i)\}^2 \le E\{F_{2i}(X_i) - EF_{1i}(X_i)\}^2$$
$$\le E\{F_{2i}(X_i) - F_{1i}(X_i)\}^2$$

and the Laplace transform of $\sup_{f \in \mathcal{F}} \nu_n(f)$ is bounded using functional Bürkholder-Davis-Gundy inequality of Proposition 5.1 for non identically distributed variables

$$L_f(t) = E \exp\{tn^{-\frac{1}{2}}(S_n - ES_n)(f)\}$$
$$\le \exp[t\{n^{-1}E(F_{2i} - F_{1i})(X_i)\}^{\frac{1}{2}}]$$
$$- t\{n^{-1}\sum_{i=1}^n E(F_{2i} - F_{1i})(X_i)\}^{\frac{1}{2}}.$$

The proof ends like in proving Bennet's inequality. □

If the Laplace transform of the variables $F_{ki}(X_i)$ is not finite, the inequality of Corollary 5.2 can be replaced by the following one, for every $\eta > 0$

$$P(\sup_{f \in \mathcal{F}} |\nu_n(f)| \ge t) \le 2 C_{\mathcal{F}} \exp\{-n\phi(\frac{t}{\sqrt{n\eta}})\}$$
$$+ P(n^{-1}\sum_{i=1}^n (F_{2i} - F_{1i})(X_i) > \eta).$$

Not only the expectation of $n^{-1}\sum_{i=1}^n (F_{2i} - F_{1i})(X_i)$ enters in the bound but its range though a value of η larger than the expectation of $n^{-1}\sum_{i=1}^n (F_{2i} - F_{1i})(X_i)$ reduces the exponential term. The second term is bounded by $\eta^{-2}n^{-1}\sum_{i=1}^n E\{(F_{2i} - F_{1i})^2(X_i)\}$.

An exponential inequality for $\sup_{f \in \mathcal{F}} |\nu_n(f)|$ is proved like the inequality (A.3) with the bound $E\{F_{2i}(X_i) - EF_{1i}(X_i)\}^2$ for the variance of the variables $f(X_i)$.

Theorem 5.2. *Let* $(X_i)_{i=1,\dots,n}$ *be independent variables satisfying the conditions of Corollary 5.2. For every* $\lambda > 0$

$$P(\sup_{f \in \mathcal{F}} |\nu_n(f)| \ge t) \le C_{\mathcal{F}} \exp\{-\frac{t^2}{2n^{-1}\sum_{i=1}^n E(F_{2i} - F_{1i})(X_i)}\}.$$

A process $(X_t)_{t \in \mathbb{R}}$ with values in a metric space $(\mathcal{X}, d)$ and with a variance function σ_t^2 which satisfies the equality in Hoefding's exponential bound $P(|X_t - X_t| > x) \le \exp\{-\frac{x^2}{2|\sigma_t^2 - \sigma_s^2|}\}$, for every $x > 0$, has a tail behaviour similar to that of a Gaussian process, or a Brownian motion if $\sigma_t^2 = t$. The Gaussian martingales also reach the bound. With a strict inequality, it is called *sub-Gaussian* (van der Vaart and Wellner, 1996). Thus, the empirical process and other martingales are sub-Gaussian. Let $(X_t)_{t \in \mathbb{R}}$ be a separable sub-Gaussian process, in their Corollary 2.2.8 van der Vaart and Wellner (1996) proved that there exists a constant K such that for every $\delta > 0$

$$E \sup_{\|t-s\| \leq \delta} |X_t - X_s| \leq K \int_0^\delta \sqrt{\log D(x, \| \cdot \|)} \, dx,$$

where $D(\varepsilon, d)$ is the packing number, defined as the maximum number of ε-separated points of the metric space $(\mathcal{X}, d)$, and it is equivalent to the covering number $N(\varepsilon, d)$ by the inequality $N(\varepsilon, d) \leq D(\varepsilon, d) \leq N(\frac{1}{2}\varepsilon, d)$. The constant $C_{\mathcal{F}}$ is larger than the constant $\int_0^\delta \sqrt{1 + \log N(x, \| \cdot \|)} \, dx$ of their inequalities which cannot be compared with the above results.

5.3 Exponential inequalities for functional martingales

On a filtered probability space $(\Omega, \mathcal{G}, (\mathcal{G}_n)_{n \geq 0}, P)$, let $X = (X_n)_n$ be in $\mathcal{M}_{0,loc}^2$ and $(A_n)_{n \geq 1}$ be a predictable process. A functional inequalities is established for the martingales

$$Y_n(f) = \sum_{k=1}^n f(A_k)(X_k - X_{k-1}) \tag{5.4}$$

defined for functions f of a class $\mathcal{F}$. The proofs are deduced from Propositions 4.8 and 4.9 for the process Y with quadratic variations

$$V_n(Y(f)) = \sum_{k=1}^n f^2(A_k)(V_k - V_{k-1})$$

and from Propositions 4.14 and 4.15 for continuous martingales.

Proposition 5.4. *On a filtered probability space* $(\Omega, \mathcal{G}, (\mathcal{G}_n)_{n \geq 0}, P)$, *let* $(X_n)_{n \geq 0}$ *be a real centered local martingale of* L_α, $\alpha \geq 2$, *with a process of quadratic variations* $(V_n(X))_{n \geq 0}$, *and let* $(A_n)_{n \geq 0}$ *be a predictable process of* L_α. *Let* $\mathcal{F}$ *be a class of measurable* L_α *functions on* $(\mathbb{R}, \mathcal{B})$. *There exists a constant* $c_\alpha > 0$ *such that for every stopping time* N *in the filtration* $(\mathcal{G}_n)_{n \geq 0}$

$$c_{\alpha,n} E\{\sup_{f \in \mathcal{F}} V_n(Y(f))\}^{\frac{\alpha}{2}} \leq E |\sup_{f \in \mathcal{F}} Y_n(f)|^\alpha \leq C_{\alpha,n} E\{\sup_{f \in \mathcal{F}} V_n(Y(f))\}^{\frac{\alpha}{2}}.$$

Proposition 4.9 applies to stopping times. For every $p \geq 2$ and for every stopping time N, there exists a constant C_p independent of N such that

$$E[\{\sup_{f \in \mathcal{F}} V_N(Y(f))\}^p] \leq E|\sup_{f \in \mathcal{F}} Y_N(f)|^{2p} \leq C_p E|\sup_{f \in \mathcal{F}} Y_N(f)|^p.$$

If $\mathcal{F}$ has an envelope F, by monotonicity of $V_N(Y(f))$ in $\mathcal{F}$ this inequality and a projection of every function of $\mathcal{F}$ on ε-nets implies $E|\sup_{f \in \mathcal{F}} Y_N(f)|^{2p} \leq C_p C_{\mathcal{F}} E|Y_N(F)|^p$. For every $\lambda > 0$, $p \leq 2$ and for stopping time N

$$P(\sup_{f \in \mathcal{F}} Y_N(f) > \lambda) \leq C_p C_{\mathcal{F}} \lambda^{-2p} E\{V_N^p(Y(F))\},$$

$$P(\sup_{f \in \mathcal{F}} Y_N^*(f) > \lambda) \leq C_p C_{\mathcal{F}} \lambda^{-2p} E\{\sum_{n=1}^{N} V_n^p(Y(F))\}.$$

Theorem 5.3. *On a filtered probability space* $(\Omega, \mathcal{G}, (\mathcal{G}_n)_{n \geq 0}, P)$, *let* $(X_n)_{n \geq 0}$ *belong to* $L_{0,loc}^p$, *for every integer* $p \geq 2$ *and let* $(A_n)_{n \geq 0}$ *be a predictable process of* L_p. *Let* $\mathcal{F}$ *be a class of measurable* $L_p(P)$ *functions on* $(\mathbb{R}, \mathcal{B})$ *with envelope* F *such that* $\sup_{f \in \mathcal{F}} Y_n(f)$ *belongs to* L_p *for every integer* p, $P(f) = 0$ *in* $\mathcal{F}$ *and* $C_{\mathcal{F}}$ *is finite. If there exists a constant* M *such that* $|f(X_i)| \leq \sigma_P(F)M$ *a.s. in* $\mathcal{F}$, *then for every* $x > 0$

$$P(\sup_{f \in \mathcal{F}} Y_n(f) \geq x) \leq C_{\mathcal{F}} \exp\{-n\phi(\frac{x}{n\sigma_P(F)M})\}$$

where $\phi(x) = (1 + x)\log(1 + x) - x$. *More generally*

$$P(\sup_{f \in \mathcal{F}} Y_n(f) \geq x) \leq C_{\mathcal{F}} \exp\{-n\phi(\frac{x}{n\sqrt{\eta}})\} + P(V_n(Y(F)) > \eta).$$

Since $V_n(Y)$ is increasing over $\mathcal{F}$, $\sup_{f \in \mathcal{F}} V_n(Y(f)) = V_n(Y(F))$. This implies the second inequality of Theorem 5.3.

Considering a local martingale $M = (M_t)_{t \geq 0}$ indexed by $\mathbb{R}_+$ and a predictable process $A = (A_t)_{t \geq 0}$, the local martingale

$$Y_t(f) = \int_0^t f(A_s) \, dM_s, \tag{5.5}$$

with predictable process $< Y_t > (f) = \int_0^t f^2(A_s) \, d < M >_s$, satisfies similar inequalities uniformly in $\mathcal{F}$.

Proposition 5.5. *On a filtered probability space* $(\Omega, \mathcal{G}, (\mathcal{G}_t)_{t \geq 0}, P)$, *let* M *be in* $\mathcal{M}_{0,loc}^\alpha$, $\alpha \geq 2$, *and let* A *be a predictable process of* L_α. *Let* $\mathcal{F}$ *be* *a class of measurable functions of $L_\alpha(P)$ on $(\mathbb{R}, \mathcal{B})$. There exist constants $C_\alpha > c_\alpha > 0$ such that for every stopping time T*

$$c_\alpha E\{\sup_{f \in \mathcal{F}} \int_0^T f^2(A_s)\, d < M >_s\}^{\frac{\alpha}{2}} \leq E|\sup_{f \in \mathcal{F}} \{\int_0^T f(A_s)\, dM_s|^\alpha\}$$

$$\leq C_\alpha E\{\sup_{f \in \mathcal{F}} \int_0^T f^2(A_s)\, d < M >_s\}^{\frac{\alpha}{2}}.$$

Like in Theorem 5.3, the local martingale Y satisfies an exponential inequality related to a bound for its predictable process and depending on the cumulated entropy function.

Theorem 5.4. *On a filtered probability space $(\Omega, \mathcal{G}, (\mathcal{G}_t)_{t \geq 0}, P)$, let M be in in $\mathcal{M}_{0,loc}^p$, $p \geq 2$ and let A be a predictable process of L_p. Let $\mathcal{F}$ be a class of measurable $L_p(P)$ functions on $(\mathbb{R}, \mathcal{B})$ with envelope F such that $\|Y_t(f)\|_{\mathcal{F}}$ belongs to L_p, for every integer p, $P(f) = 0$ in $\mathcal{F}$ and $C_{\mathcal{F}}$ is finite. If there exists a constant M such that $|Y_t(f)| \leq M \|Y_t(F)\|_{L_2}$ a.s. in $\mathcal{F}$, then for every stopping time T and for every $x > 0$*

$$P(\sup_{f \in \mathcal{F}} |Y_T(f)| \geq x) \leq C_{\mathcal{F}} E \exp\{-\phi(\frac{x}{M\|Y_T(F)\|_{L_2}})\}.$$

More generally

$$P(\sup_{f \in \mathcal{F}} |Y_T(f)| \geq t) \leq C_{\mathcal{F}} 2 \exp\{-\phi(\frac{x}{\sqrt{\eta}})\}$$

$$+ P(\int_0^T F^2(A_s)\, d < M >_s > \eta).$$

Extending Proposition 4.22, the Brownian distribution of the process

$$Y_t = \int_0^t (\int_0^s \beta_y^2\, dy)^{-\frac{1}{2}} \beta_s\, dB_s,$$

where β is a predictable process with sample paths in $\mathcal{C}_b(\mathbb{R}_+)$, allows to write a uniform tightness property.

Proposition 5.6. *On very finite interval $[S, T]$ of $\mathbb{R}_+$ and for every $x > 0$*

$$\lim_{\varepsilon \to 0} P(\sup_{S \leq s \leq t \leq T, |t-s| < \varepsilon} |Y_t - Y_s| > x) = 0.$$

Proof. The predictable compensator of Y is

$$< Y >_t = \int_0^t (\int_0^s \beta_y^2\, dy)^{-1} \beta_s^2\, ds$$

and its continuity implies that for every $\eta > 0$ and $x > 0$, there exists $\varepsilon > 0$ such that $P(\sup_{S \leq s \leq t \leq T, |t-s| < \varepsilon} < Y >_t - < Y >_s > x) \leq \eta$. The real ε defines an integer $k_\varepsilon = [\varepsilon^{-1}(T - S)]$, $k_\varepsilon + 1$ points $x_k = S + k\varepsilon$ such that $x_1 = S$ and $x_{k_\varepsilon+1} = T$, and variables

$$Z_k = \sup_{x_k \leq s \leq t \leq x_{k+1}} (< Y >_t - < Y >_s).$$

It follows that for every finite interval $[S, T]$ of $\mathbb{R}_+$ and $x > 0$

$$P(\sup_{S \leq s \leq t \leq T} < Y >_t - < Y >_s > x) \leq \sum_{k=1,\ldots,k_\varepsilon} P(Z_k > k_\varepsilon^{-1} x)$$

$$\leq k_\varepsilon \eta = O(\frac{\eta}{\varepsilon}).$$

The proof ends by using Lenglart's inequality of Proposition 1.2. $\qquad\square$

Considering the uniform metric on the space $C(\mathbb{R}_+)$, the functional variations of the process Y defined by (5.5) satisfy a property similar to the previous proposition.

Proposition 5.7. *Let $\mathcal{F} = \{\alpha \geq 0, \alpha \in D(\mathbb{R}_+)\}$. On every finite interval $[0, T]$ of $\mathbb{R}_+$*

$$\lim_{\varepsilon \to 0} \sup_{0 \leq t \leq T} P(\sup_{\alpha, \beta \in \mathcal{F}, \|\alpha - \beta\|_{[0,t]} \leq \varepsilon} |Y_t(\alpha) - Y_t(\beta)| > x) = 0.$$

Theorem 5.5. *Let $\mathcal{F}$ be a class of measurable functions on $(\mathbb{R}, \mathcal{B})$ such that there exist functions F_1 and F_2 in $\mathcal{F}$, belonging to $L_p(\mathbb{X})$ and satisfying $F_1 \leq \beta \leq F_2$ for every β of $\mathcal{F}$. Then for every stopping time T and for every $x > 0$*

$$P(\sup_{\beta \in \mathcal{F}} \sup_{t \in [0,T]} |Y_t(\beta)| \geq x) \leq 2C_{\mathcal{F}} E \int_0^T (\int_0^s F_1^2(y)\,dy)^{-1} F_2^2(s)\,ds.$$

Proof. The predictable compensator of the process $Y(\beta)$ has a uniform bound in $\mathcal{F}$

$$< Y >_t (\beta) \leq \int_0^t (\int_0^s F_1^2(y)\,dy)^{-1} F_2^2(s)\,ds.$$

Following the proof of Theorem 5.3 with the bounds of Proposition 4.14 for the moments of the process $Y(\beta)$ yields the result, with $E\sqrt{b_T}$. $\qquad\square$

Replacing the Brownian motion by a local martingale M of $\mathcal{M}_{0,loc}^p$, let Y be the process defined by

$$Y_t = (\int_0^t \beta_y^2 d < M >_y)^{-\frac{1}{2}} \int_0^t \beta_s\,dM_s,$$

where β is a process with sample paths in $\mathcal{C}_b(\mathbb{R}_+)$.

Theorem 5.6. *Let $\mathcal{F}$ be a class of measurable functions on $(\mathbb{R}, \mathcal{B})$ such that there exist functions F_1 and F_2 in $\mathcal{F}$, belonging to $L_p(\mathbb{X})$, for every $p \geq 2$ and satisfying $F_1 \leq \beta \leq F_2$ for every β of $\mathcal{F}$. For all stopping time T and $x > 0$*

$$P(\sup_{\beta \in \mathcal{F}} |Y_T(\beta)| \geq x) \leq 2C_{\mathcal{F}} E \exp\{-\phi(\frac{x}{\sqrt{a_T}})\}$$

where $a_T = \int_0^T F_2^2 \, d < M > \{\int_0^T F_1^2 \, d < M >\}^{-1}$.

The proof is the same as above, with the following uniform bounds for the predictable compensator of the process $Y(\beta)$

$$\sup_{\beta \in \mathcal{F}} < Y >_t (\beta) \leq (\int_0^t F_1^2(y) \, dy)^{-1} \int_0^t F_2^2(s) \, ds = a_t.$$

The bound is replaced by an exponential inequality by Hoeffding's inequality (Theorem A.4). Under the conditions of Theorem 5.6, for every $x > 0$

$$P(\sup_{\beta \in \mathcal{F}} Y_T(\beta) \geq x) \leq C_{\mathcal{F}} E \exp\{-\frac{x^2}{2a_T}\}.$$

5.4 Weak convergence of functional processes

Let Φ be a functional defined on the set $\mathcal{P}_{\mathcal{X}}$ of the probability distributions on $(C([0,1]), \mathcal{C})$, with values in $(C([0,1]), \mathcal{C})$ and satisfying a Lipschitz condition

$$\|\Phi(X_t) - \Phi(Y_t)\|_{[0,1]} \leq K \|X_t - Y_t\|_{[0,1]},$$

for processes X and Y of $C([0,1])$. For every $x > 0$, the empirical process of independent uniform variables on $[0,1]$ and the Brownian bridge satisfy

$$P(\|\Phi(\nu_{n,t}) - \Phi(W_t)\|_{[0,1]} > x) \leq \frac{K}{x} E \|\nu_{n,t} - W_t\|_{[0,1]}$$

and it converges to zero.

The continuously differentiability of a function $\Phi : \mathbb{R} \mapsto \mathbb{R}$ implies

$$\Phi(\nu_{n,t}) - \Phi(W_t) = (\nu_{n,t} - W_t)\Phi'(W_t) + o(\|\nu_{n,t} - W_t\|),$$

and the function is lipschitzian if there exists a constant K such that $\sup_{x \in \mathbb{R}} |\Phi'(x)|^\alpha \leq K$, the remainder term of the expansion is uniform over $[0,1]$ if the derivative of the function Φ is uniformly continuous, i.e. $\limsup_{x,x' \in \mathbb{R}, |x-x'| \to 0} |\Phi'(x) - \Phi'(x')| = 0$. Conversely, a Lipschitz function with exponent $\alpha > 1$ is continuously differentiable.

The weak convergence of the empirical processes in the space $C(\mathcal{X})$ of the continuous functions on $(\mathcal{X}, \mathcal{B})$ with the uniform metric has been proved under the condition that the class of functions has a $L_2(P)$ envelope and under conditions about the dimension of $\mathcal{F}$ that ensure the tightness of the process.

Example 5.1. On a probability space $(\Omega, \mathcal{A}, P)$, let $T > 0$ and $C > 0$ be independent real random variables with distribution functions F and G respectively and let $\delta = 1_{\{T \leq C\}}$ and $X = T \wedge C$ be the minimum variable defined on $[0, \tau)$, $\tau \leq \infty$. Let $(T_i, C_i)_{i=1,\ldots,n}$ be a vector of independent variables having the same distribution as (T, C) and let φ be a real function of $L_2(F)$. Let $\Lambda(t) = \int_0^t \{1 - F^-\}^{-1} \, dF$ and let the empirical processes

$$N_n(t) = \sum_{i=1}^{n} 1_{\{T_i \wedge C_i \leq t\}},$$

$$Y_n(t) = \sum_{i=1}^{n} 1_{\{T_i \wedge C_i \geq t\}}.$$

The difference $M_n(t) = n^{\frac{1}{2}} \int_0^t 1_{\{Y_n > 0\}} Y_n^{-1} N_n - \int_0^t 1_{\{Y_n > 0\}} \, d\Lambda$ is a local square integrable martingale in the support of I_F of F, $EM_n(t) = 0$ and

$$EM_n^2(t) = \int_{[0, t \wedge \tau]} n EY_n^{-1} \, d\Lambda$$

where $\int_{[0, t \wedge \tau]} nY_n^{-1} \, d\Lambda$ converges to $\sigma_t^2 = \int_{[0, t \wedge \tau]} \{(1 - F^-)^2 (1 - G^-)\}^{-1} \, dF$ a.s., uniformly in every compact subinterval of I_F, the predictable compensator of N_n is denoted $\widetilde{N}_n(t) = \int_0^t 1_{\{Y_n > 0\}} Y_n \, d\Lambda$. Let $\mathcal{H}$ be a family of functions with envelope H such that $\sigma_t^2(h) = \sup_{h \in \mathcal{H}} \int_0^t h^2 \{(1 - F^-)^2 (1 - G^-)\}^{-1} \, dF$ is finite. Applying Proposition 5.5, for every $\alpha \geq 2$ and $t < \tau$

$$E| \sup_{h \in \mathcal{H}} \int_0^t h \, dM_n|^\alpha \leq C_{\alpha, t} E\{ \int_0^t 1_{\{Y_n > 0\}} H^2 Y_n^{-1} d\Lambda \}^{\frac{\alpha}{2}}.$$

If the constant of entropy dimension $C_{\mathcal{H}}$ is finite and if there exist m_n such that $N_n(t) \leq m_n \int_0^t Y_n \, d\Lambda$ in I_F, Theorem 5.3 entails that for every stopping time T

$$P(\sup_{h \in \mathcal{H}} \sup_{t \leq T} \int_0^t h \, dM_n \geq x) \leq k_0 C_{\mathcal{H}} \exp\{-\phi(\frac{x}{\sqrt{n}\eta})\}$$

$$+ P(\int_0^{T \wedge \tau} H^2 1_{\{Y_n > 0\}} Y_n^{-1} \, d\Lambda > \eta), \quad n \geq 1.$$

For every $\varepsilon > 0$, t and η can be chosen sufficiently large to ensure that the upper bound is smaller than ε. This inequality implies the tightness of the sequence $(\sup_{h \in \mathcal{H}} \sup_{t \leq T} \int_0^t h \, dM_n)_{n \geq 1}$, and therefore its convergence to a centered Gaussian variable with variance $\int_0^T H_t^2 d\sigma_t^2$.

Example 5.2. Let $0 < S < T$ and $C > 0$ be real random variables on a probability space $(\Omega, \mathcal{A}, P)$, with C independent of S and T, and let $\delta = 1_{\{T \leq C\}}$ and $\delta' = 1_{\{S \leq C\}}$. Let $X = T - S$ and $\widetilde{X} = \delta' \{X \wedge (C - S)\}$ which takes the values X if $\delta = 1$, $C - S$ if $\delta' = 1$ and $\delta = 0$, and zero if $\delta' = 0$. Let $\tau > 0$ be included in the support $I_{T \wedge C}$ of the distribution of the variable $T \wedge C$ and let Y be the indicator process

$$Y(x) = 1_{\{T \wedge C \geq S + x\}}, \quad x \leq \tau.$$

We consider the empirical process related to the right-censored variables $(S, T \wedge C)$ defined from a vector of independent variables $(S_i, T_i, C_i)_{i=1,\ldots,n}$ having the same distribution as (S, T, C). Counting processes are defined for (s, x) in I_τ by

$$N_n(x) = \sum_{i=1}^n N_i(x) = \sum_{i=1}^n \delta_i 1_{\{T_i \leq S_i + x\}},$$

$$Y_n(x) = \sum_{i=1}^n Y_i(x) = \sum_{i=1}^n 1_{\{T_i \wedge C_i \geq S_i + x\}},$$

$$\widetilde{N}_n(x) = \sum_{i=1}^n \widetilde{N}_i(x) = \sum_{i=1}^n \int_0^x Y_i(y) \lambda_{Y|S}(y; S_i) \, dy,$$

where $\lambda_{X|S}(x; s) = \lim_{\varepsilon \downarrow 0} \frac{1}{\varepsilon} P(x \leq X < x + \varepsilon | X \geq x, S = s)$ is the hazard function of X conditionally on $S = s$. Summing the weighted differences $N_i - \widetilde{N}_i$ yields

$$M_n(x) = n^{-\frac{1}{2}} \left\{ \int_0^x 1_{\{Y_n > 0\}} Y_n^{-1} \, dN_n - \sum_{i=1}^n \int_0^x 1_{\{Y_n(y) > 0\}} \lambda_{Y|S}(y; S_i) \, dy \right\}$$

is a local square integrable martingale with respect to the filtration generated by $((N_i(t), Y_i(t))_{t \geq S_i})_{i=1,\ldots,n}$, then

$$EM_n^2(x) = \int_{[0, x \wedge \tau]} E \frac{\lambda_{Y|S}(y; S)}{n^{-1} Y_n(y)} \, dy := E\widetilde{N}_n(x).$$

Let $\mathcal{H}$ be a family of functions h defined in $I_{T \wedge C}$, with an envelope H such that the integral

$$E \int_0^x \frac{H^2(s+y)}{P(T \geq (s+y) \wedge C)} \lambda_{Y|S}(y; s) \, dy \tag{5.6}$$

is finite for every (s, x) with $s + x$ in $I_{T \wedge C}$. For every h of $\mathcal{H}$

$$E\{\delta h(S + X)1_{\{T \leq S+x\}}1_{\{T \leq S+x|S\}}$$
$$= \int_{[0, x \wedge \tau]} E\{h(S + y)Y(y)|S = s\}\lambda_{Y|S}(y; s)\, dy.$$

For every function h of $\mathcal{H}$, the process $W_n(s, x) = \int_{[0, x \wedge \tau]} h(s + y)\, dM_n(y)$, (s, x) in $I_{T \wedge C}$, converges weakly to a centered Gaussian process W_h, as an empirical process in the class of functions $\mathcal{F}_h = \{\delta h(s + y)1_{\{s+y \leq s+x\}} - \int_0^x E\{h(S + u)Y(s, u)|S = s\}\lambda_{X|S}(u; s)\, du, (s, x) \in I_{T \wedge C}$, at fixed h. For every integer $p \geq 2$ and for every x in $I_{T \wedge C}$

$$E|\sup_{h \in \mathcal{H}} \int_0^x h\, dM_n|^{2p} \leq C_{\mathcal{H}} E\{\sup_{h \in \mathcal{H}} \int_0^x h^2\, d\widetilde{N}_n\}^p$$

and this inequality extends to a uniform inequality in subintervals $[0, T]$ of $I_{T \wedge C}$

$$E|\sup_{h \in \mathcal{H}} \sup_{x \geq T} \int_0^x h\, dM_n|^{2p} \leq C_{\mathcal{H}} E\{\int_0^T h^2\, d\widetilde{N}_n\}^p.$$

This inequality implies Chernov and Bennet inequalities for the variable $\sup_{h \in \mathcal{H}} \sup_{x \geq T} \int_0^x h\, dM_n$, by Theorem 5.3, hence its tightness and therefore its weak convergence.

Extending the process to a class of functions $\mathcal{H}$ defined by (5.6) and having a finite entropy integral $C_{\mathcal{H}}$, the weak convergence of the process is also uniform on $\mathcal{H}$.

5.5 Differentiable functionals of empirical processes

Let Φ be a function defined on the subset $\mathcal{P}_{\mathcal{X}}$ of the probability distributions on $(\mathcal{X}, \mathcal{B})$ and with values in a normed vector space and let $(B_2, \mathcal{B}_2)$ be the Borel σ-algebra. The function Φ is supposed to be differentiable in the following sense.

Definition 5.1. Let B_1 and B_2 be normed vector spaces provided with the Borel σ-algebra, a measurable map $\Phi : E \in B_1 \to B_2$ is *differentiable* at $x \in B_1$, *tangentially to a separable subspace* C of B_1 if there exists a linear map $d\phi(x) : B_1 \to B_2$ such that for all sequences $(t_n)_n$ in $\mathbb{R}$ and $(h_n)_n$ in B_1, such that t_n tends to zero, $x + t_n h_n$ belongs to E and h_n converges to a limit h in C as n tends to infinity, then

$$\lim_{n \to \infty} \|\frac{\phi(x + t_n h_n) - \phi(x)}{t_n} - d\phi(x).h\| = 0.$$

A measurable map $\phi : E \subset B_1 \to B_2$ is *continuously differentiable* at $x \in E$ along a sequence $(x_n)_n \in E$ that converges to x and *tangentially to a separable subspace* C of B_1 if there exists a linear map $d\phi(x) : B_1 \to B_2$, continuous on E and such that for all sequences $(t_n)_n$ in $\mathbb{R}$ and $(h)_n$ in B_1, such that t_n tends to zero, $x_n + t_n h_n$ in E and h_n converges to h in C as n tends to infinity

$$\lim_{n \to \infty} \left\| \frac{\phi(x_n + t_n h_n) - \phi(x_n)}{t_n} - d\phi(x).h \right\| = 0.$$

In Pons (1986) and Pons and Turckheim (1991), we established the following results.

Proposition 5.8. *Let P be a probability on $(\mathcal{X}, \mathcal{B})$ and let $\mathcal{F}$ be a functional subset of $\mathcal{L}_P^2$ with an envelope $F \in \mathcal{L}_P^2$ and such that $\int_0^1 (\log D_F^{(2)}(x, \mathcal{F}))^{\frac{1}{2}} dx < \infty$. Let Φ be a map defined from a subset $\mathcal{P}_\mathcal{X}$ of $\ell^\infty(\mathcal{F})$ to B_2, differentiable at P tangentially to $\mathcal{U}_b(\mathcal{F}, P)$, then $\sqrt{n}\{\Phi(\widehat{P}_n) - \Phi(P)\}$ converges weakly to $d\Phi(P).G_P$.*

It is a consequence of the assumption of differentiability and of Pollard's central limit theorem.

The differentiability of a functional is applied to nonparametric models. Here, it is used to prove a triangular version of the $\sqrt{n}$-consistency of $\Phi(\widehat{P}_n)$ to $\Phi(P)$, for a differentiable function Φ. It relies on the existence of a sequence of Brownian bridges G_{P_n} related to P_n and converging to the Brownian bridge G_P as P_n converges to P, uniformly on $\mathcal{F}$.

Proposition 5.9. *Let P_n and P be probabilities on $(\mathcal{X}, \mathcal{B})$ and let $\mathcal{F}$ be a subset of $\bigcap_n \mathcal{L}_{P_n}^2 \bigcap \mathcal{L}_P^2$ with an envelope F in $\bigcap_n \mathcal{L}_{P_n}^2 \bigcap \mathcal{L}_P^2$. Suppose that $\bigcap_n \mathcal{L}_{P_n}^2 \bigcap \mathcal{L}_P^2$ has a finite entropy dimension, $\lim \|P_n - P\|_\mathcal{F} = 0$ and $\lim \|P_n - P\|_{\mathcal{F}^2} = 0$. Then, for every n, there exist uniformly continuous versions of the Brownian bridges for G_P and G_{P_n}, defined on the same space and such that for every $\varepsilon > 0$, $\lim_n P\{\|G_{P_n}^{(n)} - G_P^{(n)}\|_\mathcal{F} > \varepsilon\} = 0$.*

Proof. Let $\mathcal{F}_{\sigma_n}$ be a σ_n-net of $\mathcal{F}$ adapted to P and let π be the projection $\mathcal{F} \to \mathcal{F}_{\sigma_n}$. Since $\sigma_{P_n}^2(f - \pi(f)) \leq \sigma_P^2(f - \pi(f)) + 4\|P_n - P\|_{\mathcal{F}^2}$, $\|G_P(f - \pi(f))\|_\mathcal{F}$ and $|G_{P_n}(f - \pi(f))\|_\mathcal{F}$ tend to zero with σ_n. The restrictions of G_P and G_{P_n} to $\mathcal{F}_{\sigma_n}$ are Gaussian variables with dimension $k_n \leq C\sigma_n^{-d}$, for n large enough, if d is the entropy dimension of $\mathcal{F}$. Strassen's theorem implies the existence of a probability space where a Gaussian variable (X, Y) is defined as having the same margins as the restrictions of G_{P_n} and G_P to $\mathcal{F}_{\sigma_n}$ and such that $P\{\|X - Y\|_\infty > \Pi\} < \Pi$, where Π is the Prohorov distance for the uniform norm on $\mathbb{R}^{k_n}$. The existence of uniformly continuous versions of the Brownian bridge G_{P_n} and G_P defined on the same probability space is deduced from Berkes and Philipp's lemma (1979). Finally, from Dehling (1983), Π has a bound depending on k_n and δ_n as follows

$$\Pi \leq C\delta_n^{\frac{1}{2}} k_n^{\frac{1}{6}} (1 + |\log \frac{k_n}{\delta_n}|^{\frac{1}{2}}),$$

$$\delta_n \leq k_n^2 \sup_{f,g \in \mathcal{F}_{\sigma_n}} |\mathrm{cov}_{P_n}(f,g) - \mathrm{cov}_P(f,g)|.$$

Since $|\mathrm{cov}_{P_n}(f,g) - \mathrm{cov}_P(f,g)| \leq \|P_n - P\|_{\mathcal{F}^2} + 2\|P_n - P\|_{\mathcal{F}} \sup_n \int F \, dP_n$, by choosing σ_n that converges to zero with a rate such that $\lim_n \delta_n^{\frac{1}{2}} k_n^{\frac{1}{6}} = 0$, Π tends to 0 as $n \to \infty$. $\qquad\square$

Proposition 5.10. *Let $\mathcal{F}$ be a family of finite entropy dimension, let P_n and P be probabilities on $(\mathcal{X}, \mathcal{B})$ such that $\lim_n \|P_n - P\|_{\mathcal{F}} = 0$, $\lim_n \|P_n - P\|_{\mathcal{F}^2} = 0$, $F \in \mathcal{L}_P^{2+\delta}$ and $F \in \mathcal{L}_{P_n}^{2+\delta}$ for every n, with $\delta > 0$. Then, for every n, there exists a uniformly continuous version of the Brownian bridge G_P defined on the same space as ν_{n,P_n} and such that for every $\varepsilon > 0$, $P\left\{\|\nu_{n,P_n} - G_P^{(n)}\|_{\mathcal{F}} > \varepsilon\right\}$ tends to zero.*

Let Φ be a map defined from a subset $\mathcal{P}_\mathcal{X}$ of $\ell^\infty(\mathcal{F})$ to B_2, differentiable at P along the sequence $(P_n)_n$ and tangentially to $\mathcal{U}_b(\mathcal{F}, P)$, then

$$\sqrt{n}\left(\Phi(\widehat{P}_n) - \Phi(P_n)\right) \xrightarrow{\mathcal{D}} d\Phi(P).G_P.$$

Proof. This is a consequence of Massart (1986) that ensures for every n, the existence of a uniformly continuous Brownian bridge related to P_n and such that

$$P\{\|\nu_{n,P_n} - G_{P_n}^{(n)}\|_{\mathcal{F}} \geq \alpha_n\} \leq \beta_n,$$

with α_n and β_n depending only on the entropy dimension of $\mathcal{F}$ and converging to zero.

Let $\mathcal{F}_{\sigma_n, P_n}$ be a (σ_n, P_n)-net of $\mathcal{F}$ and let $\nu_{n,P_n}(\sigma_n)$ be the restriction of ν_{n,P_n} to $\mathcal{F}_{\sigma_n, P_n}$. By Proposition 5.9 and Berkes and Philipp's lemma, there exists a distribution on $\ell^\infty(\mathcal{F}_{\sigma_n, P_n}) \times \mathcal{U}_b(\mathcal{F}, P_n) \times \mathcal{U}_b(\mathcal{F}, P)$ having $\nu_{n,P_n}(\sigma_n)$ and G_P as margins on $\ell^\infty(\mathcal{F}_{\sigma_n, P_n})$ and $\mathcal{U}_b(\mathcal{F}, P)$. Skorohod's lemma implies the existence of a uniformly continuous version of G_P defined on Ω and such that $P\left\{\|\nu_{n,P_n}(\sigma_n) - G_P^{(n)}\|_{\mathcal{F}} > \varepsilon\right\}$ tends to zero, then the bounds for the variations of ν_{n,P_n} end the proof. $\qquad\square$

Corollary 5.3. *Let P be a probability distribution on $(\mathcal{X}, \mathcal{B})$ and let $\mathcal{F}$ be a subset of a family $\mathcal{L}_P^2$ with finite entropy dimension, such that F belongs to $\mathcal{L}_P^{2+\delta}$, for some $\delta > 0$. Let Φ be a map defined from a subset $\mathcal{P}_\mathcal{X}$ of $\ell^\infty(\mathcal{F})$ to B_2, differentiable at P tangentially to $\mathcal{U}_b(\mathcal{F}, P)$ along every probability sequence $(P_n)_n$ such that $\lim_n \|P_n - P\|_\mathcal{F} = 0$, $\lim_n \|P_n - P\|_{\mathcal{F}^2} = 0$, and F belongs to $\bigcap_n \mathcal{L}_{P_n}^{2+\delta}$. Then for a.e. $(X_1, \ldots, X_n)$*

$$\sqrt{n}\left(\Phi(P_n^*) - \Phi(\widehat{P}_n)\right) \xrightarrow{\mathcal{D}} d\Phi(P).G_P$$

where P_n^* is the empirical distribution of an i.i.d. sequence of variables with distribution function P_n.

As an application, let $[0, \tau]$ be a subset of $\mathbb{R}_+^2$, let H be a continuous distribution function on $\mathbb{R}_+^2$ and Λ be defined on $\mathbb{R}_+^2$ by $\Lambda(t) = \int_{]0,t]} \bar{H}^{-1} \, dH$, for $t = (t_1, t_2)$ such that $\bar{H}(t) = \int_{[t_1, \infty[\times [t_2, \infty[} dH > 0$. For t in $\mathbb{R}_+^2$, let Q_t be the quadrant larger than t and T_t be the triangle under the diagonal and lower than t

$$Q_t = \left\{(u, v) : u \in \mathbb{R}_+^2, v \in \mathbb{R}_+^2, u \geq t, v \geq t\right\},$$

$$T_t = \left\{(u, v) : u \in \mathbb{R}_+^2, v \in \mathbb{R}_+^2, u \leq v, u \leq t\right\},$$

then $\mathcal{Q}_\tau = \{Q_t, t \in [0, \tau]\}$ and $\mathcal{T}_\tau = \{T_t, t \in [0, \tau]\}$ are Vapnik-Cervonenkis classes, i.e. $\|\widehat{P}_n - P\|_{\mathcal{Q}_\tau}$ and $\|\widehat{P}_n - P\|_{\mathcal{T}_\tau}$ converge a.s. to zero as n tends to infinity. Let $\Lambda = \Phi(P)$, where P is the product probability distribution $P_H \times P_G$ related to continuous distribution functions H and G

$$\Phi(P)(t) = \int_{\mathbb{R}_+^4} \frac{1}{P(Q_u)} 1_{T_t}(u, v) \, dP(u, v).$$

On the set $E_n = \{\widehat{P}_n(Q_\tau) > 0\}$ having a probability that tends to 1, $\widehat{\Lambda}_n = \Phi(\widehat{P}_n)$. The process $W_{n,P} = \sqrt{n}(\widehat{\Lambda}_n - \Lambda)$ is written as

$$W_{n,P}(t) = \nu_{n,P}(g_t) - \int \frac{\nu_{n,P}(Q_u)}{\widehat{P}_n(Q_u)P(Q_u)} 1_{T_t}(u, v) \, d\widehat{P}_n(u, v),$$

with $g_t(u, v) = \frac{1}{P(Q_u)} 1_{T_t}(u, v)$. Similarly, let

$$g_{n,t}(u, v) = \frac{1}{P_n(Q_u)} 1_{T_t}(u, v),$$

$\mathcal{F}_n = \{g_{n,t}, g_t, 1_{Q_t}, 1_{T_t}; t \in [0, \tau]\}$ and $\mathcal{F} = \bigcup_n \mathcal{F}_n$.

Proposition 5.11. *Let P be a probability on $\mathbb{R}_+^4$ and let W be the Gaussian process defined by*

$$W(t) = G_P(g_t) - \int \frac{G_P(Q_u)}{P^2(Q_u)} 1_{T_t} \, dP(u, v),$$

then $W_{n,P} = \sqrt{n}\{\Phi(\widehat{P}_n) - \Phi(P)\} = \sqrt{n}(\widehat{\Lambda}_n) - \Lambda)$ converges weakly to W on every interval $[0, \tau]$ such that $\Lambda(\tau)$ and the variance of $W(\tau)$ are finite.

Its proof relies on the differentiability of the functional Φ (Pons, 1986). The functional Φ is also differentiable in the sense of Definition 5.1 under the next conditions.

Proposition 5.12. *Let P_n and P be probability distributions on $\mathbb{R}^4_+$ such that $P(Q_\tau) > 0$, $n^\alpha \sup_{u \leq \tau} |P_n(Q_u) - P(Q_u)| \leq M$, for constants $\alpha > 0$ and $M > 0$, and such that $\lim_n \|P_n - P\|_{\mathcal{F}} = 0$ and $\lim_n \|P_n - P\|_{\mathcal{F}^2} = 0$. Then Φ defining Λ as $\Phi(P)$ on $D([0,\tau])$ is differentiable at P along $(P_n)_n$ and tangentially to $\mathcal{U}_b(\mathcal{F}, P)$.*

The condition $n^\alpha \sup_{u \leq \tau} |P_n(Q_u) - P(Q_u)| \leq M$ entails that $\mathcal{F}$ has a finite entropy dimension, moreover the envelope F is finite under the condition $P(Q_\tau) > 0$. Let P_n^* the empirical distribution of an i.i.d. sequence of variables with the empirical distribution function $\widehat{P}_n$.

Proposition 5.13. *Under the conditions of Proposition 5.10, the process $W_{n,P_n} = \sqrt{n}\{\Phi(\widehat{P}_n) - \Phi(P_n)\}$ converges weakly to W under P_n and the process $W_n^* = \sqrt{n}\{\Phi(\widehat{P}_n^*) - \Phi(\widehat{P}_n)\}$ converges weakly to W under $\widehat{P}_n$, conditionally on the random vector $(X_i, \delta_i)_{i \leq n}$.*

5.6 Regression functions and biased length

On a probability space $(\Omega, \mathcal{F}, P)$, let (X, Y) be a random variable with values in a separable and complete metric space $(\mathcal{X}_1 \times \mathcal{X}_2, \mathcal{B})$. For every x in $\mathcal{X}_1$, a regression function is defined by the conditional mean of Y given $X \leq x$

$$m(x) = E(Y|X \leq x) = \frac{E(Y 1_{\{X \leq x\}})}{P(X \leq x)}.$$

Its empirical version is defined from a sequence of independent random variables distributed like (X, Y)

$$\widehat{m}_n(x) = \frac{\sum_{i=1}^n Y_i 1_{\{X_i \leq x\}}}{\sum_{i=1}^n 1_{\{X_i \leq x\}}}, \quad x \in \mathcal{X}_1, \tag{5.7}$$

where the denominator is the empirical distribution $\widehat{F}_{X,n}(x)$, with expectation $F(x) = P(X \leq x)$, and the numerator is an empirical mean process denoted $\mu_n(x)$ with expectation $\mu(x) = E(Y 1_{\{X \leq x\}})$.

Proposition 5.14. *If Y belongs to $L_4(\mathcal{X}_2)$, for every x in a subset of I_X of $\mathcal{X}_1$ such that there exist constants for which $0 < M_1 < m(x) < M_2$, then*

$E\widehat{m}_n(x) = m(x) + O(n^{-\frac{1}{2}})$ and

$$var\, \widehat{m}_n(x) = n^{-1} F_X^{-1}(x)\{E(Y^2|X \leq x) - m^2(x)\} + o(1),$$

$$n^{\frac{1}{2}}(\widehat{m}_n - m) = F_X^{-1}\{n^{\frac{1}{2}}(\mu_n - \mu) - m\,\nu_{X,n}\} + r_n$$

where $\sup_{I_X} r_n = o_{L_2}(1)$, *as n tends to infinity.*

Proof. Let $A_n(x) = n^{-1} \sum_{i=1}^{n} Y_i 1_{\{X_i \leq x\}}$ be the numerator of $\widehat{m}_n$ and let $\mu(x) = E(Y 1_{\{X \leq x\}})$ be its expectation. Under the condition, there exists a strictly positive $k(x)$ such that for every x in I_X, $k(x) \leq \widehat{F}_{Xn}(x)$ if n is large enough. For the mean of $\widehat{m}_n(x)$

$$E\widehat{m}_n(x) - m(x) = E\frac{\mu_n - \mu}{\widehat{F}_{Xn}}(x) - m(x)E\frac{\widehat{F}_{Xn} - F_X}{\widehat{F}_{Xn}}(x)$$

$$\leq m(x)E\frac{(\widehat{F}_{Xn} - F_X)^2}{F_X\widehat{F}_{Xn}}(x) + E\frac{|(\mu_n - \mu)(\widehat{F}_{Xn} - F_X)|}{F_X\widehat{F}_{Xn}}(x)$$

$$\leq m(x)\frac{\|\widehat{F}_{Xn}(x) - F_X(x)\|_2^2}{F_X(x)k_2(x)} + \|\mu_n(x) - \mu(x)\|_2 \frac{\|\widehat{F}_{Xn}(x) - F_X(x)\|_2}{F_X(x)k_2(x)}$$

with the L_2-norm, then $\|\widehat{F}_{Xn} - F_X\|_2 = O(n^{-\frac{1}{2}})$ and $\|\mu_n - \mu\|_2 = O(n^{-\frac{1}{2}})$. For its variance, $var\widehat{m}_n(x) = E\{\widehat{m}_n(x) - m(x)\}^2 + O(n^{-1})$ and the first term develops as $F_X^2(x)E\{\widehat{m}_n(x) - m(x)\}^2 - \{E\widehat{m}_n(x)\}^2\{var\widehat{F}_{Xn}(x) - 2E\widehat{m}_n(x)cov(\mu_n(x), \widehat{F}_{Xn}(x))\} + var\mu_n(x) + o(n^{-1}) = O(n^{-1})$. $\square$

Replacing the variable Y by $f(Y)$, for a function f belonging to a class of functions $\mathcal{F}$, let

$$m_f(x) = E\{f(Y)|X \leq x\}$$

be a regression function indexed by f, the empirical regression function becomes

$$\widehat{m}_{f,n}(x) = \frac{\sum_{i=1}^{n} f(Y_i)1_{\{X_i \leq x\}}}{\sum_{i=1}^{n} 1_{\{X_i \leq x\}}}, \quad x \in \mathcal{X}_1,$$

as n tends to infinity, the variance of the normalized process

$$\zeta_{f,n} = n^{-\frac{1}{2}}(\widehat{m}_{f,n} - m_f)$$

is approximated by

$$\sigma_f^2(x) = var\zeta_{f,n}(x) = F_X^{-1}(x)\{E(f^2(Y)|X \leq x) - m_f^2(x)\} := \sigma_{F_1,F_2}^2(x).$$

If $F_1 \leq f \leq F_2$, then $\sigma_f^2(x) \leq F_X^{-1}(x)\{E(F_2^2(Y)|X \leq x) - m_{F_1}^2(x)\}$.

Proposition 5.15. *Let $\mathcal{F}$ be a class of measurable functions on $(\mathbb{X}_2, \mathcal{X}_2)$ such that $C_{\mathcal{F}}$ is finite and there exist envelopes $F_1 \leq f \leq F_2$ belonging to $L_p(\mathbb{X})$. Under the conditions of Proposition 5.14, for every x of I_X*

$$\lim_{n \to \infty} P(\sup_{f \in \mathcal{F}} |n^{-\frac{1}{2}}(\widehat{m}_{f,n} - m_f)(x)| \geq t) \leq 2C_{\mathcal{F}} \frac{\sigma_{F_1,F_2}^2(x)}{t^2}.$$

The odd moments of order $p \geq 2$ of the process $\zeta_{f,n}$ are $o(1)$ and its even moments are $O(1)$, from its expansion of Proposition 5.14, like those of the empirical process ν_n. From Proposition 5.1, for every $p \geq 2$ and for every x of I_X

$$\lim_{n \to \infty} P(\sup_{f \in \mathcal{F}} |\zeta_{f,n}(x)| \geq t) \leq 2C_{\mathcal{F}} \frac{\sigma_{F_1, F_2}^p(x)}{t^p}.$$

Biased length variables appear in processes observed on random intervals (Cox, 1960). Let Y be a positive random variable sampled at a uniform and independent random time variable U on $[0,1]$. The variable Y is not directly observed and only a biased length variable $X = YU$ is observed, therefore $F_Y \leq F_X$ and $EX = \frac{1}{2}EY$. The variable $U = (Y^{-1}X) \wedge 1$ has a uniform distribution on $[0,1]$ and its mean is $\int_0^\infty \{x \int_x^\infty y^{-1} \, dF_Y(y) + F_Y(x)\} \, dF_X(x) = \frac{1}{2}$.

Lemma 5.1. *The distribution function of X and Y are defined for every positive x by*

$$F_X(x) = E(xY^{-1} \wedge 1) = F_Y(x) + x \int_x^\infty y^{-1} \, dF_Y(y), \qquad (5.8)$$

$$F_Y(y) = 1 - E(Xy^{-1} \wedge 1) = F_X(y) - y^{-1} \int_0^y x \, dF_X(x). \qquad (5.9)$$

Proof. Let $x > 0$, the distribution function of $X = UY$ is defined by

$$F_X(x) = \int_0^1 P(Y \leq u^{-1}x) \, du = \int_0^1 F_Y(u^{-1}x) \, du$$
$$= F_Y(x) + E(Y^{-1}x 1_{\{x < Y\}}) = E(xY^{-1} \wedge 1).$$

The distribution of $Y = U^{-1}X$ is written

$$F_Y(y) = \int_0^1 P(X \leq uy) \, du = \int_0^1 F_X(uy) \, du = y^{-1} \int_0^y F_X(x) \, dx$$
$$= F_X(x) - y^{-1} \int_0^y x \, dF_X(x). \qquad \square$$

The expected mean lifetime function is defined as

$$m(y) = E\{X 1_{X \leq y}\}.$$

It is related to the distribution functions of the variables X and Y by (5.9) in Lemma 5.1. For every $y > 0$, $m(y) = y\{F_X(y) - F_Y(y)\}$.

From the observation of n independent and identically distributed random variables X_i distributed like X, we define the empirical distribution function $\widehat{F}_{X,n}(x) = n^{-1}\sum_{i=1}^n 1_{\{X_i \leq x\}}$ of F_X and the empirical version of the function $m(y)$

$$\widehat{m}_n(y) = n^{-1}\sum_{i=1}^n X_i 1_{\{X_i \leq y\}}.$$

By plugging in (5.9), $\widehat{F}_{X,n}$ and $\widehat{m}_n$ define the empirical distribution function $\widehat{F}_{Y,n}$ of the unobserved variable Y

$$\widehat{F}_{Y,n}(y) = n^{-1}\sum_{i=1}^n (1 - \frac{X_i}{y})1_{\{X_i \leq y\}}.$$

The variance of the empirical process related to $\widehat{F}_{Y,n}$ is

$$\sigma_Y^2(y) = \{F_Y(1-F_Y)\}(y) + E(\frac{X^2}{y^2}1_{\{X \leq y\}}) - \frac{m_X(y)}{y} \leq \{F_Y(1-F_Y)\}(y).$$

Proposition 5.16. *The estimator $\widehat{F}_{Y,n}$ converges uniformly to F_Y in probability and $n^{\frac{1}{2}}(\widehat{F}_{Y,n} - F_Y)$ converges weakly to a centered Gaussian variable with variance function σ_Y^2.*

From the inequality (1.12), for all $t > 0$ and $y > 0$

$$P(|n^{-\frac{1}{2}}\{\widehat{F}_{Y,n}(y) - F_Y(y)\})| \geq t) \leq 2\frac{\sigma_Y^2(y)}{t^2} \leq \frac{2F_Y(y)\{1 - F_Y(y)\}}{t^2}.$$

There exists a constant C such that for every $t > 0$

$$P(\sup_{y>0}|n^{-\frac{1}{2}}\{\widehat{F}_{Y,n}(y) - F_Y(y)\})| \geq t) \leq \frac{C}{t^2},$$

since $\sigma_Y^2(y) \leq \{F_Y(1-F_Y)\}(y)F_Y(y) \leq \frac{1}{4}$ for every y.

A continuous multiplicative mixture model is more generally defined for a real variable U having a non uniform distribution function on $[0,1]$. Let F_U denote its distribution function and let F_Y be the distribution function of Y. The distribution functions of X and Y are

$$F_X(x) = \int F_U(xy^{-1})\, dF_Y(y),$$

$$F_Y(y) = \int_0^1 F_X(uy)\, dF_U(u)$$

$$= \int_0^y \{1 - F_U(y^{-1}x)\}\, dF_X(x)$$

$$= F_X(y) - E_X\{1_{\{X \leq y\}}F_U(y^{-1}X)\} \tag{5.10}$$

and the conditional density of Y given X is

$$f_{Y|X}(y;x) = \frac{f_Y(y)f_U(y^{-1}x)}{f_X(x)}.$$

The empirical distribution of Y is deduced from (5.10) in the form

$$\widehat{F}_{Y,n}(y) = n^{-1}\sum_{i=1}^{n}1_{\{X_i\leq y\}}\{1 - F_U(y^{-1}X_i)\}.$$

The variance of the empirical process related to $\widehat{F}_{Y,n}$ is

$$\sigma^2_{F_U}(y) = F_Y(y)\{1 - F_Y(y)\} + E\{F_U^2(y^{-1}X)1_{\{X\leq y\}}\} - m_{F_U}(y)$$
$$\leq F_Y(y)\{1 - F_Y(y)\},$$

where the expected mean lifetime is now $m_{F_U}(y) = E_X\{1_{\{X\leq y\}}F_U(y^{-1}X)\}$. Applying the inequality (1.12), for every $t > 0$

$$P(|n^{-\frac{1}{2}}\{\widehat{F}_{Y,n}(y) - F_Y(y)\})| \geq t) \leq \frac{2F_Y(y)\{1 - F_Y(y)\}}{t^2} \leq \frac{1}{2t^2}.$$

The process $n^{-\frac{1}{2}}(\widehat{F}_{Y,n} - F_Y)$ is bounded in probability in $\mathbb{R}$ endowed with the uniform metric, and it converges weakly to a Gaussian process with mean zero and variance function $\sigma^2_{F_U}$.

Let $p \geq 2$ be an integer and let $\mathcal{F}_U$ be a class of distribution functions on $[0,1]$ with an envelope F of $L_p([0,1])$, from Proposition 5.14

$$P(\sup_{y>0}|n^{-\frac{1}{2}}\{\widehat{F}_{Y,n}(y) - F_Y(y)\})| \geq t) \leq C\frac{F_U^p(1)}{t^p},$$

$$P(\sup_{F_U\in\mathcal{F}}\sup_{y>0}|n^{-\frac{1}{2}}\{\widehat{F}_{Y,n}(y) - F_Y(y)\})| \geq t) \leq C\frac{F^p(1)}{t^p}, \ t > 0.$$

Another biased length model is defined by the limiting density, as t tends to infinity, of the variations $X(t) = S_{N(t)+1} - t$ between t and the sum of random number of independent and identically distributed random variables ξ_k having the distribution function G, $S_{N(t)+1} = \sum_{k=1}^{N(t)+1}\xi_k$, with the random number $N(t) = \sum_{i=1}^{\infty}1_{\{S_i\leq t\}}$. This limiting density only depends on the distribution of the variables ξ_i in the form

$$f_X(x) = \mu^{-1}\{1 - G(x)\}$$

where $\mu = \{f_X(0)\}^{-1}$ (Feller, 1971). This is equivalent to

$$G(x) = 1 - f_X(x)\{f_X(0)\}^{-1}.$$

Let F_Y be the limiting density of $S_{N(t)+1} - S_{N(t)}$. The distribution functions F_X and F_Y are

$$F_X(x) = \mu^{-1} \int_0^x \{1 - G(y)\} \, dy$$
$$= \mu^{-1} x \{1 - G(x) + x^{-1} E_G(\xi 1_{\{\xi \leq x\}})\},$$
$$F_Y(y) = F_X(y) - y^{-1} E(X 1_{\{X \leq y\}})$$

and the expected mean lifetime distribution function for X is

$$m_X(y) = (2\mu)^{-1} [y^2 \{1 - G(y)\} - E_G(\xi^2 1_{\{\xi \leq x\}})].$$

The empirical versions of the functions m, G, F_X and F_Y are all easily calculated from a sample $(\xi_i)_{i \leq n}$.

5.7 Regression functions for processes

Let $X = (X_n)_{n \geq 0}$ be an adapted process of $L_2(P, (\mathcal{F}_n)_{n \geq 0})$, with values in a separable and complete metric space $\mathcal{X}_1$, and let $Y = (Y_n)_{n \geq 0}$ be an adapted real process. We assume that the processes have independent increments $(X_n - X_{n-1})_{n \geq 0}$, with a common distribution function F_X with a density f_X and there exists a function $m > 0$ of $C_1(\mathcal{X}_1)$, such that

$$E(Y_n - Y_{n-1} | X_n - X_{n-1} \leq x) = m(x), \quad n \geq 1.$$

This implies that $(Y_n - Y_{n-1})_{n \geq 0}$ is a sequence of independent variables with a common distribution function F_Y, with a density f_Y. The empirical version of the function m is

$$\widehat{m}_n(x) = \frac{\sum_{i=1}^n (Y_i - Y_{i-1}) 1_{\{X_i - X_{i-1} \leq x\}}}{\sum_{i=1}^n 1_{\{X_i - X_{i-1} \leq x\}}}, \quad x \in \mathcal{X}_1. \tag{5.11}$$

The empirical means $n^{-1} \sum_{i=1}^n (1_{\{X_i - X_{i-1} \leq x\}}, (Y_i - Y_{i-1}) 1_{\{X_i - X_{i-1} \leq x\}})$ converge a.s. uniformly to the expectation of the variables $F_X(x)(1, m(x))$, therefore $\lim_{n \to \infty} \sup_{\mathcal{X}_1} \|\widehat{m}_n - m\| = 0$, a.s. and Propositions 5.14 and 5.15 are satisfied for the processes X and Y.

Let $(X, Y) = (X_n, Y_n)_{n \geq 0}$ be an ergodic sequence of $L_2(P, (\mathcal{F}_n)_n)$ with values in $\mathcal{X}^2$, there exists an invariant measure π on $\mathcal{X}$ such that for every continuous and bounded function φ on $\mathcal{X}^2$

$$\frac{1}{n} \sum_{k=1}^n \varphi(X_k, Y_k, X_{k-1}, Y_{k-1}) \to \int_{\mathcal{X}} \varphi(z_k, z) F_{X_k, Y_k | X_{k-1}, Y_{k-1}}(dz_k, z)) \, d\pi(z)$$

and the estimator

$$\widehat{m}_n(x) = \frac{\sum_{i=1}^{n} Y_i 1_{\{X_i \leq x\}}}{\sum_{i=1}^{n} 1_{\{X_i \leq x\}}}, \ x \in \mathcal{X}_1,$$

converges in probability to $m(x)$, uniformly in $\mathcal{X}_1$. Under the condition (4.3) and a φ-mixing assumption, it converges in distribution to a centered Gaussian process with variance function

$$\sigma_m^2(x) = F_X^{-1}(x)\{E(Y^2|X \leq x) - m^2(x)\}.$$

Let $\mathcal{H}$ be a class of functions, the transformed variables $h(Y_i)$ define a functional empirical regression

$$\widehat{m}_n(h, x) = \frac{\sum_{i=1}^{n} h(Y_i) 1_{\{X_i \leq x\}}}{\sum_{i=1}^{n} 1_{\{X_i \leq x\}}}, \ x \in \mathcal{X}_1,$$

under the condition (4.3), it converges to $m(f, x) = E\{h(Y)|X = x\}$ defined as a mean with respect to the invariant measure. If the class $\mathcal{H}$ has a finite constant $C_{\mathcal{H}}$ and an envelope H, the convergence is uniform over $\mathcal{H}$ and the process $\sup_{h \in \mathcal{H}} \sup_{x \in I} |\widehat{m}_n(f, x) - m(h, x)|$ converges to zero in every real interval where m has lower and upper bounds.

Chapter 6

Inequalities for Processes

6.1 Introduction

The stationary covariance function of a centered process $(X_t)_{t>0}$ is

$$R_X(t) = E\{X(s)X(s+t)\},$$

for all s and $t > 0$. It is extended to every real t with the covariance $R(-t) = R(t)$ for $t < 0$. Its variance satisfies $EX^2(s) = EX^2(0) = R_X(0)$, for every $s > 0$. By the Cauchy-Schwarz inequality

$$E\|X_t X_s\| \le E\|X_0\|^2, \quad R_X(t) \le R_X(0).$$

For a Gaussian process, there is equivalence between the stationarity of the distribution and the stationarity of its mean and covariance functions. For a stationary process

$$E\{X(s+t) - X(s)\}^2 = 2\{R(0) - R_X(t)\},$$

and a necessary and sufficient condition for the continuity of the process is the continuity of the mean and covariance functions. Thus, the standard Brownian motion is continuous and stationary, but its covariance function has no derivative at zero.

A stationary process X with mean function $x = EX$ has the mean process

$$\bar{X}(T) = \frac{1}{T} \int_{[0,T]} X(t)\, dt, \ T > 0,$$

with expectation $\bar{x} = \bar{X}$. For every $T > 0$

$$E\frac{1}{T} \int_{[0,T]} \{X(t) - \bar{x}(t)\}^2\, dt = R(0),$$

153

$$E[\frac{1}{T}\int_{[0,T]}\{X(t)-\bar{x}(t)\}\,dt]^2 = \frac{R(0)}{T} + \frac{1}{T^2}\int_{[0,T]^2} cov\{X(s)X(t)\}\,ds\,dt$$

$$= \frac{R(0)}{T} + \frac{1}{T^2}\int_{[0,T]^2} R_{|t-s|}\,ds\,dt,$$

this is the variance of the mean process $\bar{X}_T$ and the first term of its expression tends to zero as T tends to infinity. The Bienaymé-Chebychev implies

$$P(\sup_{t\in[0,T]} (\bar{X}-\bar{x})(t) > x) \le \frac{1}{x^2 T^2}\int_{[0,T]^2} R_{|t-s|}\,ds\,dt + \frac{R(0)}{x^2 T^2}.$$

Stochastic inequalities between processes have been established from inequalities between the covariance function of real valued stationary processes (Slepian, 1962). They are generalized in the next section to $\mathbb{R}^n$ and to processes which are not Gaussian. In Section 6.3, sufficient conditions are established for an infinite time in random ruin models, in the mean models, and the probabilities of ruin in random intervals of the processes are defined. Stochastic orders between several models are deduced.

6.2 Stationary processes

The distribution of a Gaussian vector with mean zero is determined by its variance matrix $\Sigma_X = (E(X_i - EX_i)(X_j - EX_j))_{i,j=1,\ldots,n}$. For Gaussian variables $X = (X_1,\ldots,X_n)$ and $Y = (Y_1,\ldots,Y_n)$ having the same variances and such that $E(X_iX_j) \ge E(Y_iY_j)$ for every $i \ne j$, the quadratic form $t^T\Sigma_X t - t^T\Sigma_Y t = \sum_{i\ne j} t^2(\sigma_{X,ij}^2 - \sigma_{Y,ij}^2)$, t in $\mathbb{R}^n$, is positive, the densities of X and Y are therefore ordered and satisfy $f_X(t) \le f_Y(t)$ which is equivalent to Slepian's lemma (1962).

For all real numbers $u_1,\ldots,u_n > 0$

$$P(X_i \le u_i, i=1,\ldots,n) \ge P(Y_i \le u_i, i=1,\ldots,n).$$

Let $(X_t)_{t\ge 0}$ and $(Y_t)_{t\ge 0}$ be centered Gaussian processes with stationary covariance functions R_X and R_Y. By passage to the limit in Slepian's lemma, the inequality $R_X \ge R_Y$ in an interval $[0,T_0]$ implies

$$P(\sup_{t\in[0,T]} X_t \le c) \ge P(\sup_{t\in[0,T]} Y_t \le c)$$

for all $c > 0$ and T in $[0,T_0]$.

Theorem 6.1 (Slepian, 1962). *For all times $S \geq 0$ and $T \geq 0$, the inequality $R(t) \geq 0$ in $[0, T + S]$ implies*

$$P(\sup_{t \in [0,T+S]} X_t \leq c) \geq P(\sup_{t \in [0,T]} X_t \leq c)P(\sup_{t \in [0,S]} X_t \leq c).$$

Proof. For all times $S \geq 0$ and $T \geq 0$, the variables $\sup_{t \in [0,T]} X_t$ and $\sup_{t \in [0,S]} X_t$ are independent. By stationarity, $(\max_{t=t_1,\ldots,t_{k+m} \in [0,T+S]} X_t)$ and $\max(\max_{t=t_1,\ldots,t_k \in [0,T]} X_t, \max_{t_{k+1},\ldots,t_{k+m} \in [0,S]} X_t)$ have the same distribution and the covariance matrix of $((X_{t_j})_{j=1,\ldots,k}, (X_{t_j})_{j=k+1,\ldots,k+m})$ is diagonal by blocks, hence the result is an application of Slepian's Lemma (1962), by passage to the limit as the partition of $[0, T + S]$ increases. $\square$

A centered Gaussian process with values in $\mathbb{R}^n$ also satisfies the inequality of Theorem 6.1 with the Euclidean norm

$$P(\sup_{t \in [0,T+S]} \|X_t\|_{2,n} \leq c) \geq P(\sup_{t \in [0,T]} \|X_t\|_{2,n} \leq c)P(\sup_{t \in [0,S]} \|X_t\|_{2,n} \leq c).$$

It can be extended to balls of $\mathbb{R}^n$ centered at zero and with radius $r > 0$, B_r. If X_t belongs to B_r, its norm is bounded by $\sqrt{n}r$ and reciprocally, therefore we have the following.

Theorem 6.2. *For every $r > 0$, and for all times $S \geq 0$ and $T \geq 0$*

$$P(\sup_{t \in [0,T+S]} X_t \in B_r) \geq P(\sup_{t \in [0,T]} X_t \in B_r)P(\sup_{t \in [0,S]} X_t \in B_r).$$

Let τ_T be the time when a stationary Gaussian process X reaches its supremum on the interval $[0, T]$, by stationarity $\sup_{t \in [0,T+S]} X_t = X_{\tau_{S+T}}$ has the same distribution as the maximum of $\sup_{t \in [0,T]} X_t = X_{\tau_T}$ and $\sup_{t \in [0,S]} X_t = X_{\tau_S}$ therefore τ_{S+T} and $\tau_S \wedge \tau_T$ have the same distribution. Applying Lévy's arcsine law (1.21)

$$P(\tau_T \leq t) = P(\sup_{s \in [0,t]} X_s \geq \sup_{s \in [t,T]} B_s) = \frac{2}{\pi} \arcsin \sqrt{\frac{R(t)}{R(T)}}.$$

The proof is the same as for Proposition 4.21.

The process of the predictable quadratic variations of an increasing process X of $L_2(P)$, with a stationary covariance function R, satisfies

$$E < X >_t^2 = R(t)$$

by the martingale property of the centered process $M = X - < X >$. This implies

$$E(M_{t+s} - M_t)^2 = E(M_{t+s}^2 - M_t^2)$$
$$= E(X_{t+s}^2 - < X >_{t+s}^2) - E(X_s^2 - < X >_s^2)$$

therefore $E(M_{t+s} - M_t)^2 = 0$.

The variations of the martingale have the quadratic mean

$$E\{(M_{t+s} - M_t)^2|\mathcal{F}_t\} = E\{M_{t+s}^2|\mathcal{F}_t\} - M_t^2$$
$$= E\{(X_{t+s}^2 - X_t^2|\mathcal{F}_t\} + 2X_t < X >_t$$
$$- E\{< X >_{t+s}^2 - < X >_t^2 |\mathcal{F}_t\}$$

and the process of the predictable quadratic variations of M^2 is written

$$< M^2 >_t = E\{X_t - < X >_t)^2|\mathcal{F}_t^-\} = E(X_t^2|\mathcal{F}_t^-) - < X >_t^2.$$

The Slepian theorems extend to non Gaussian processes.

Theorem 6.3. *Let X and Y be processes with stationary moment functions $m_{X,k}$ and $m_{Y,k}$ such that for every $k \geq 1$, $m_{X,k} \leq m_{Y,k}$ in an interval $[0, T_0]$. For all $c > 0$ and T in $[0, T_0]$*

$$P(\sup_{t \in [0,T]} X_t \leq c) \geq P(\sup_{t \in [0,T]} Y_t \leq c).$$

Proof. The assumption of ordered moments implies the inequality of the Laplace transforms $L_{X_t} \leq L_{Y_t}$, for every t in $[0, T_0]$, and the theorem is a consequence of Chernov's theorem for the processes X and Y. It applies for example to point processes with independent increments. $\square$

Theorem 6.4. *Let X be a process with independent increments and with stationary moment functions. For all stopping times $S \geq 0$ and $T \geq 0$, the inequality $m_{X,k} \geq 0$, for every $k \geq 1$, in $[0, T + S]$ implies*

$$P(\sup_{t \in [0,T+S]} X_t \leq c) \geq P(\sup_{t \in [0,T]} X_t \leq c)P(\sup_{t \in [0,S]} X_t \leq c).$$

The proof is similar to the proof of Theorems 6.1 and 6.3, using the Laplace transforms.

6.3   Ruin models

A marked point process $(T_k, X_k)_{k \geq 1}$ is defined by positive random times T_k, with $T_0 = 0$, and by positive jump size X_k of the process at T_k, for $k \geq 0$, it is written as

$$N_X(t) = \sum_{k \geq 1} X_k 1_{\{T_k \leq t\}}, t \geq 0. \tag{6.1}$$

Under the conditions of a process N with independent increments and an independent sequence of mutually independent variables $(X_k)_{k \geq 1}$, the process N_X has independent increments and its Laplace transform is

$$L_{N_X(t)}(\lambda) = e^{\lambda N_X(t)} = \prod_{k \geq 1} L_{X_k}(\lambda) P(T_k \leq t)$$

$$= E \prod_{k \geq 1} L_{X_k}(\lambda) F_k(t - T_{k-1}).$$

If the inter-arrival times of the point process are identically distributed, with density f, N is a renewal process and

$$L_{N_X(t)}(\lambda) = \prod_{k \geq 1} L_{X_k}(\lambda) P(T_k \leq t) = \prod_{k \geq 1} L_{X_k}(\lambda) \sum_{k \geq 1} F * f^{*(k-1)}(t),$$

where the density of T_{k-1} is the convolution $f^{*(k-1)}$.

In the Sparre Andersen model (Thorin, 1970), a process decreases from an initial model by the random arrival of claims at times T_k and the inter-arrival variables $S_k = T_k - T_{k-1}$ are identically distributed with a distribution function F and a density f in $\mathbb{R}_+$, hence the density of T_k is the convolution f^{*k}.

The process is written as

$$Y_t = a + ct - N_X(t), \ t \geq 0,$$

with $N_X(0) = 0$, $a > 0$ and $b > 0$. The time of ruin is the random variable

$$T_a = \inf_{t \geq 0} \{Y_t < 0 | Y_0 = a\},$$

this is a stopping time for the process Y and its probability of occurrence is $P(T_a < \infty) = P(Y_{T_a} < 0 | Y_0 = a)$. Let $\mu = ES_k$, for every $k \geq 1$, the mean of $T_{N(t)}$ is $\mu^{-1} EN(t)$.

Lemma 6.1. *If* $0 < \mu X_k \leq c$ *a.s., for every* $k \geq 1$*, the time of ruin is a.s. infinite on the set* $\{\mu^{-1} N(t) - T_{N(t)} < c^{-1} a, \forall t > 0\}$.

Proof. The process Y is written as

$$Y_t = a + c(t - T_{N(t)}) + \sum_{k \geq 1} (c\mu^{-1} - X_k) 1_{\{T_k \leq t\}} + c\{T_{N(t)} - \mu^{-1} N(t)\}$$

where $t - T_{N(t)} \geq 0$ is the current time since the last occurrence of an event of the process N_X and $U(t) = \sum_{k \geq 1} (c - X_k) 1_{\{T_k \leq t\}}$ is strictly positive under the assumption for the variables X_k. Therefore $Y_t > 0$ if the last term is positive. $\square$

The probability of $\{\mu^{-1}N(t) - T_{N(t)} \geq c^{-1}a, \forall t > 0\}$ is

$$P(\sup_{t \geq 0}\{\mu^{-1}N(t) - T_{N(t)}\} \geq c^{-1}a) \leq c^2 a^{-2}\{\mu^{-1}N(t) - T_{N(t)}\}^2$$

where $ET_{N(t)} = \mu^{-1}EN(t)$. It follows that a large constant a ensures a large probability for an infinite time of ruin.

The mean of the current time after the last event of the process N is $t - T_{N(t)} > 0$ is $t - \mu^{-1}EN(t)$, moreover $EN(t) = \sum_{k \geq 1} F*f^{*(k-1)}(t)$, where the convolutions of the distributions are $F^{*k}(t) = \int_0^\infty F(t-s)f^{*(k-1)}(s)\,ds$, $k \leq 2$. Therefore, for every $t > 0$

$$\sum_{k \geq 1} F * f^{*(k-1)}(t) < \mu t. \tag{6.2}$$

The mean of Y_t is $EY_t = a + ct - EN_X(t)$ and it remains strictly positive up to $t_0 = \arg\min\{t : \sum_{k \geq 1} EX_k\, EF(t - T_{k-1}) \geq a + ct\}$.

Lemma 6.2. *Let N be a renewal process with a mean inter-event time μ, independent of a sequence of independent and identically distributed variables $(X_k)_{k \geq 0}$ with mean μ_X, and let $c < \mu_X \mu$ be a strictly positive constant. The mean time of ruin is then $t_0 > (\mu_X \mu - c)^{-1}a$.*

Proof. From (6.2), if the variables X_k have the same mean μ_X

$$EN_X(t) = \mu_X EN(t) < \mu_X \mu t.$$

The bound for t_0 is deduced from the equation of the model

$$EY(t_0) = a + ct_0 - EN_X(t_0) \leq 0. \qquad \square$$

Assuming that t_0 is finite, the question is to characterize the smallest values of t such that $Y(t) = a + ct - \sum_{k=1}^{N(t)} X_k 1_{\{T_k \leq t\}}$ is negative. Let

$$\Lambda(t) = \int_0^t \frac{dF}{1 - F}$$

be the cumulative hazard function of the variables S_k. By the Bienaymé-Chebychev inequality, for every $\varepsilon > 0$

$$P(Y(t) < -\varepsilon) = P(\sum_{k \geq 1} 1_{\{T_k \leq t\}} = n)P(\sum_{k=1}^n X_k 1_{\{T_k \leq t\}} > a + ct - \varepsilon)$$

$$\leq \frac{\mu_X}{a + ct - \varepsilon}e^{-\Lambda(t)} \sum_{n \geq 1} \frac{\Lambda^n(t)}{n!} \sum_{k=1}^n P(T_k \leq t).$$

Taking the limit as ε tends to zero, we obtain the next result.

Proposition 6.1. *Under the conditions of Lemma 6.2, for every $t > 0$*

$$P(Y(t) < 0) \le \frac{\mu_X}{a + ct} e^{-\Lambda(t)} \sum_{n \ge 1} \frac{\Lambda^n(t)}{n!} \sum_{k=1}^{n} F^{*k}(t).$$

The probability that the time of ruin belongs to a random interval $]T_n, T_{n+1}]$ is $P_n = P(Y(T_k) > 0, k = 1, \ldots, n, Y(T_{n+1}) \ge 0)$. Let also $\widetilde{P}_n = P(Y(T_k) > 0, k = 1, \ldots, n+1) = \widetilde{P}_{n-1} - P_n$. For $n = 1$, the independence properties of the model imply

$$P_1 = P(X_1 < a + cT_1, X_1 + X_2 \ge a + cT_2)$$
$$= P(X_1 < a + cT_1)P(X_2 \ge c(T_2 - T_1))$$
$$= \{\int_0^\infty F_X(a + cs)\,dF(s)\}\{1 - \int_0^\infty F_X(cs)\,dF(s)\},$$
$$\widetilde{P}_1 = \{\int_0^\infty F_X(a + cs)\,dF(s)\}\{\int_0^\infty F_X(cs)\,dF(s)\},$$

they involve additive and multiplicative convolutions. If F has a density f, $\widetilde{P}_1 \le c^{-2}\|F_X\|_p^2\|f\|_{p'}^2$ and $P_1 \le c^{-2}\|F_X\|_p\|1 - F_X\|_p\|f\|_{p'}^2$, for all conjugate numbers $p \ge 1$ and p'. For $n = 2$, the same arguments entail

$$P_2 = \widetilde{P}_1 P(X_3 > cS_3)$$
$$= \{\int_0^\infty F_X(a + cs)\,dF(s)\}\{\int_0^\infty F_X(cs)\,dF(s)\}\{1 - \int_0^\infty F_X(cs)\,dF(s)\},$$

and

$$\widetilde{P}_2 = \widetilde{P}_1 - P_2 = \{\int_0^\infty F_X(a + cs)\,dF(s)\}\{\int_0^\infty F_X(cs)\,dF(s)\}^2.$$

Proposition 6.2. *The probability of ruin in the interval $]T_n, T_{n+1}]$ is*

$$P_n = \widetilde{P}_{n-1} P(X_{n+1} > cS_{n+1})$$
$$= \{\int_0^\infty F_X(a + cs)\,dF(s)\}\{\int_0^\infty F_X(cs)\,dF(s)\}^{n-1}$$
$$\times \{1 - \int_0^\infty F_X(cs)\,dF(s)\}.$$

For all conjugate numbers $p \ge 1$ and p'

$$P_n \le c^{-(n+1)}\|F_X\|_p^n\|1 - F_X\|_p\|f\|_{p'}^{n+1}.$$

The result is proved recursively, from the equality $\widetilde{P}_n = \widetilde{P}_{n-1} - P_n$.

Proposition 6.3. *For every integer $n \geq 1$, the time of ruin T_a belongs to the interval $]T_n, T_{n+1}]$ with probability P_n and the interval with the greatest probability is $]T_1, T_2]$, except if $\int_0^\infty F_X(cs)\,dF(s) = \frac{1}{2}$. In that case, the probabilities that T_a belongs to $]0, T_1]$ or $]T_1, T_2]$ are equal and larger than the probabilities of the other intervals.*

The model has been generalized to allow variations by adding a stochastic diffusion model to a parametric trend $y(t)$ minus $N_X(t)$

$$dY(t) = dy(t) + bY(t)\,dt + \sigma(t)\,dB(t) - dN_X(t),$$

where B is the standard Brownian motion and $Y(0) = y(0)$. Its mean $m_Y(t) = EY(t)$ satisfies $m_Y(t) = y(t) + b\int_0^t m(s)\,ds - EN_X(t)$. The solution of the mean equation is $m_Y(t) = y(t) + (e^{bt} - 1) - EN_X(t)$, denoted $\psi(t) - EN_X(t)$ and Lemma 6.2 applies replacing the linear trend by $\psi(t)$ in the sufficient condition.

Using the solution (4.17) of the stochastic diffusion model

$$Y(t) = \psi(t) + \int_0^t \sigma(s)\,dB(s) - N_X(t). \tag{6.3}$$

Under the condition that the function ψ belongs to $C_1(\mathbb{R}_+)$, $\psi(T_n) - \psi(T_{n-1}) = (T_n - T_{n-1})\psi'(T_{n-1} + \theta(T_n - T_{n-1}))$, with θ in $]0, 1[$. For every $k \geq 1$, the variable $Z_k = \int_{T_{k-1}}^{T_k} \sigma(s)\,dB(s)$ has a normal distribution with variance $v(T_{k-1}, T_k) = \int_{T_{k-1}}^{T_k} \sigma^2(s)\,ds$ conditionally on (T_{k-1}, T_k). The probabilites P_n are therefore convolutions

$$P\Big(X_1 < \psi(T_1) + \int_0^{T_1} \sigma\,dB\Big) = \int_{\mathbb{R}} \int_0^\infty F(\psi(t) + z)f(t)f_{\mathcal{N}_{v(t)}}(z)\,dt\,dz.$$

The probability that the time of ruin belongs to the interval $]T_1, T_2]$ is

$$P_1 = P\Big(X_1 < \psi(T_1) + \int_0^{T_1} \sigma\,dB, X_2 > \psi(T_2) - \psi(T_1) + \int_{T_1}^{T_2} \sigma\,dB\Big)$$

$$= \int_{\mathbb{R}} \int_0^\infty \Big\{1 - \int_{\mathbb{R}} \int_0^\infty F_X(s + \psi'(t + \theta s + z))f(s)f_{\mathcal{N}_{v(t, s+t)}}(z)\,ds\,dz\Big\}$$

$$F_X(\psi(t + z_1))f(t)f_{\mathcal{N}_{v(0, t)}}(z_1)\,dt\,dz_1.$$

The probability that the time of ruin belongs to the interval $]T_n, T_{n+1}]$ is calculated using the notation $s_k = t_k - t_{k-1}$ in the integrals

$$P_n = P(X_1 < \psi(T_1) + \int_0^{T_1} \sigma \, dB, \ldots, X_n < \psi(T_n) - \psi(T_{n-1})$$

$$+ \int_{T_{n-1}}^{T_n} \sigma \, dB, X_{n+1} > \psi(T_{n+1}) - \psi(T_n) + \int_{T_n}^{T_{n+1}} \sigma \, dB)$$

$$= \int_{\mathbb{R}^n} \int_{\mathbb{R}_+^n} \prod_{k=1}^n F_X(cs_k + z_k) f(s_k) f_{\mathcal{N}_{v(t_{k-1}, t_k)}}(z_k) \, ds_k \, dz_k$$

$$\{1 - \int_{\mathbb{R}} \int_0^\infty F_X(s + \psi'(t + \theta s + z) f(s) f_{\mathcal{N}_{v(t, s+t)}}(z) \, ds_{n+1} \, dz_{n+1}\}$$

$$ds_1 \ldots ds_n \, dz_1 \, dz_n.$$

The process Y_{T_a} is negative and it stops at this value. Adding a stochastic noise to the Sparre Andersen model is therefore necessary to avoid a large probability of an early end of the process as described by Propositions 6.2 and 6.3.

The independence of the variables $Y(T_n) - Y(T_{n-1})$ can also be replaced by an assumption of a positive martingale sequence $(X_k)_{k\geq 0}$ with respect to the filtration $(\mathcal{F}_n)_n$ generated by the process N_X, or by a submartingale (respectively supermartingale) sequence, without an explicit expression for the model also denoted $\psi(t)$. Adding a diffusion noise to the Sparre Andersen model, the process $Y = \psi + \int_0^\cdot \sigma \, dB - N_X$ is a martingale under this assumption.

With a renewal process $(T_k)_{k\geq 1}$ and an independent martingale $(X_k)_{k\geq 1}$, the process N_X has the conditional means

$$E\{N_X(t)|\mathcal{F}_n\} = \sum_{k=1}^n X_k 1_{\{T_k \leq t\}} + \sum_{k>n} E\{X_k 1_{\{\sum_{j=n+1}^k S_j \leq t - T_n\}}|\mathcal{F}_n\}$$

$$= \sum_{k=1}^n X_k 1_{\{T_k \leq t\}} + \sum_{k>n} X_n \int_0^{t-T_n} f^{*(k-n)}(s) \, ds.$$

If the martingale has the property that X_n is $\mathcal{F}_{n-1}$-measurable, N_X has the predictable compensator

$$\tilde{N}_X(t) = \int_0^t (\sum_{k\geq 1} X_k 1_{\{T_k \geq s\}}) \, d\Lambda(s).$$

The expression of mean of Y is not modified by the assumption of the dependence between the variables X_k, it is also equal to

$$EY(t) = \psi(t) - EX_0 \sum_{k\geq 1} \int_0^\infty \int_0^t \{1 - F(s - t)\} f^{*(k-1)}(t) \, d\Lambda(s) \, dt$$

if X_n is $\mathcal{F}_{n-1}$-measurable for every n. Lemmas 6.1 and 6.2 are still valid.

6.4 Comparison of models

The weighted processes Y and Z defined (4.5) on a filtered probability space $(\Omega, \mathcal{F}, (\mathcal{F}_n)_{n \geq 0}, P)$, are generalized to an adapted sequence $U = (U_n)_{n \geq 0}$, with adapted weighting variables $(A_n)_n$. Let

$$Y_n = \sum_{k=1}^{n} A_k (U_k - U_{k-1}), \qquad (6.4)$$

$$Z_n = \sum_{k=1}^{n} A_k^{-1} (U_k - U_{k-1}).$$

By the Cauchy inequality

$$E|Y_n| \leq E\{\sum_{k=1}^{n} E(A_k^2 | \mathcal{F}_{k-1}) V_n(U)\}^{\frac{1}{2}} \qquad (6.5)$$

$$\leq [E\{\sum_{k=1}^{n} E(A_k^2 | \mathcal{F}_{k-1})\}]^{\frac{1}{2}} \{EV_n(U)\}^{\frac{1}{2}}$$

and Z_n satisfies the same kind of inequality. The bounds are increasing series and the inequalities hold for non increasing processes Y and Z.

Let U be a local martingale of $L_2(P)$ and A be a predictable process of $L_2(P)$. The sequences $(Y_n)_{n \geq 0}$ and $(Z_n)_{n \geq 0}$ are local martingales of $L_2(P)$ and their means are zero. Their L_1-means have smaller increasing bounds defined as $u_n = \{E \sum_{k=1}^{n} A_k^2 (V_k - V_{k-1})(U)\}^{\frac{1}{2}}$ for $E|Y_n|$ and $v_n = \{E \sum_{k=1}^{n} A_k^{-2} (V_k - V_{k-1})(U)\}^{\frac{1}{2}}$ for $E|Z_n|$. The Kolmogorov inequality (4.7) applies with the random weights A_k and A_k^{-1}. The same inequalities are true with the bounds (6.5) instead of the u_n and v_n, for general adapted processes U and A of $L_2(P)$.

Let us consider ruin models where the random series (6.4) replace the deterministic trend ψ. They are more realistic for applications in other fields where the process N_X is sequentially corrected by inputs. The variables U_k may be defined as the random times of the process N_X or other processes depending on them, they are not necessarily local martingales and their means are supposed to be different from zero. Their weights may be predictable or not. The model becomes

$$Y(t) = a + c \sum_{n \geq 1} A_n (U_n - U_{n-1}) 1_{\{T_n \leq t\}} - N_X(t) \qquad (6.6)$$

and the time of ruin is a.s. infinite if for every $n \geq 1$

$$0 \leq X_n \leq \frac{a}{nc} + A_n (U_n - U_{n-1}), \text{ a.s.}$$

The mean of the process is everywhere positive if $EN_X(t)$ is sufficiently large with respect to the bound of the inequality (6.5).

Lemma 6.3. *A sufficient condition for a finite mean time of ruin t_0 is*

$$a + cE(\sum_{k=1}^{n} A_k^2 1_{\{T_k \leq t_0\}})^{\frac{1}{2}} \{EV_n(U)\}^{\frac{1}{2}} \leq EN_X(t_0).$$

It comes from the Cauchy inequality for the sum $E(\sum_{k=1}^{n} A_k(U_k - U_{k-1})1_{\{T_k \leq t_0\}}) \leq \{E(\sum_{k=1}^{n} A_k^2 1_{\{T_k \leq t_0\}})EV_n(U)\}^{\frac{1}{2}}$ and $EY(t_0) < 0$ is satisfied if this bound is lower than $c^{-1}\{EN_X(t) - a\}$. The probabilities P_n of ruin in an interval $]T_n, T_{n+1}]$ are written as

$$P(T_a \in]0, T_1]) = P(Y(T_1) \leq 0) = P(X_1 \geq a + cA_1U_1),$$
$$P(T_a \in]T_1, T_2]) = P(X_1 < a + cA_1U_1, X_2 > cA_2(U_2 - U_1),$$

$$P(T_a \in]T_1, T_2]) = P(X_1 < a + cA_1U_1, X_2 > cA_2(U_2 - U_1),$$
$$P(T_a \in]T_n, T_{n+1}]) = P(\{X_1 < a + cA_1U_1\} \cap [\cap_{k=1,\dots,n}\{X_1 + \cdots + X_k < a$$
$$+ c(A_1U_1 + \cdots + A_n(U_n - U_{n-1}))\}]$$
$$\cap \{X_{n+1} \geq cA_{n+1}(U_{n+1} - U_n)\}).$$

Under the assumption of sequences of independent variables $(X_n)_n$, $(A_n)_n$ and $(U_{n+1} - U_n))_n$, the probabilities are factorized and they are calculated recursively as in Proposition 6.2, by the formula

$$P_n = \widetilde{P}_{n-1} - \widetilde{P}_n = \widetilde{P}_{n-1}\{1 - P(X_{n+1} \geq cA_{n+1}(U_{n+1} - U_n))\}.$$

Under the condition of a nonlinear trend function ψ, the assumption of independent sequences cannot be satisfied and the probability P_n is written as a convolution of $n + 1$ marginal probabilities. The discrete model (6.6) does not include the stochastic diffusion but it can be extended to this model by adding the integral $A_{n+1} \int_{T_n}^{t} \sigma \, dB$ to the n-th term of the sum in (6.6) where $U_n = \int_0^{T_n} \sigma \, dB$.

The expression of the diffusion in (6.3) does not vary with the past of the process and it can be generalized in the form

$$W(t) = \sum_{n \geq 1} \int_{T_{n-1}}^{t \wedge T_n} \sigma(A_n(s)) \, dB(s)$$

where the integral is the stochastic integral for a predictable sequence $(A_n)_n$. For every sequence of adapted stopping times

$$E\{W(t)|\mathcal{F}_n\} = \sum_{k=1,\dots,n} \int_{T_{n-1}}^{t \wedge T_n} \sigma(A_n(s)) \, dB(s),$$

it is therefore a local martingale with mean zero. Under the assumption of a sequence $(A_n)_n$ independent of the Brownian motion, its variance is

$$var W(t) = \int_0^t E\{\sum_{n \geq 1} 1_{[T_{n-1}, T_n]}(t) \sigma^2(A_n(s))\} \, ds.$$

6.5   Moments of the processes at T_a

From Wald's equations (A.1) for a sequence of independent and identically distributed variables $(X_n)_{n \leq 0}$, such that $EX_i = 0$ and $var X_i = \sigma^2$, and a stopping time T with a finite mean, $ES_T^2 = \sigma^2 T$.

Lemma 6.4. *Let φ be a real function defined on $\mathbb{N}$. For every stopping time T on a filtered probability space $(\Omega, \mathcal{F}, (\mathcal{F}_n)_{n \geq 0}, P)$*

$$\sum_{k \geq 1} \varphi(k) P(T \geq k) = E \sum_{k=1}^T \varphi(k).$$

Proposition 6.4. *On a probability space $(\Omega, \mathcal{F}, P)$, let $(\mathcal{J}_n)_{n \geq 0}$ be a filtration and let N_X be the process defined by (6.1) with an $(\mathcal{F}_n)_{n \geq 0}$-adapted sequence of independent variables $(X_n)_{n \geq 0}$ and an independent renewal process. For every integer valued stopping time T with respect to $(\mathcal{F}_n)_{n \geq 0}$, $EN_X(T) = EX \, EN(T)$ and*

$$var N_X(T) = E(X^2) E[\sum_{k=1}^T \{N(k) - N(k-1)\}^2 - \{EN(T)\}^2].$$

Proof. Since the process Y is only decreasing at the jump times of N_X, T is a stopping time with integer values. For every $n \geq 1$

$$EN_X(T \wedge n) = EX \, [EN(T \wedge (n-1))$$
$$+ E\{N(n) - N(T \wedge (n-1))\} P(T \geq n)]$$
$$= EX \sum_{k=1}^n E\{N(k) - N(k-1)\} P(T \geq k)$$

since X_n is independent of $\{T \geq n\}$ and $N_X(T \wedge (n-1))$. Moreover, the mean of T is $ET = \sum_{k \geq 1} P(T \geq k)$ and by Lemma 6.4, for every process φ on $\mathbb{N}$

$$E \sum_{k \geq 1} \varphi(k) P(T \geq k) = E \sum_{k=1}^T \varphi(k),$$

which implies

$$EN_X(T) = EX \sum_{k \geq 1} E\{N(k) - N(T \wedge (k-1))\} P(T \geq k)$$

$$= EX\, E \sum_{k=1}^{T} \{N(k) - N(T \wedge (k-1))\} = EX\, EN(T).$$

The variance develops as the squared sum

$$EN_X^2(T \wedge n) = E[N_X(T \wedge (n-1))$$
$$+ \{N_X(n) - N_X(T \wedge (n-1))\} 1_{\{T \geq n\}}]^2$$
$$= EN_X^2(T \wedge (n-1))$$
$$+ E\{N_X(n) - N_X(T \wedge (n-1))\}^2 P(T \geq n)$$
$$= \sum_{k \geq 1} E[\{N_X(n) - N_X(T \wedge (n-1))\}^2 1_{\{T \geq n\}}].$$

Using again Lemma 6.4, we have

$$\sum_{k \geq 1} E[\{N_X(k) - N_X(T \wedge (k-1))\}^2 1_{\{T \geq k\}}]$$

$$= E \sum_{k=1}^{T} \{N_X(k) - N_X(T \wedge (k-1))\}^2$$

$$= E(X^2)\, E \sum_{k=1}^{T} \{N(k) - N(k-1)\}^2. \qquad \square$$

Corollary 6.1. *In the Sparre Andersen model*

$$\frac{ET_a}{EN(T_a)} \geq \frac{a}{cEN(T_a)} + \frac{EX}{c} \geq \frac{EX}{c}.$$

Other consequences of Proposition 6.4 are deduced from the Bienaymé-Chebychev inequality. For every $y < a$

$$P(Y(T_a) < y) = P(N_X(T_a) - cT_a > a - y)$$
$$\leq \frac{EN_X^2(T_a) - 2T_a N_X(T_a) + c^2 T_a^2}{(a-y)^2},$$

where $EN_X^2(T_a)$ is given in Proposition 6.4, $N(0) = 0$ and, by Lemma 6.4, $E\{T_a N_X(T_a)\} = EX\, E \sum_{k=0}^{T_a - 1} \{N(T_a) - N(k)\}$. Since $N_X(T_a) \geq a + cT_a$, for every $t > 0$

$$P(T_a > t) \leq P(N_X(T_a) > a + ct) \leq \frac{EX\, EN(T_a)}{a + ct}.$$

The moments of higher order are calculated in the same way, by an expansion of

$$EN_X^p(T_a \wedge n) = E[N_X(T_a \wedge (n-1))$$
$$+\{N_X(n) - N_X(T_a \wedge (n-1))\}1_{\{T_a \geq n\}}]^p.$$

The centered moments of the process $Y(T_a)$ in the Sparre Andersen model are equal to those of $N_X(T_a)$ and to those of $Y(T_a)$ in the model (6.3) with a stochastic diffusion. In model (6.6), $EY(T_a) = a + c\sum_{n \geq 1}, \{A_n(U_n - U_{n-1}) - X_n\}1_{\{T_n \leq T_a\}}$ has the same form and its centered moments are similar if the sequence of variables $(A_n, U_n - U_{n-1}, X_n)_{n \geq 1}$ are independent and identically distributed, writing $E\{A_n(U_n - U_{n-1} - X_n\}$ instead of EX in the expression of EN_X.

If the assumption of independent and identically distributed variables is replaced by an assumption of martingales, conditional means are added to the above expressions.

6.6 Empirical process in mixture distributions

Section 3.4 presents the general form of continuous mixture of densities with some examples. With a location parameter lying in an interval I, the density of the observed variable is the convolution $f_X(x) = \int_I f_{X|W}(x - w)f_W(w)\,dw = f_{X|W} * f_W(x)$ and the Fourier transform of f_W is the ratio of the transforms for f_X and $f_{X|W}$.

In a generalized exponential model, the conditional density $f_{X|Y}$ of the variable X, given a dependent real variable Y, is expressed as

$$f_{X|Y}(x; Y) = \exp\{T(x, Y) - b(Y)\},$$

where $b(Y) = \log \int \exp\{T(x, Y)\}\,dx$ is the normalization function of the conditional density and the $T(X, Y)$ only depends on the variables. The model is extended to a semi-parametric conditional density

$$f_{X|Y,\eta}(x; Y) = \exp\{\eta^T T(x, Y) - b(\eta, Y)\}$$

with $b(\eta, Y) = \log \int \exp\{\eta^T T(x, Y)\}\,dx$. When Y is a hidden real variable with distribution function F_Y, the distribution of the variable X has the density $f_\eta(x) = E_Y f_{X|Y,\eta}(x; Y)$ where E_Y is the unknown mean with respect to F_Y.

Let $E_{X|Y}$ be the conditional expectation with respect to the probability distribution of X conditionally on Y. When the distribution function F_Y

is known, the distribution function of X is parametric with parameter η. The derivative with respect to η of $\log f_\eta$ is $f_\eta^{-1} \dot{f}_\eta$ and

$$\dot{f}_\eta(x) = E_Y[\{T(x,Y) - E_{X|Y}T(X,Y)\}f_{X|Y,\eta}(x;Y)] \qquad (6.7)$$

with $\dot{b}_\eta(\eta, Y) = E_{X|Y}T(X,Y)$. For the sample $(X_i)_{1 \leq i \leq n}$ of the variable X, the maximum likelihood estimator of η is a solution of the score equation $\dot{l}_n(\eta) = \sum_{i=1}^n f^{-1}(X_i; \eta) \dot{f}_\eta(X_i) = 0$ and its asymptotic behaviour follows the classical theory of the parametric maximum likelihood.

With a nonparametric mixing distribution, the distribution function F_Y is approximated by a countable function

$$F_{Yn}(y) = \sum_{k=1}^{K_n} p_{nk} 1_{\{y_{nk} \leq y\}},$$

where the probabilities

$$p_{nk} = f_{Yn}(y_{nk}) \qquad (6.8)$$

sum up to 1, and the distribution function F_X is written as

$$F_X(x) = \sum_{k=1}^{K_n} p_{nk} F_{X|Y,\eta}(x; y_{nk}),$$

with a parametric distribution of X conditionally on Y, $F_{X|Y,\eta}$. The empirical distribution function of the observed variable X is denoted $\widehat{F}_{Xn}$. Let $f_{X|Y,\eta}(x; y)$ be the density of $F_{X|Y,\eta}$, it is supposed to be twice continuously differentiable with respect to the parameter η and with respect to y, with first derivative with respect to η, $\dot{f}_{X|Y,\eta}$, satisfying (6.7), and $f^{(1)}_{X|Y,\eta}(x; y)$, with respect to y.

Proposition 6.5. *Under the constraint $\sum_{k=1}^{K_n} p_{nk} = 1$, the maximum likelihood estimators of the probabilities p_{nk} are*

$$\widehat{p}_{nK_n} = \frac{\int f_{Y|X,\widehat{\eta}_n}(\widehat{y}_{nK_n}; x)\, d\widehat{F}_n(x)}{\sum_{k=1}^{K_n} \int \widehat{f}_{X,\widehat{\eta}_n}(\widehat{y}_{nk}; x)\, d\widehat{F}_n(x)},$$

$$\widehat{p}_{nk} = \frac{\int f_{Y|X,\widehat{\eta}_n}(\widehat{y}_{nk}; x)\, d\widehat{F}_n(x)}{\sum_{k=1}^{K_n} \int f_{Y|X,\widehat{\eta}_n}(\widehat{y}_{nk}; x)\, d\widehat{F}_n(x)}, \quad k = 1, \ldots, K_n - 1,$$

where $\widehat{\eta}_n$ and $(\widehat{y}_{nk})_{k \leq K_n}$ are the maximum likelihood estimators of the parameters η and $(y_{nk})_{k \leq K_n}$.

Proof.   The values of $(p_{nk})_{k \leq K_n}$ that maximize

$$L_n = \prod_{i=1}^{n} \log \sum_{k=1}^{K_n} p_{nk} f_{X|Y,\eta}(X_i; y_{nk})$$

under the constraint $\sum_{k=1}^{K_n} p_{nk} = 1$ are solutions of the equations

$$0 = \sum_{i=1}^{n} \frac{\sum_{k=1}^{K_n} p_{nk} \dot{f}_{X|Y,\eta}(X_i; y_{K_n,n})}{f_{X,\eta}(X_i)},$$

$$0 = \sum_{i=1}^{n} \frac{\sum_{k=1}^{K_n} p_{nk} f_{X|Y,\eta,k}^{(1)}(X_i; y_{nk})}{f_{X,\eta}(X_i)},$$

$$\sum_{i=1}^{n} \frac{f_{X|Y,\eta}(X_i; y_{K_n})}{f_{X,\eta}(X_i)} = \sum_{i=1}^{n} \frac{f_{X|Y,\eta}(X_i; y_{nk})}{f_{X,\eta}(X_i)}, \quad k = 1, \ldots, K_n - 1,$$

where $f_{X,\eta}(X_i) = \sum_{k=1}^{K_n} p_{nk} f_{X|Y,\eta}(X_i; y_{nk})$ and

$$\frac{f_{X|Y,\eta}(X_i; y_{nk})}{f_{X,\eta}(X_i)} = \frac{f_{Y|X,\eta}(y_{nk}, X_i)}{f_{Y,\eta}(y_{nk})}.$$

By the definition (6.8) of the probabilities p_{nk}, the last equations are therefore equivalent to

$$\frac{\sum_{i=1}^{n} f_{Y|X,\eta}(X_i; y_{nK_n})}{p_{nK_n}} = \frac{\sum_{i=1}^{n} f_{Y|X,\eta}(X_i; y_{nk})}{p_{nk}}, \quad k = 1, \ldots, K_n - 1,$$

using the constraint yields

$$p_{nK_n} = \frac{\sum_{i=1}^{n} f_{Y|X,\eta}(X_i; y_{nK_n})}{\sum_{k=1}^{K_n} \sum_{i=1}^{n} f_{Y|X,\eta}(X_i; y_{nk})}$$

and the expression of p_{nk} is similar.   □

The maximum likelihood estimators of the parameters η and $(y_{nk})_{k \leq K_n}$ are the values of η and $(y_{nk})_{k \leq K_n}$ that maximize $L_n(\widehat{p}_n, y_n, \eta)$, they are solutions of the equations

$$0 = \sum_{i=1}^{n} \frac{\sum_{k=1}^{K_n} \widehat{p}_{nk} \dot{f}_{X|Y,\eta}(X_i; y_{K_n,n})}{f_{X,\eta}(X_i)},$$

$$0 = \sum_{i=1}^{n} \frac{\widehat{p}_{nk} f_{X|Y,\eta,k}^{(1)}(X_i; y_{nk})}{f_{X,\eta}(X_i)}, \quad k = 1, \ldots, K_n.$$

By (6.7), the first equation is written

$$\int f_{X|\eta}^{-1}(x) \sum_{k=1}^{K_n} \widehat{p}_{nk} \{T(x, y_{nk}) - E_{X|Y} T(X, y_{nk})\} f_{X|Y,\eta}(x, y_{nk}) \, d\widehat{F}_n(x) = 0$$

and the second equation is equivalent to

$$\int f_{X|\eta}^{-1}(x) \sum_{k=1}^{K_n} \widehat{p}_{nk}\{T_y^{(1)}(x, y_{nk}) - E_{X|Y}T_y^{(1)}(X, y_{nk})\}f_{X|Y,\eta}(x, y_{nk}) \, dx = 0.$$

If η is a vector of dimension d, $K_n + p$ parameters are estimated by these equations are they converge to the true parameter values as n tends to infinity with $K_n = o(n)$, by the classical theory of the maximum likelihood estimation.

6.7 Integral inequalities in the plane

Pitt (1977) proved that for all symmetric convex subsets A and B of $\mathbb{R}^2$, the density of a Gaussian variable X in $\mathbb{R}^2$ satisfies

$$\int_{A\cap B} f(x) \, dx \geq \int_A f(x) \, dx \int_B f(x) \, dx,$$

which is equivalent to

$$P(X \in A \cap B) \geq P(X \in A)P(X \in B).$$

Note that the assumptions imply that the intersection of A and B is not empty. Removing the conditions and applying the Hölder inequality to indicator functions 1_A and 1_B implies that for every density function

$$\int_{A\cap B} f(x) \, dx \leq \{\int_A f(x) \, dx \int_B f(x) \, dx\}^{\frac{1}{2}}$$

or $P(X \in A \cap B) \leq \{P(X \in A)P(X \in B)\}^{\frac{1}{2}}$.

These inequalities entail that for every positive function h of $L_1(\mu_{\mathcal{N}})$, where $\mu_{\mathcal{N}}$ is a Gaussian distribution, the inequalities apply to the integral of h with respect to the measure $\mu_{\mathcal{N}}$ restricted to symmetric subsets of $\mathbb{R}^2$

$$\int_{A\cap B} h(x) \, d\mu_{\mathcal{N}}(x) \geq \int_A h(x) \, d\mu_{\mathcal{N}}(x) \int_B h(x) \, d\mu_{\mathcal{N}}(x),$$

$$\int_{A\cap B} h(x) \, d\mu_{\mathcal{N}}(x) \leq \{\int_A h(x) \, d\mu_{\mathcal{N}}(x) \int_B h(x) \, d\mu_{\mathcal{N}}(x)\}^{\frac{1}{2}}.$$

Moreover, for every positive probability measure μ and every subsets of $\mathbb{R}^2$, positive functions of $L_1(\mu_{\mathcal{N}})$ satisfy

$$\int_{A\cup B} h(x) \, d\mu(x) + \int_{A\cap B} h(x) \, d\mu(x) \geq \int_A h(x) \, d\mu(x) + \int_B h(x) \, d\mu(x),$$

$$\int_{A\cup B} h(x) \, d\mu(x) \leq \int_A h(x) \, d\mu(x) + \int_B h(x) \, d\mu(x),$$

with equality if and only if $A \cap B$ is empty.

The L_p-distance of non intersecting subsets A and B of $\mathbb{R}^n$ is

$$d_p(A, B) = \inf_{x \in A} \inf_{y \in B} \{ \sum_{i=}^{n} |x_i - y_i|^p \}^{\frac{1}{p}}, \ p \geq 1.$$

For independent random subsets A and B of a probability space $(\Omega, \mathcal{A}, P)$, there exist independent variables X and Y defined in $(\Omega, \mathcal{A}, P)$ and with values in $\mathbb{R}^n$ such that

$$P(d_p(A, B) > t) \leq t^{-1} E \inf_{X \in A, Y \in B} \| X - Y \|_p.$$

With a centered Gaussian probability, the bound is

$$t^{-1}(2\pi)^{-\frac{n}{2}} \{ \det(\Sigma_1)\det(\Sigma_2) \}^{-\frac{1}{2}} \int_{A \times B} \| x - y \|_p e^{-\frac{1}{2} x^t \Sigma_1^{-1} x} e^{-\frac{1}{2} y^t \Sigma_2^{-1} y} \, dx \, dy.$$

Under the constraint that the distance between sets A and B_α is at least equal to α

$$E \inf_{X \in A, Y \in B_\alpha} \| X - Y \|_p = \alpha P(A \times B_\alpha)$$

and $P(d_p(A, B_\alpha) > t) \leq t^{-1} \alpha P(A \times B_\alpha)$.

Let A and B subsets of $\mathbb{R}^n$ and let X in A, applying this inequality conditionally on X gives $P(d_p(X, B) > t | X = x) \leq t^{-1} E \inf_{Y \in B} \| x - Y \|_p$. Under a Gaussian probability P on $\mathbb{R}$, the paths of the process $d(x, B)$ stay a.s. inside frontiers $(2\sigma^2(x)h_x)^{\frac{1}{2}}$ determined by Proposition 4.23, with the variance $Ed^2(x, B) = \int_{\mathbb{R}} (x - s)^2 dP_Y(s) = \sigma^2(x)$ for some Y that achieves the minimum over B and with a function h such that h^{-1} belongs to $L_1(P)$.

6.8   Spatial point processes

A Poisson process N indexed by $\mathbb{R}_+$, with intensity λ, has a covariance function

$$R(s, t) = cov(N_s, N_t) = var N_{s \wedge t} = \lambda(t \wedge s)$$

and the higher moments of the Poisson process are $E\{(N_t - \lambda t)^k\} = \lambda t$, for every integer $k \geq 2$.

Lemma 6.5. *Let $t_1, t_2, \ldots, t_k$ be positive reals, the crossed-moments of the Poisson process in $\mathbb{R}_+$ are*

$$E\{(N_{t_1} - \lambda t_1) \cdots (N_{t_k} - \lambda t_k)\} = \lambda t_{m_k},$$

where $t_{m_k} = \min(t_1, \ldots, t_k)$ and $k \geq 2$.

Proof. Let $0 < t_1 < t_2 < t_3$

$$\nu_3(t_1, t_2, t_3) = E\{(N_{t_1} - \lambda t_1)(N_{t_2} - \lambda t_2)(N_{t_3} - \lambda t_3)\}$$
$$= E\{(N_{t_1} - \lambda t_1)(N_{t_2} - \lambda t_2)^2\}$$
$$= \lambda t_1 + E[(N_{t_1} - \lambda t_1)\{(N_{t_2} - \lambda t_2) - (N_{t_1} - \lambda t_1)\}$$
$$\{(N_{t_2} - \lambda t_2) + (N_{t_1} - \lambda t_1)\}]$$
$$= \lambda t_1 + E[(N_{t_1} - \lambda t_1)\{(N_{t_2} - \lambda t_2) - (N_{t_1} - \lambda t_1)\}^2]$$
$$= \lambda t_1.$$

The result for any k is deduced by induction. $\qquad\square$

Let us consider two Poisson processes with intensities λ_1 and $\lambda_2 = \lambda_1 + x$, $x > 0$, then $P_{\lambda_1}(N_t = k)P_{\lambda_2}^{-1}(N_t = k) = e^{-xt}(1 + \lambda_2^{-1}xt)^k$, it is increasing with respect to x if $k > [xt + \lambda_2]$ and decreasing otherwise, therefore the distributions of Poisson processes cannot be ordered.

A heterogeneous Poisson process with a cumulative intensity Λ has the moments $E[\{N_t - \Lambda(t)\}^k] = \Lambda(t)$, for every integer $k \geq 2$ and

$$E[\{N_{t_1} - \Lambda(t_1)\} \cdots \{N_{t_k} - \Lambda(t_k)\}] = \Lambda(t_{m_k}).$$

Some characteristics of the first two moments of point processes in the plane are introduced in Section 4.9 and they are based on martingales properties. Spatial point processes are characterized by their moments on the balls. In $\mathbb{R}^d$, let $r > 0$ and $B_r(x)$ be the ball of radius r centered at x, its volume is $|B_r(x)| = c_d r^d$ for every center x. The k-th moment of N, $k \geq 2$, is defined by the values of the process in k balls of radius r. For every $x = (x_1, \ldots, x_k)$ in $\mathbb{R}^{kd}$, let k balls $B_r(x_j)$ with a non empty intersection $\cap_{j=1,\ldots,k} B_r(x_j)$,

$$\nu_{k,r}(x) = \frac{1}{(c_d r^d)^{\frac{k}{2}}} E\{N(B_r(x_1)) \cdots N(B_r(x_k))\}.$$

For a spatial stationary process N in $\mathbb{R}^d$, it is invariant by translation and defined for $(k - 1)$ location parameters

$$\nu_{k,r}(x) = \frac{1}{(c_d r^d)^{\frac{k}{2}}} E\{N(B_r(x_1 - x_k)) \cdots N(B_r(0))\} = \nu_{k,r}(x_1 - x_k, \ldots, 0).$$

The second moments of a process with independent increments are

$$E\{N(B_r(x_1))N(B_r(x_2))\} = EN(B_r(x_1)) \, EN(B_r(x_2))$$
$$+ var N(B_r(x_1) \cap B_r(x_2)).$$

For a stationary process in $\mathbb{R}^d$, this equality becomes

$$
\begin{aligned}
E\{N(B_r(0))N(B_r(x_2 - x_1))\} = {} & EN(B_r(0))\, EN(B_r(x_2 - x_1)) \\
& + varN(B_r(0) \cap B_r(x_2 - x_1)). \quad (6.9)
\end{aligned}
$$

A Poisson process N with a cumulative intensity Λ in $\mathbb{R}^d$ has integer values and its distribution on balls is the distribution of a Poisson variable having as parameter the measure with respect to Λ of the balls

$$
P\{N(B_r(x)) = k\} = e^{-\Lambda(B_r(x))}\frac{\Lambda^k(B_r(x))}{k!}
$$

and its moments are $EN^k(B_r(x)) = \Lambda(B_r(x))$, $k \geq 1$. Its crossed-moments in balls are given by Lemma 6.5, according to the number of balls intersecting with each ball. Let $\varepsilon > 0$ and consider an ε-net in $\mathbb{R}^d$, balls of radius r and centered at the points of the ε-net have intersections by pairs with their nearest neighbours. Balls having a radius r such that $\varepsilon \geq 2r$ are disjoint, then for every k

$$
E[\{N_{B_r(x_1)} - \Lambda(B_r(x_1))\} \cdots \{N_{B_r(x_k)} - \Lambda(B_r(x_k))\}] = 0.
$$

Let r be in the interval $I_1(\varepsilon) =]\frac{\varepsilon}{2}, \frac{\varepsilon}{\sqrt{2}}]$ and let $\mathcal{V}_\varepsilon(x)$ be the set of the centers of the nearest balls of $B_r(x_1)$, the values of the counting process on non-overlapping subsets of the balls are independent and they have the same value in the pairwise intersections of the balls, therefore

$$
\begin{aligned}
& E[\{N_{B_r(x_1)} - \Lambda(B_r(x_1))\} \cdots \{N_{B_r(x_k)} - \Lambda(B_r(x_k))\}] \\
& = \sum_{i=1}^{k} \sum_{x_j \in \mathcal{V}_\varepsilon(x_i)} varN(B_r(x_i) \cap B_r(x_j)).
\end{aligned}
$$

If $\sqrt{2}r \geq \varepsilon$, the number $K_r(x)$ of balls intersecting a ball $B_r(x)$ increases and they are mutually intersecting. In $\mathbb{R}^2$, each ball with a radius r belonging to $I_1(\varepsilon)$ has 4 nearest neighbours (Fig. 6.1). There are $K_r(x) = 8$ balls centered in an ε-net around x, with radius r belonging to the interval $I_2(\varepsilon) =]\frac{\varepsilon}{\sqrt{2}}, \varepsilon]$, (Fig. 6.2). Under the same condition in $\mathbb{R}^3$, a ball $B_r(x)$ has 6 intersecting balls centered in the ε-net if r belongs to $I_1(\varepsilon)$ and $K_r(x) = 24$ intersecting balls centered in the ε-net if r belongs to $I_2(\varepsilon)$.

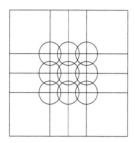
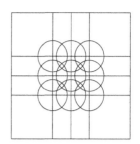

Fig. 6.1   Intersections of balls with ra-
dius $r = 3\varepsilon$, in the interval $I_1(\varepsilon)$, and
centered in an ε-net.

Fig. 6.2   Intersections of balls with ra-
dius $r = .8\varepsilon$, in the interval $I_2(\varepsilon)$, and
centered in an ε-net.

Proposition 6.6. *Let N be a heterogeneous Poisson process in $\mathbb{R}^d$, with cumulative intensity Λ and let r be in the interval $I_2(\varepsilon)$. For every $k \geq 2$*

$$E[\{N_{B_r(x_1)} - \Lambda(B_r(x_1))\} \cdots \{N_{B_r(x_k)} - \Lambda(B_r(x_k))\}]$$

$$= \sum_{i=1}^{k} \sum_{j_i=1}^{K_r(x_i)} [\Lambda(B_r(x_i) \cap B_r(x_{j_i}))$$

$$- \sum_{k_i \neq j_i, k_i=1}^{K_r(x_i)} \{\Lambda(B_r(x_i) \cap B_r(x_{j_i}) \cap B_r(x_{k_i}))$$

$$+ \sum_{l_i \neq j_i, k_i, l_i=1}^{K_r(x_i)} \Lambda(B_r(x_i) \cap B_r(x_{j_i}) \cap B_r(x_{k_i}) \cap B_r(x_{l_i}))\}].$$

Proof.   Let $E_k = E[\{N_{B_r(x_1)} - \Lambda(B_r(x_1))\} \cdots \{N_{B_r(x_k)} - \Lambda(B_r(x_k))\}]$, it is expanded as

$$E_k = \sum_{i=1}^{k} \sum_{j_i=1}^{K_r(x_i)} \{E[\{N_{B_r(x_i) \cap B_r(x_{j_i})} - \Lambda(B_r(x_i) \cap B_r(x_{j_i}))\}^2]$$

$$- \sum_{k_i \neq j_i, k_i=1}^{K_r(x_i)} (E[\{N_{B_r(x_i) \cap B_r(x_{j_i}) \cap B_r(x_{k_i})}$$

$$- \Lambda(B_r(x_i) \cap B_r(x_{j_i}) \cap B_r(x_{k_i}))\}^3]$$

$$+ \sum_{l_i \neq j_i, k_i, l_i=1}^{K_r(x_i)} E[\{N_{B_r(x_i) \cap B_r(x_{j_i}) \cap B_r(x_{k_i}) \cap B_r(x_{l_i})}$$

$$- \Lambda(B_r(x_i) \cap B_r(x_{j_i}) \cap B_r(x_{k_i}) \cap B_r(x_{l_i}))\}^4)\}.$$

The result is deduced from the expression $EN^k(A) = \Lambda(A)$ for every Borel set of $\mathbb{R}^d$ and for every $k \geq 2$. $\square$

With r in the interval $I_3(\varepsilon) = \,]\varepsilon, \sqrt{2}\varepsilon]$, the moments calculated in Proposition 6.6 have additional terms including intersections of orders 5 to 8, as in Fig. 6.3.

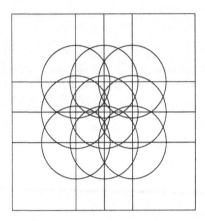

Fig. 6.3   Intersections of balls with radius $r = 1.2\,\varepsilon$, in the interval $I_3(\varepsilon) = \,]\varepsilon, \sqrt{2}\varepsilon]$, and centered in an ε-net.

With a larger radius, the number $K_r(x)$ increases and the k-order moment $E[\{N_{B_r(x_1)} - \Lambda(B_r(x_1))\} \cdots \{N_{B_r(x_k)} - \Lambda(B_r(x_k))\}]$ is a sum of moments of higher order, up to k as ε tends to zero. In the interval $I_4(\varepsilon) = \,]\sqrt{2}\varepsilon, 2\varepsilon]$, the eight nearest balls of $B_r(x)$ are intersecting by pair. The moments calculated in Proposition 6.6 must include intersections of order larger than 8 (Fig. 6.4).

A stationary process with independent increments in $\mathbb{R}^d$ has its finite dimensional distributions defined by the distribution of the vector $\{N_{B_r(x_1)} - \Lambda(B_r(x_1)), \ldots, N_{B_r(x_k)} - \Lambda(B_r(x_k))\}$ for every $(x_1, \ldots, x_k)$ and for every $r > 0$. It splits into the variations of $N - \Lambda$ on intersection of balls according to the number of intersecting balls in each subset. Under the assumption of stationarity, it is written as the sum of the values of $N - \Lambda$ in independent subsets having the same pattern, they are therefore identically distributed. It follows that every x in $\mathbb{R}^{kd}$, the variable

$$X_{k,r}(x) = (c_d r^d)^{-\frac{k}{2}} \{N_{B_r(x_1-x_k)} \cdots N_{B_r(0)} - \nu_{k,r}(x)\}$$

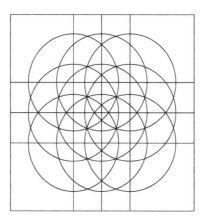

Fig. 6.4   Intersections of balls with radius $r = 1.6\,\varepsilon$, in the interval $I_4(\varepsilon) = \,]\sqrt{2}\varepsilon, 2\varepsilon]$, and centered in an ε-net.

converges weakly as r tends to infinity to a normal variable with variance

$$\sigma^2(x) = \lim_{r\to\infty}(c_d r^d)^{-k} E\{N^2_{B_r(x_1-x_k)}\cdots N^2_{B_r(0)} - \nu^2_{k,r}(x)\}$$

and for all pairs of distinct points (x_i, x_j) in $\mathbb{R}^{2d}$, the covariance between $N_{B_r(x_i)}$ and $N_{B_r(x_j)}$ is defined by $varN_{B_r(0)\cap B_r(x_i-x_j)}$, from (6.9). The covariances differ only by the differences of the centers of the balls and $cov(N_{B_r(x_i)}, N_{B_r(x_j)}) = 0$ if $|x_i - x_j| > 2r$.
The variables $X_{k,r}(x)$, where $\varepsilon = |x_i - x_j| > 2r$, satisfy

$$P(\|X_{k,r}(x)\|_k > a) = \frac{1}{a^2 c_d r^d} \sum_{i=1}^{k} varN_{B_r(x_i-x_k)}$$

$$= \frac{1}{a^2 c_d r^d} \sum_{i=1}^{k} \Lambda(B_r(x_i - x_k)),$$

for all $a > 0$ and $k \geq 2$. The bound is modified according to the domain of r with respect to ε. For all $a > 0$ and r in $\mathbb{R}_+$

$$P\{\sup_{x\in\mathbb{R}^d}(c_d r^d)^{-\frac{1}{2}}|N_{B_r(x)} - \Lambda(B_r(x))| > a\}$$

$$= P\{(c_d r^d)^{-\frac{1}{2}}|N_{B_r(0)} - \Lambda(B_r(0))| > a\} \leq \frac{\Lambda(B_r(0))}{a^2 c_d r^d}.$$

The function $\Lambda(B_r(0))$ is increasing with r and under the condition of the convergence of $r^{-d}\Lambda(B_r(0))$ to a limit λ

$$\lim_{r \to \infty} P\{\sup_{x \in \mathbb{R}^d}(c_d r^d)^{-\frac{1}{2}}|N_{B_r(x)} - \Lambda(B_r(x))| > a\} \leq \frac{\lambda}{a^2 c_d}. \tag{6.10}$$

Proposition 6.7. *The spatial process* $(c_d r^d)^{-\frac{1}{2}}\{N_{B_r(x)} - \Lambda(B_r(x)\}_{x \in \mathbb{R}^d}$ *converges weakly to a centered Gaussian process with variance function* $(c_d r^d)^{-1}\Lambda(B_r(x))$.

Equation (6.10) with a sufficiently large proves the tightness of the process and the weak convergence of its finite dimensional distributions as r tends to infinity proves the result.

Schladitz and Baddeley (2000) give the expression of second and third order characteristics of stationary point processes, with explicit formulae for the Poisson with intensity λ and for other parametric processes. For $d = 2$ it is defined as the expected mean number of points of the process contained in $B_r(0)$ and not farther than 1

$$T_2(r) = (2\lambda^2|B_r(0)|^2)^{-1}E\int_{B_r^2(0)} 1_{\{\|x-y\| \leq 1\}} \, dN(x) \, dN(y)$$

$$= (2\lambda^2|B_r(0)|^2)^{-1}E\sum_{X_i \in B_r(0)} N_{B_r(0) \cap B_1(X_i)},$$

or with a correction to avoid multiple countings of the centers of the balls

$$\mu_2(r) = (2\lambda^2|B_r(0)|^2)^{-1}EN_2(r),$$

$$N_2(r) = \sum_{X_i \in B_r(0)} N_{\{B_r(0) \cap B_1(X_i)\} \setminus \{X_i\}}.$$

For $d = 3$, the third order mean number of points is

$$T_3(r) = \frac{1}{2\lambda^3|B_r(0)|^3}E\int_{B_r^3(0)} 1_{\{\|x-y\| \leq 1\}}1_{\{\|y-z\| \leq 1\}}$$

$$1_{\{\|x-z\| \leq 1\}} \, dN(x) \, dN(y) \, dN(z)$$

and it equals

$$T_3(r) = \frac{1}{2\lambda^3|B_r(0)|^3}E\int_{B_r^2(0)} N_{B_1(x) \cap B_1(y) \cap B_r(0)} \, dN_{B_1(x)}(y) \, dN(x)$$

$$= \frac{1}{2\lambda^3|B_r(0)|^3}E\sum_{X_i \in B_r(0)}\sum_{X_j \in B_1(X_i) \cap B_r(0)} N_{B_1(X_i) \cap B_1(X_j) \cap B_r(0)}.$$

With a correction for the centers of the balls, it becomes

$$\mu_3(r) = \frac{1}{2\lambda^3 |B_r(0)|^3} E N_3(r),$$

$$N_3(r) = \sum_i [1_{B_r(0)}(X_i) \sum_{j \neq i} \{1_{\{B_1(X_i) \cap B_r(0)\}\}}(X_j)$$

$$\sum_{k \neq j, i} 1_{\{B_1(X_i) \cap B_1(X_j) \cap B_r(0)\}}(X_k)\}].$$

The functions $T_2(r)$ and $T_3(r)$ are mean convolutions of stochastic measures and they cannot be compared to the mean products $E \prod_{i=1}^{k} N_{B_r(x_i)}$ studied previously. Higher order measures are easily defined in the same way. The processes $N_2(r)$ and $N_3(r)$ cannot be split into sums of the counting processes in disjoint subsets and a condition of mixing, as r increases, is not relevant except for sparse processes.

The same measures can be calculated for a heterogeneous Poisson process with a continuous cumulative intensity measure Λ in $\mathbb{R}^d$. The conditional means of the spatial counting processes

$$A_r = \sum_i 1_{B_r(0)}(X_i),$$

$$C_r(X_i) = \sum_{j \neq i} 1_{\{B_1(X_i) \cap B_r(0)\}\}}(X_j),$$

$$D_r(X_i, X_j) = \sum_{k \neq j, i} 1_{\{B_1(X_i) \cap B_1(X_j) \cap B_r(0)\}}(X_k)$$

have the values

$$a_r = \sum_i P\{X_i \in B_r(0)\} = \Lambda(B_r(0)),$$

$$c_r(X_i) = \sum_{j \neq i} P\{X_j \in B_1(X_i) \cap B_r(0)|X_i\}$$
$$= \Lambda(B_1(X_i) \cap B_r(0)),$$

$$d_r(X_i, X_j) = \sum_{k \neq j, i} P\{X_k \in B_1(X_i) \cap B_1(X_j) \cap B_r(0)|X_i, X_j\}$$
$$= \Lambda(B_1(X_i) \cap B_1(X_j) \cap B_r(0)),$$

therefore

$$\mu_2(r) = \frac{1}{2\Lambda^2(B_r(0))} \int_{B_r(0)} \Lambda(B_r(0) \cap B_1(x)) \, d\Lambda(x),$$

$$\mu_3(r) = \frac{1}{2\Lambda^3(B_r(0))} \int_{B_r^2(0)} \Lambda(B_1(x) \cap B_1(y) \cap B_r(0)) \, d\Lambda(y) \, d\Lambda(x).$$

The normalized processes

$$W_{r,A} = \{2\lambda^2(B_r(0))\}^{-1}(A_r - a_r)\{B_r(0)\},$$
$$W_{r,C} = \{2\lambda^2(B_r(0))\}^{-1}(C_r - c_r)\{B_r(0)\},$$
$$W_{r,D} = \{2\lambda^2(B_r(0))\}^{-1}(C_r - c_r)\{B_r(0)\}$$

converge weakly to centered Gaussian processes having variances given by Proposition 6.7. This implies the weak convergence of the normalized and centered processes

$$\{\{2\lambda^2(B_r(0))\}^{-1}\int_{B_r(0)}\{C_r(x,dy)\,A_r(dx) - \mu_2(r)\},$$

$$\{\{2\lambda^3(B_r(0))\}^{-1}\int_{B_r^3(0)}\{D_r(x,y)\,C_r(x,dy)\,A_r(dx) - \mu_3(r)\}$$

to the Gaussian process $\int_{B_r(0)} W_C\,dA + \int_{B_r(0)} C\,dW_A$ and $\int_{B_r^2(0)} W_D\,dC\,dA + \int_{B_r(0)} D\,dW_C\,dA + \int_{B_r(0)} D\,dC\,dW_A$, respectively.

Chapter 7

Inequalities in Complex Spaces

7.1 Introduction

Trigonometric series with real coefficients a_k and b_k are written as

$$S(t) = a_0 + \sum_{k=1}^{\infty} \{a_k \cos(k\omega t) + b_k \sin(n\omega t)\},$$

they are periodic functions with period $T = 2\pi\omega^{-1}$. Let f be a periodic and locally integrable function with period T, then it develops as a trigonometric series with coefficients

$$a_k = \frac{2}{T} \int_{-\frac{T}{2}}^{\frac{T}{2}} f(x) \cos(\frac{2\pi}{T} kx) \, dx, \ k \neq 0,$$

$$b_k = \frac{2}{T} \int_{-\frac{T}{2}}^{\frac{T}{2}} f(x) \sin(\frac{2\pi}{T} kx) \, dx,$$

$$a_0 = \frac{1}{T} \int_{-\frac{T}{2}}^{\frac{T}{2}} f(x) \, dx.$$

A periodic odd function $f : [-T, T] \to \mathbb{R}$ develops as a sine series and a periodic odd function develops as a cosine series. The Fourier representation of a periodic function f with period $T = 2\pi\omega^{-1}$ and coefficients λ_k is

$$f(x) = \sum_{k=-\infty}^{\infty} \lambda_k e^{i\omega kx}, \tag{7.1}$$

$$\lambda_k = \frac{1}{T} \int_{-\frac{T}{2}}^{\frac{T}{2}} e^{-i\omega kt} f(t) \, dt.$$

Parseval's equality for the L_2 norm of the Fourier transform is

$$\|f\|_2 = \{\sum_{k \in \mathbb{Z}} |\lambda_k|^2\}^{\frac{1}{2}}. \tag{7.2}$$

179

This is a consequence of the Fourier representation of the function with a period T and of the integral $(2T)^{-1} \int_{-T}^{T} e^{i(m-n)wx} \, dx = 1_{\{m=n\}}$.

The expansions of functions as trigonometric series have produced noticable equalities. For example, for every integer m and for x in $[0, \frac{\pi}{2}]$

$$\int_0^x \frac{\sin(mx)}{\sin x} \, dx = \mathcal{R}e\{\int_0^x e^{i(m-1)x} \frac{1 - e^{-2imx}}{1 - e^{-2ix}} \, dx\}$$

$$= \sum_{k=0}^{m} \frac{\sin(m - 2k)x)}{m - 2k},$$

with the notation $\frac{\sin 0}{0} = 0$. Therefore

$$\frac{1}{2} \int_0^{\frac{\pi}{2}} \frac{\sin(mx)}{\sin x} \, dx = 0, \quad \text{for every even } m,$$

$$= 1 - \frac{1}{3} + \frac{1}{5} - \frac{1}{7} + \cdots + \frac{1}{m}, \quad \text{for every odd } m.$$

So the integral is finite, with

$$\lim_{m \to \infty} \int_0^{\frac{\pi}{2}} \frac{\sin(2m + 1)x}{\sin x} \, dx = \frac{\pi}{2},$$

using the limit of the series proved in Section 3.1.

Applying functional equations to trigonometric series allows us to solve some of them. For example, the equation $f(x + y) - f^2(x) - f^2(y) + 1 = 0$, where the function $f : \mathbb{R} \mapsto [-1, 1]$ satisfies $f(0) = 1$, $f^{(1)}(0) = 0$ and $f^{(2)}(0) = -2a$, has a solution f defined by $f(x) = \cos(\sqrt{2a}x)$ if and only if $y = x + \frac{k\pi}{\sqrt{2a}}$, k in $\mathbb{Z}$. The functions u that are solutions of the differential equations

$$\frac{d^2 u}{dx^2} \pm \alpha^2 u = 0,$$

$$x \frac{d^2 u}{dx^2} + 2 \frac{du}{dx} - k(k + 1)u = 0,$$

and defined as sums of power functions or as a trigonometric series satisfy polynomial equations depending on the coefficients of the series. They can be solved iteratively, which yield explicit expressions of the solutions. For Fourier's equation

$$\frac{\partial^2 v}{\partial x^2} + \frac{\partial^2 v}{\partial y^2} + \frac{\partial^2 v}{\partial z^2} = 0,$$

expansions of the solution in Fourier series yield explicit solutions (Fourier, 1822). This is not true for all equations and other expansions have been defined such as the development in series defined by the projection of the solution on other functional basis. Legendre's basis of polynomials provides the solution of Legendre's equations for modeling the curve of comets. Several classes of polynomials have been defined for the expansion of solutions of differential equations (Legendre, 1805 Fourier, 1822 Byerly, 1893).

In $\mathbb{C}$, the series are defined by two real sequences

$$u_n = v_n + iw_n = \rho_n(\cos\theta_n + i\sin\theta_n),$$

where $(v_n)_{n\geq 0}$ and $(w_n)_{n\geq 0}$ are real series, $(\rho_n)_{n\geq 0}$ and $(\theta_n)_{n\geq 0}$ are defined by the change of variables

$$\rho_n = (u_n^2 + v_n^2)^{\frac{1}{2}}$$

and θ_n is defined modulo $2k\pi$ if $u_n > 0$ and modulo $(2k+1)\pi$ if $u_n < 0$ by

$$\theta_n = \arctan(w_n v_n^{-1}).$$

Reciprocally, the trigonometric functions are expressions of the formulæ

$$e^{i\theta_n} = \cos\theta_n + i\sin\theta_n \text{ and } e^{-i\theta_n} = \cos\theta_n - i\sin\theta_n,$$

the logarithm of a complex number $u_n = v_n + iw_n$ is

$$\log u_n = \log\rho_n + i(\theta_n \pm 2k\pi), \quad \text{if } v_n > 0,$$

$$= \log\rho_n + i(\theta_n \pm (2k+1)\pi), \quad \text{if } v_n < 0.$$

For all real x and integer m, the equality $e^{mix} = (e^{ix})^m$ implies

$$\cos(mx) + i\sin(mx) = (\cos x + i\sin x)^m.$$

A necessary and sufficient condition for the convergence of $\sum_{k=0}^{\infty} u_k$ is the convergence of both series $\sum_{k=0}^{\infty} v_k$ and $\sum_{k=0}^{\infty} w_k$ or, equivalently, the convergence of $\sum_{k=0}^{\infty} \rho_k$. Conditions for their convergence is $\rho_{n+1}\rho_n^{-1} < 1$ for every n larger some integer n_0 or $\rho_n^{\frac{1}{n}} < 1$ for every n larger some integer n_0. If one of these ratios remains larger than 1, the series diverges. Inequalities for complex series are inequalities in the two-dimensional space $\mathbb{C}$, for the real series $(v_n)_{n\geq 0}$ and $(w_n)_{n\geq 0}$ of $u_n = v_n + iw_n$ and they entail inequalities for ρ_n and θ_n.

Cauchy's inequality is written with the scalar product in $\mathbb{C}$, where for all complex numbers z_1 and z_2

$$\|z_1 + z_2\|^2 \geq 4(z_1, \bar{z}_2),$$

with the scalar product (1.1), it is equivalent to

$$0 \leq \|z_1 + z_2\|^2 - 4 < z_1, \bar{z}_2 > = \|z_1\|^2 + \|z_2\|^2 - 2 < z_1, \bar{z}_2 > = \|z_1 - z_2\|^2.$$

The geometric equalities (1.2) are deduced from this definition. The Cauchy-Schwarz and the Minkowski inequalities are still valid.

7.2 Polynomials

The algebraic numbers are the roots of polynomials with integer coefficients. Dividing by the coefficient of the higher degree of the variable, this extends to rational coefficients.

The class of polynomials in x with coefficients in $\mathbb{C}$ is denoted by $\mathbb{C}[x]$. Gauss-d'Alembert's theorem states that every polynomial of $\mathbb{C}[x]$ has at least a complex root. Let P_k be a polynomial of degree k in $\mathbb{C}[x]$, having a complex root z_0, it is written in the form $P_k(z) = (z - z_0)P_{k-1}(z)$ and a necessary condition ensuring that P_k has at least two complex roots is: P_{k-1} belongs to $\mathbb{C}[x]$. These conditions are not necessarily satisfied.

The roots of the equation $x^n = a + ib$ are the n values $x_{n,k} = \rho^{\frac{1}{n}} e^{\frac{2k\pi}{n}}$, with $k = 1, \ldots, n$ if $a > 0$, and $x_n = \rho^{\frac{1}{n}} e^{\frac{(2k+1)\pi}{n}}$, with $k = 0, \ldots, n-1$ if $a < 0$, and $\rho = (a^2+b^2)^{\frac{1}{2}}$. Writing $x^n - 1 = (x-1)(x^{n-1}+x^{n-2}+\cdots+x+1)$, it follows that the $n-1$ roots of the equation $(x^{n-1}+x^{n-2}+\cdots+x+1) = 0$ are $x_{n,k} = \rho^{\frac{1}{n}} e^{\frac{2k\pi}{n}}$, for $k = 1, \ldots, n-1$ and they belong to $\mathbb{C}$. The cubic root of i is $\sqrt[3]{i} = -i$, its fifth root is $\sqrt[5]{i} = i$, $\sqrt[7]{i} = -i$, etc., $\sqrt[3]{-i} = i$, $\sqrt[5]{-i} = -i$, $\sqrt[7]{-i} = i$, etc.

Proposition 7.1. *The odd roots of i and $-i$ are cyclic in $\mathbb{C}$*

$$^{2k+1}\!\sqrt{i} = -i, \quad \text{if } k \text{ is odd,} \qquad ^{4k+1}\!\sqrt{i} = i,$$
$$^{2k+1}\!\sqrt{-i} = i, \quad \text{if } k \text{ is odd,} \qquad ^{4k+1}\!\sqrt{-i} = -i.$$

The even roots of i belong to $\mathbb{C}$, in particular

$$\sqrt{i} = \pm\frac{1+i}{\sqrt{2}},$$

$$\sqrt[4]{i} = \pm\frac{\sqrt{\sqrt{2}+1}+i\sqrt{\sqrt{2}-1}}{\sqrt{2\sqrt{2}}},$$

$$\sqrt[6]{i} = \pm\frac{1-i}{\sqrt{2}}.$$

Proof. Writing $\sqrt[4]{i} = a + ib$ implies $a^2 - b^2 + 2iab = \sqrt{i}$, hence $\sqrt{2}(a^2 + b^2) = 1$ and $2\sqrt{2}ab = 1$, then b^2 is a zero of $P(x) = 8x^4 + 4\sqrt{2}x^2 - 1 = 0$ in $\mathbb{R}[x^2]$ and P has one positive real root. Let $z = a + ib = \sqrt[6]{i}$, a and b are deduced from $z^2 = -i$ which implies $a = -b$ and $2ab = -2a^2 = -1$. $\square$

As a consequence, all integer roots of -1 belong to $\mathbb{C}$. It follows that $\sqrt[x]{-1}$ belongs to $\mathbb{C}$, for every real x.

A polynomial of $\mathbb{R}[x]$ having a complex root z has not necessarily the root $\bar{z}$ as proved by the next example. Bernoulli (1742) stated that the polynomial $x^4 - 4x^3 + 2x^2 + 4x + 4$ has four complex roots $1 \pm \sqrt{2 \pm \sqrt{-3}}$ (Bradley, d'Antonio and Sandifer, 2007). The same method applies for the factorization of 4-th degree polynomials of $\mathbb{R}[x]$ of the form

$$\alpha\{x^4 - 4ax^3 + 2x^2(3a^2 - b) - 4ax(a^2 - b) + (a^2 - b)^2 + c\}$$
$$= \alpha \prod (x - a \pm \sqrt{b \pm \sqrt{-c}})$$

with real numbers a, b and c. The coefficients of x^2 and x of the polynomial are not free and this factorization is not general. It can be extended as a product of four complex roots of the form $\pm a \pm \sqrt{\pm b \pm \sqrt{c}}$ with complex constants but the 4-th degree polynomials of $\mathbb{R}[x]$ are not always in this form.

A third degree polynomial of $\mathbb{R}[x]$ with real roots a, b and c has the form $P(x) = \alpha(x - a)(x - b)(x - c)$ or $P(x) = \alpha(x^3 - Sx^2 + S_2X - P)$ with $S = a + b + c$, $S_2 = ab + bc + ac$, $P = abc$. If $S^2 - 3S_2 \geq 0$, $P(x)$ has three real roots, then a root is between $x_1 = \frac{1}{3}\{S - (S^2 - 3S_2)^{\frac{1}{2}}\}$ and $x_2 = \frac{1}{3}\{S + (S^2 - 3S_2)^{\frac{1}{2}}\}$, another one is smaller than x_1 and the third one is larger than x_2. If $P(x)$ has a double root a, its factorization is easily calculated by solving the equations $S = 2a + b$, $S_2 = a^2 + 2ab$, $P = a^2b$.

If $S^2 - 3S_2 < 0$, $P(x)$ has only one real root and two conjugate complex roots. A polynomial $P(x) = (x - a - \sqrt{b})(x - a + \sqrt{b})(x - c) = x^3 - (2a + c)x^2 + (2ac + a^2 - b)x - (a^2 - b)c$ with real or complex roots can be factorized from the specific expressions of the sums and of the product. There is no general method for the factorization of the polynomials of degree larger than two.

7.3 Fourier and Hermite transforms

A complex function from $\mathbb{R}$ to $\mathbb{C}$ has the form $F(x) = f(x) + ig(x)$ where f and g are real functions defined on the same subset of $\mathbb{R}$. Its complex conjugate is $\bar{F}(x) = f(x) - ig(x)$. Its Euclidean norm is

$$\|F(x)\| = F(x)\bar{F}(x) = (f^2(x) + g^2(x))^{\frac{1}{2}}$$

and its norm $\ell_2(\mu, \mathbb{R})$ is $\|F\|_2 = \{\int_{\mathbb{R}}(f^2 + g^2)\,d\mu\}^{\frac{1}{2}}$.

Let f and g be functions from $\mathbb{C}$ to $\mathbb{C}$ provided with the product Lebesgue, the norm of f is $\|f\|_2 = \int_{\mathbb{C}} f(z)\bar{f}(z)\,d\mu(z)$ and the scalar product of f and g is

$$< f,g > = \int_{\mathbb{C}} f(z)\bar{g}(z)\,d\mu(z) = \int_{\mathbb{C}} g(z)\bar{f}(z)\,d\mu(z),$$

the conjugate of $f\bar{g}$ is $\bar{f}g$. Cauchy's inequality applies and for all functions f and $g : \mathbb{C} \to \mathbb{C}$

$$2\{\int_{\mathbb{C}} f(z)\bar{g}(z)\,d\mu(z)\}^{\frac{1}{2}} \leq \|f + g\|_2 \leq \|f\|_2 + \|g\|_2.$$

The Fourier transform of a convolution is the product of the Fourier transforms.

A necessary and sufficient condition for the differentiability of a complex function $f(x + iy) = f_1(x + iy) + if_2(x + iy)$ on $\mathbb{C}$ is Cauchy's condition

$$\frac{\partial f_1}{\partial x} = \frac{\partial f_2}{\partial y}, \qquad \frac{\partial f_2}{\partial x} = -\frac{\partial f_1}{\partial y}$$

then the derivative of f at $z = (x + iy)$ is

$$f^{(1)}(x + iy) = \frac{\partial f_1}{\partial x}(z) + i\frac{\partial f_2}{\partial x} = \frac{\partial f_2}{\partial y} - i\frac{\partial f_1}{\partial y}.$$

Lipschitz's condition can be extended to a complex function f on $\mathbb{C}$: for all z and z', the m-th order derivative of f satisfies

$$\|f^{(m)}(z) - f^{(m)}(z')\|_2 \leq k\|z - z'\|_2$$

for a constant k. The condition is equivalent to

$$[\{\frac{\partial^m f_1(z)}{\partial x^m} - \frac{\partial^m f_1(z')}{\partial x^m}\}^2 + \{\frac{\partial^m f_2(z)}{\partial x^m} - \frac{\partial^m f_2(z')}{\partial x^m}\}^2]^{\frac{1}{2}}$$
$$= [\{\frac{\partial^m f_1(z)}{\partial y^m} - \frac{\partial^m f_1(z')}{\partial y^m}\}^2 + \{\frac{\partial^m f_2(z)}{\partial y^m} - \frac{\partial^m f_2(z')}{\partial y^m}\}^2]^{\frac{1}{2}}$$
$$\leq k\|z - z'\|_2.$$

Let X be a real random variable with distribution function F, its Fourier transform is the function with values in $\mathbb{C}$

$$\varphi_X(t) = \int_{-\pi}^{\pi} e^{-itx}\,dF(x).$$

Let f be an even function having a finite Fourier transform $\mathcal{F}f$, then all odd moments of f are zero and φ_X develops as

$$\varphi_X(t) = \sum_{k \geq 0} \frac{(-t^2)^k}{(2k)!} E(X^{2k}).$$

Conversely, all even moments of an odd function are zero and, when it is finite, its Fourier transform develops as

$$\varphi_X(t) = i \sum_{k \geq 1} (-1)^{k+1} \frac{t^{2k+1}}{(2k+1)!} E(X^{2k+1}).$$

The derivative of a differentiable function f with period $T = 2\pi w^{-1}$ is written as $f'(x) = \sum_{k \in \mathbb{Z}} \lambda_k(f') e^{i\omega k x}$ where the coefficients are obtained by integration by parts

$$\lambda_k(f) = \frac{1}{T} \int_{-\frac{T}{2}}^{\frac{T}{2}} e^{-i\frac{2\pi}{T}kt} f(t)\, dt = \frac{1}{2ki\pi} \int_{-\frac{T}{2}}^{\frac{T}{2}} e^{-i\frac{2\pi}{T}kt} f'(t)\, dt = \frac{1}{iwk} \lambda_k(f')$$

hence $\lambda_k(f') = iwk\lambda_k(f)$ and its norm $L_2([0,T])$ satisfies

$$\|f'\|_2 = 2\pi T^{-1} \{\sum_{k \in \mathbb{Z}} k^2 \lambda_k^2\}^{\frac{1}{2}}.$$

If f is periodic and belongs to $C_1[0,T] \cap L_2([0,T])$, the coefficients $\lambda_k(f)$ tend to zero as k tends to infinity and the approximation

$$f_n(x) = \sum_{k=1}^{n} \lambda_k e^{iwkt}$$

of the function f by its Fourier series converges in L_2, $\lim_{n \to \infty} \|\widehat{f_n} - f\|_2 = 0$.

If φ_X is $L_1(\mathbb{C})$ and F has a density, it equates to the inverse of the Fourier transform at every continuity point x of f

$$f(x) = \int_{\mathbb{R}} e^{iwtx} \varphi_X(t)\, dt = \frac{1}{T} \int_{-\frac{T}{2}}^{\frac{T}{2}} \int_{\mathbb{R}} e^{iw(t-s)x}\, dF(s)\, dt,$$

by Proposition A.4, and the inverse Fourier series has the value $\frac{1}{2}\{f(x^+) + f(x^-)\}$ at every point x where the function f is discontinuous.

Example 7.1. The Fourier transform of the normal density is the function defined on $\mathbb{R}$ by $\varphi(x) = e^{-\frac{x^2}{2}}$. The inverse of the Fourier transform of the normal density f is

$$f(x) = \int_{\mathbb{R}} e^{itx} e^{-\frac{x^2}{2}}\, dt = \int_{\mathbb{R}} \int_{\mathbb{R}} e^{i(t-s)w} f(s)\, ds\, dt.$$

The transform of a normal variable with mean μ and variance σ^2 is $e^{ix\mu} e^{-\frac{x^2\sigma^2}{2}}$.

Example 7.2. The function $(ax)^{-1}\sin(ax)$, with a constant $a > 0$, is the Fourier transform of the uniform density on $[-a, a]$. Using the inverse transform yields

$$\frac{1}{2\pi}\int_{-\pi}^{\pi}\frac{\sin(ax)}{ax}e^{-itx}\,dx = \frac{1}{2a}\frac{1}{2\pi}\int_{-\pi}^{\pi}\int_{-a}^{a}e^{-itx}e^{its}\,dx\,ds$$

$$= \frac{1}{at^2\pi}\sin(t\pi)\sin(ta)$$

and it differs from $f(t) = a^{-1}1_{]-a,a[}(t)$.

Let f and g be periodic and integrable functions defined from $[-T, T]$ to $\mathbb{R}$ or $\mathbb{C}$, the expression of their Fourier transform (7.1) provides a scalar product for f and g

$$T^{-1}\int_{-T}^{T}f(x)\bar{g}(x)\,dx = \sum_{n\geq 0}\lambda_n(f)\bar{\lambda}_n(g).$$

Parseval's equality (7.2) for the Fourier transform $\widehat{f}$ of a function f of L_2 is a consequence of this expression

$$\|f\|_2 = \|\widehat{f}\|_2 = \sum_{n\geq 0}\lambda_n^2(f).$$

The operator T_g is defined as $T_g(f) = \int_{-T}^{T}f(x)g(x)\,dx$ for periodic integrable functions f and g on $[-T, T]$. Let g be a periodic function on $[-T, T]$ with Fourier transform $\widehat{g}$, it defines the operator $\widehat{T}_g = T_{\widehat{g}}$. Developing the functions f and g as a Fourier series S_f and S_g yields

$$\int f(x)\bar{S}_g(x)\,dx = \sum_{k\geq 0}a_k(f)\bar{a}_k(g) + \sum_{k\geq 0}b_k(f)\bar{b}_k(g) = \int S_f(x)\bar{g}(x)\,dx,$$

from Fubini's theorem. The scalar product of a function and the Fourier transform of another one is a symmetric operator

$$\widehat{T}_g(f) = T_{\widehat{g}}(f) = T_g(\widehat{f}). \tag{7.3}$$

With the normal density $f_{\mathcal{N}}$ and a function g on $[-T, T]$, Equation (7.3) becomes

$$\widehat{T}_g(f_{\mathcal{N}}) = T_{\widehat{g}}(f_{\mathcal{N}}) = \frac{1}{2T}\int_{-T}^{T}g(x)e^{-\frac{x^2}{2}}\,dx = \frac{1}{2T}\int_{-T}^{T}g(x)\widehat{f}(x)\,dx. \tag{7.4}$$

Corollary 7.1. *Let g be a function of $L_2([-T, T])$, a normal variable X satisfies*

$$E\{g(X)1_{|X|\leq T}\} = \frac{T}{\sqrt{2\pi}}\int_{-\infty}^{\infty}\widehat{g}(x)f_{\mathcal{N}}(x)\,dx,$$

in particular

$$P(|X| \le a) = \frac{2}{\sqrt{2\pi}} E\{\frac{\sin(aX)}{X}\} = 1 - 2e^{-\frac{a^2}{2}}.$$

Proof. This is a consequence of the equality (7.3) with the normal and uniform densities (Examples 7.1 and 7.2)

$$\frac{1}{\sqrt{2\pi}} \int_{\mathbb{R}} \frac{\sin(ax)}{ax} e^{-\frac{x^2}{2}} \, dx = \frac{1}{2a} \int_{-a}^{a} e^{-\frac{x^2}{2}} \, dx = \frac{\sqrt{2\pi}}{2a} P(|X| \le a)$$

for a variable X with the normal distribution $\mathcal{N}(0,1)$. Chernov's theorem for the normal variable X implies $1 - 2e^{-\frac{a^2}{2}} = P(|X| \le a)$. $\square$

Hermite's polynomials are real functions defined $\mathbb{R}$ from the derivatives of the normal density function $f_{\mathcal{N}}$ as

$$H_k(t) = (-1)^k \frac{d^k e^{-\frac{t^2}{2}}}{dt^k} e^{\frac{t^2}{2}} = \frac{d^k f_{\mathcal{N}}(x)}{dx^k} \{f_{\mathcal{N}}(x)\}^{-1}, \tag{7.5}$$

they are symmetric. Let $\varphi(t) = e^{\frac{-t^2}{2}}$ be Fourier's transform of the normal density, by the inversion formula

$$\frac{d^k f_{\mathcal{N}}(x)}{dx^k} = H_k(x) f_{\mathcal{N}}(x) = \frac{1}{2\pi} \int_{-\pi}^{\pi} (it)^k e^{itx} \varphi(t) \, dt, \tag{7.6}$$

or by the real part of this integral. Hermite's polynomials are recursively defined by $H_0 = 1$ and $H_{k+1} = xH_k - H_k'$, for every $k > 2$, hence H_k is a polynomial of degree k, with higher term x^k. With this definition, the functions H_k are orthogonal in $(L_2(\mathbb{R}), \mu_{\mathcal{N}})$ with the normal distribution $\mu_{\mathcal{N}}$, and the inequalities of the Hilbert spaces apply to the basis $(H_k)_{k \ge 0}$. From (7.5), their norm $c_k = \|H_k\|_2$ is defined by

$$c_k^2 = \int_{\mathbb{R}} f_{\mathcal{N}}(x) H_k^2(x) \, dx = \sqrt{2\pi} \int_{\mathbb{R}} \{f_{\mathcal{N}}^{(k)}(x)\}^2 e^{\frac{t^2}{2}} \, dx$$

and it is calculated from the even moments of the normal distribution, $c_k = (k!)^{\frac{1}{2}}$ (Section A.4). The Hermite transform $H_f = \sum_{k \ge 0} a_k(f) H_k$ of a function f has the coefficients

$$a_k(f) = \frac{1}{c_k} \int_{\mathbb{R}} f(x) H_k(x) f_{\mathcal{N}}(x) \, dx = \frac{1}{c_k} \int_{\mathbb{R}} f(x) f_{\mathcal{N}}^{(k)}(x) \, dx$$

if these integrals are finite. The scalar product of the Hermite expansions of functions f and g in $(L_2(\mathbb{R}), \mu_{\mathcal{N}})$ is written as $E\{H_f(X) H_g(X)\}$, with a normal variable X, and it develops as $\int_{\mathbb{R}} H_f H_g \, d\mu_{\mathcal{N}} = \sum_{k \ge 0} k! \, a_k(f) a_k(g)$.

A condition for the convergence of the transform H_f of a function f is the convergence of the series

$$\|H_f\|_2^2 = \sum_{k \geq 1} k! \, a_k^2(f).$$

For every $k \geq 0$

$$\int_{\mathbb{R}} f(x) c_k^{-1} H_k(x) f_{\mathcal{N}}(x) \, dx = \int_{\mathbb{R}} c_k^{-2} H_f(x) H_k(x) f_{\mathcal{N}}(x) \, dx.$$

A function f is then equal to its normalized Hermite expansion.

The generating functions related to the polynomials H_k provide translations of the normal density and therefore convolutions.

Proposition 7.2. *For every w such that $|w| < 1$*

$$f_{\mathcal{N}}(x + w) = \sum_{k \geq 0} H_k(x) f_{\mathcal{N}}(x) \frac{w^k}{k!},$$

$$\sum_{k \neq j \geq 0} H_k(t) H_j(t) e^{-t^2} \frac{w^{j+k}}{(j+k)!} = \sum_{k \neq j \geq 0} \frac{d^{j+k} e^{-\frac{t^2}{2}}}{dt^{j+k}} \frac{w^{j+k}}{(j+k)!}$$

$$= \sqrt{2\pi} f_{\mathcal{N}}(t + w).$$

For every function g of $(L_2(\mathbb{R}), \mu_{\mathcal{N}})$

$$\sum_{k \geq 0} \int_{\mathbb{R}} g(x) H_k(x) f_{\mathcal{N}}(x) \frac{w^k}{k!} \, dx = \int_{\mathbb{R}} g(x - w) f_{\mathcal{N}}(x) \, dx.$$

A normal variable X then satisfies

$$Eg(X - w) = \sum_{k \geq 0} E\{g(X) H_k(X)\} \frac{w^k}{k!}$$

$$= \sum_{k \neq j \geq 0} E\{g(X) H_k(X) H_j(X) e^{\frac{-X^2}{2}}\} \frac{w^{j+k}}{(j+k)!}$$

$$= \sqrt{2\pi} \sum_{k \neq j \geq 0} E\{g(X) f_{\mathcal{N}}^{(k)}(X) f_{\mathcal{N}}^{(j)}(X)\} \frac{w^{j+k}}{(j+k)!}$$

for every function g such that $g(X)$ has a finite variance.

The derivatives of the Hermite transform of a function g of $(L_2(\mathbb{R}), \mu)$ are written as $g^{(j)} = \sum_{k \geq 0} a_k H_k^{(j)}$ where $H_k^{(1)}(x) = x H_k(x) - H_{k+1}(x)$ and the derivative of order j of H_k is a polynomial of degree $k+j$ recursively written in terms of $H_k, \ldots, H_{k+j}$ in the form $H_k^{(j)}(x) = \sum_{i=0}^{j} P_{ik}(x) H_{k+i}(x)$, where P_{ik} is a polynomial of degree $j - i$.

Proposition 7.3. *The Fourier transform of the Hermite functions H_k has the expansion*

$$\widehat{H}_k(t) = \sum_{j \geq k+1} \frac{(-it)^{j-k}}{(j-k)!} e^{\frac{t^2}{2}} \int_{\mathbb{R}} H_j(x + it)\, dx, \ k \geq 1.$$

Proof. Applying Proposition 7.2 to a function f of $(L_2(\mathbb{R}), \mu_{\mathcal{N}})$, its derivatives are written as

$$f^{(k)}(x + w) = \sum_{j \geq k+1} f^{(j)}(x) \frac{w^{j-k}}{(j-k)!} = \sum_{j \geq k+1} H_j(x) f_{\mathcal{N}}(x) \frac{w^{j-k}}{(j-k)!}$$

and the Fourier transform of H_k defined by (7.5) is

$$\widehat{H}_k(t) = \int_{\mathbb{R}} H_k(x) e^{-itx}\, dx = \sqrt{2\pi} e^{\frac{t^2}{2}} \int_{\mathbb{R}} f^{(k)}(x) e^{\frac{(x-it)^2}{2}}\, dx$$

$$= e^{\frac{t^2}{2}} \int f^{(k)}(y + it) f_{\mathcal{N}}^{-1}(y)\, dy$$

where the last integral stands for $y = x - it$ at fixed t. From the expansion of the derivative $f^{(k)}(y + it)$, it is also written as

$$\sum_{j \geq k+1} \frac{(it)^{j-k}}{(j-k)!} \int_{\mathbb{R}} H_j(x - it)\, dx$$

and it is a complex function of t. □

The Hermite series of odd functions are zero, by symmetry of the polynomials H_k. The Hermite transform of uniform variables is also zero.

Theorem 7.1. *The Hermite transform of the normal density is the series*

$$H_{f_{\mathcal{N}}}(x) = \sum_{k \geq 0} a_{2k}(f_{\mathcal{N}}) H_{2k}(x),$$

its coefficients are strictly positive

$$a_{2k}(f_{\mathcal{N}}) = \frac{1}{\sqrt{(2k)!}} \int_{\mathbb{R}} \{f_{\mathcal{N}}^{(2k)}(x)\}^2\, dx.$$

Proof. The Hermite transform of the normal density is defined by the integrals $\int_{\mathbb{R}} f_{\mathcal{N}}^2(x) H_k(x)\, dx = a_k(f_{\mathcal{N}}) c_k$, for every $k \geq 0$. Integrating by parts the expansion of the function $H_{f_N}(x) = \sum_{k \geq 0} a_k(f_N) H_k(x)$, where $f_N H_k$ satisfies (7.6), yields

$$
\begin{aligned}
a_{2k+1}(f_N) &= \frac{1}{c_{2k+1}} \int_{\mathbb{R}} f_N^{(2k+1)}(x) f_N(x)\, dx \\
&= \frac{(-1)^k}{c_{2k+1}} \int_{\mathbb{R}} f_N^{(k)}(x) f_N^{(k+1)}(x)\, dx = 0, \\
a_{2k}(f_N) &= \frac{1}{c_{2k}} \int_{\mathbb{R}} H_{2k}(x) f_N^2(x)\, dx = \frac{1}{c_{2k}} \int_{\mathbb{R}} f_N^{(2k)}(x) f_N(x)\, dx \\
&= \frac{1}{c_{2k}} \int_{\mathbb{R}} \{ f_N^{(k)}(x) \}^2\, dx.
\end{aligned}
$$

$\square$

Corollary 7.2. *The norm of* H_{f_N} *is finite.*

The norm of H_{f_N} is expanded as

$$
\| f_N \|_{L_{2,\mu_N}} = \sum_{k \geq 1} c_{2k}^2 a_{2k}^2(f_N) = \sum_{k \geq 1} \int_{\mathbb{R}} \{ f_N^{(k)}(x) \}^2\, dx.
$$

The integrals $\sum_{k \geq 1} \int_{|x| > A} \{ f_N^{(k)}(x) \}^2\, dx$ and $\int_{|x| > A} (1 - x^2)^{-1} e^{-x^2}\, dx$ are equivalent as A tends to infinity and they converge to a finite limit.

7.4 Inequalities for the transforms

The Fourier transform is extended to a bounded linear operator on $L_p(\mathbb{R})^n$, $p \geq 1$. From the equivalence between the norms of the spaces L_p, for all conjugate integers p and p', there exist constants $k_1 \leq p$ and $k_2 \geq (p')^{-1}$ such that for every function f of L_p

$$
k_1 \| f \|_p \leq \| \widehat{f} \|_{p'} \leq k_2 \| f \|_p
$$

and the Fourier transform converges in L_p for every integer $p \geq 1$.

The Hermite expansion $H_f = \sum_{k \geq 0} a_k(f) H_k$ of a function f of L_{2,μ_N} satisfies the following inequality due to the Hölder inequality.

Proposition 7.4. *Let X be a normal variable, functions f and g such that $Ef^2(X)$ and $Eg^2(X)$ are finite satisfy*

$$
E\{ H_f(X) H_g(X) \} \leq \left(\sum_{k \geq 0} [E\{ f(X) H_k(X) \}]^2 \right) \left(\sum_{k \geq 0} [E\{ g(X) H_k(X) \}]^2 \right)
$$

and $EH_k(X) = 0$ for every $k \geq 0$.

The normalized Hermite polynomials are denoted

$$h_k = \frac{H_k}{\sqrt{k!}}.$$

The functions h_k are equivalent to $(k!)^{-\frac{1}{2}} x^k$ as x tends to infinity. At zero, the polynomials converge, $H_{2k+1}(x) \sim 0$ and

$$\frac{H_{2k}}{(2k)!} \sim_{x \to 0} \frac{1}{2k}.$$

The expansion of a function f in this basis is denoted by h_f and $H_f \equiv h_f$ for every function f. The order of $\|h_{f_N}\|_{L_2,\mu_N}$ is

$$\lim_{A \to \infty} \int_{|x|>A} \sum_{k \geq 0} \frac{x^{2k}}{(2k)!} e^{-x^2} < \infty.$$

The error $R_n(f) = f - S_n(f)$ in estimating a function f by the sum of the first n terms of its Fourier expansion is $R_n(f) = \sum_{k>n} \lambda_k^2(f)$. In the Hermite expansion, we consider norms $L_p(\mu_N)$ of the error of the partial sum $S_n(f; H) = \sum_{k \leq n} a_k(f) H_k(x)$.

Lemma 7.1. *The coefficient a_k of the expansion H_f for a function f belonging to L_{2,μ_N} has a norm $\|a_k\|_{L_2,\mu_N} < 1$ if $\|f\|_{L_2,\mu_N} < \|H_k\|_{L_2,\mu_N}$.*

Proposition 7.5. *The partial sums $S_n(f; h)$ of the Hermite expansion h_f in the orthonormal basis $(h_k)_{k \geq 0}$ satisfy*

$$\|n^{-1} S_n(f; h)\|_{L_2,\mu_N} < \|f\|_{L_2,\mu_N}$$

and for all conjugate integers $p > 1$ and p'

$$\|n^{-1} S_n(f; h)\|_{L_2,\mu_N} < \|f\|_{L_p,\mu_N} n^{-1} \sum_{k>n} \|h_k\|_{L_{p'},\mu_N}.$$

Proof. The Hermite expansion h_f in the orthonormal basis of polynomials satisfies $\|h_f\|_{L_2,\mu_N}^2 = \sum_{k \geq 0} \{E(fh_k)(X)\}^2$ and for every $k \geq 1$, the Hölder inequality implies $E|f(X)h_k(X)| \leq \|f\|_{L_2,\mu_N}$, with equality if and only if $f = h_k$. □

Proposition 7.6. *For every function f of $L_2(\mu_N)$, the Hermite sum $S_n(f; H)$ has an error*

$$\|R_n(f)\|_{L_2,\mu_N} \leq \|f\|_2 \|R_n(f_N)\|_{L_2,\mu_N}$$

and it tends to zero as n tends to infinity.

Proof. The squared norm of $R_n(f)$ is $\|R_n(f)\|_2^2 = \sum_{k>n} c_k^2(f) a_k^2(f)$ with

$$a_k(f) = \frac{1}{c_k(f)} \int_{\mathbb{R}} f(x) f_{\mathcal{N}}^{(k)}(x) \, dx,$$

$$c_k^2 a_k^2(f) \leq \{ \int_{\mathbb{R}} f^2(x) \, dx \} \{ \int_{\mathbb{R}} f_{\mathcal{N}}^{(k)2}(x) \, dx \} = c_{2k} a_{2k}(f_{\mathcal{N}}) \|f\|_2^2.$$

From Corollary 7.2, the Hermite transform of the normal density converges therefore $\|R_n(f_{\mathcal{N}})\|_{L_{2,\mu_{\mathcal{N}}}}$ converges to zero as n tends to infinity. $\qquad\square$

A function f having a converging Hermite expansion can then be approximated by finite sums of this expansion, for n large enough.

7.5 Inequalities in $\mathbb{C}$

A map f from a subset $\mathcal{D}$ of $\mathbb{C}$ to $\mathbb{C}$ is *analytic* at z_0 if there exist an open disc $D = D(r, z_0)$ centered at z_0 and with radius r, and a convergent series such that for every z in D

$$f(z) = \sum_{k=0}^{\infty} a_k (z - z_0)^k.$$

This expansion is unique and all derivatives of f are analytic. Let f_n be the partial sum $f_n(z) = \sum_{k=0}^{n} a_k (z - z_0)^k$, the coefficients are defined by $a_0 = f(z_0)$ and

$$a_k = \lim_{\|z-z_0\| \to 0} \frac{\|f(z) - f_{k-1}(z)\|}{\|z - z_0\|^k} = \frac{f^{(k)}(z_0)}{k!}.$$

A holomorph function f on a disc $D(r, z_0)$ with frontier the circle $C = C(r, z_0)$ has the derivatives

$$f^{(n)}(x) = \frac{n!}{2\pi} \int_C \frac{f(z)}{(z - z_0)^{n+1}} \, dz.$$

In $\mathbb{C}^2$, a function $f(x, y)$ holomorph on circles $C(r_1, z_1) \times C(r_2, z_2)$ has derivatives

$$\frac{\partial^{n+m} f(x, y)}{\partial x^n \partial y^m} = \frac{n! m!}{(2i\pi)^2} \int_C \frac{f(x, y)}{(x - z_1)^{n+1}(y - z_2)^{m+1}} \, dx \, dy$$

$$= \frac{n! m!}{4\pi^2 r_1^n r_2^m} \int_{[0,2\pi]^2} f(z_1 + r_1 e^{i\theta}, z_2 + r_2 e^{i\theta}) e^{-ni\varphi} e^{-mi\theta} \, d\varphi \, d\theta,$$

with integrals on the circles $C_1 = C(r_1, z_1)$ and $C_2 = C(r_2, z_2)$. Then

$$\left| \frac{\partial}{\partial x^n} \frac{\partial}{\partial y^m} f(x, y) \right| < \frac{n! m! \|f\|}{r_1^n r_2^m}.$$

Cauchy's theorem states that the solution of a canonical system of differential equations in $\mathbb{R}^k$

$$\frac{dx_m}{dx_1} = \varphi_m(x_1, \ldots, x_k),\ m = 2, \ldots, k$$

with a set of holomorph functions $(\varphi_1, \ldots, \varphi_{k-1})$ on discs with radius r and centers $a_1, \ldots, a_k$, is a holomorph function of $(x_1, \ldots, a_k)$ in the interior of a disc centered at $a = (a_1, \ldots, a_k)$.

7.6 Complex spaces of higher dimensions

The bijection between the spaces $\mathbb{R}^2$ and $\mathbb{C}$ defined by $h(x, y) = x + iy$ is defined by the square root i of -1 in the Euclidean metric of $\mathbb{R}^2$. In $\mathbb{R}^3$, -1 has two square roots denoted by i and j. The map

$$X = (x, y, z) \mapsto t = x + iy + jz$$

defines a bijection between $\mathbb{R}^3$ and a complex space denoted $\mathbb{C}_2$ (with two roots of -1). The complex conjugate of $t = x + iy + jz$ is $x - iy - jz$ and the space $\mathbb{C}_2$ is a vector space endowed with the Euclidean norm defined as the scalar product of complex conjugates

$$\|t\| = \{(x + iy + jz)(x - iy - jz)\}^{\frac{1}{2}} = \{x^2 + y^2 + z^2\}^{\frac{1}{2}}.$$

The spherical coordinates of a point defined on a sphere with radius r by angles θ in $[0, 2\pi]$ and φ in $[0, \pi]$ are $X = r(\cos\varphi\cos\theta, \cos\varphi\sin\theta, \sin\varphi)$, the norm of X is r while the norm of its projection in the horizontal plane is $r\{(\cos\varphi\cos\theta)^2 + (\cos\varphi\sin\theta)^2\}^{\frac{1}{2}} = r\cos\varphi$.

Let $\rho_{xy} = \|x + iy\|_2$ and $\rho_{xz} = \|x + jz\|_2$, then $t = x - iy - jz$ is also written as

$$t = \rho_{xy}\rho_{xz}e^{i\theta}e^{j\varphi},\ \theta \in [0, 2\pi], \varphi \in [0, 2\pi],$$

$$\theta = \arctan\frac{x}{y},\ y \neq 0,$$

$$\varphi = \arctan\frac{x}{z},\ z \neq 0,$$

and $\theta = 0$ if $y = 0$, $\varphi = 0$ if $z = 0$. The product $u = e^{i\theta}e^{j\varphi}$ belongs to $\mathbb{C}^{\otimes 2}$ and its expansion using trigonometric functions is $u = \cos\varphi\cos\theta + ij\sin\varphi\sin\theta + i\cos\varphi\sin\theta + j\sin\varphi\cos\theta$, where $\cos\varphi(\cos\theta + i\sin\theta)$ is the projection of $e^{i\theta}e^{j\varphi}$ in the horizontal plane and $\sin\varphi(\cos\theta + i\sin\theta)$ is its projection in a vertical section of the sphere. By orthogonality, the squared norm of u can be written as the sum of the squares $\cos^2\varphi\|\cos\theta + i\sin\theta\|^2 +$

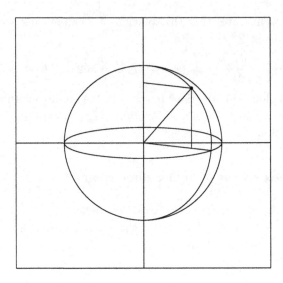

Fig. 7.1　Spherical representation in $\mathbb{R}^3$ and projection of a point in the plane.

$\sin^2 \varphi \| \cos \theta + i \sin \theta) \|^2 = 1$. The projections on the orthogonal spaces have a similar form whereas it is not satisfied for the coordinates in the spherical representation.

The equality $t = \bar{t}$ implies that t belongs to $\mathbb{R}$. The real scalar product of x and y in $\mathbb{C}_2$ is defined from the norm by (1.1)

$$(t, t') = \frac{1}{2} \{ (t + t')(\overline{t + t'}) - t\bar{t} - t'\bar{t}' \},$$

the inequalities for the norms and the geometric equalities (1.2) are still true. It differs from the complex scalar product

$$t\bar{t}' = xx' + yy' + zz' + i(xy' - x'y) + j(xz' - x'z) - ij(yz' + y'z)$$

where only $xx' + yy' + zz' + i(xy' - x'y) + j(xz' - x'z)$ belongs to $\mathbb{C}_2$ and ij belongs to $\mathbb{C}^2$ and satisfies $(ij)^2 = 1$. It follows that the space $\mathbb{C}_2$ is not a Hilbert space like $\mathbb{C}$.

Let f be a function defined from $\mathbb{R}^3$ to $\mathbb{R}$, its Fourier transform is defined from $\mathbb{R}^3$ to $\mathbb{C}_2$ by

$$\widehat{f}(s, t) = \sum_{k=0}^{\infty} \sum_{l=0}^{\infty} \lambda_{kl} e^{iks} e^{jlt}, \qquad (7.7)$$

$$\lambda_{kl} = \frac{1}{(2\pi)^2} \int_{-\pi}^{\pi} \int_{-\pi}^{\pi} e^{-ikx} e^{-jly} f(x, y) \, dx \, dy$$

and the inverse transform of $\widehat{f}$ is

$$f(x, y) = \int_{\mathbb{R}^2} e^{itx} e^{isy} \widehat{f}(s, t) \, ds \, dt.$$

If $f(x, y) = f_1(x) f_2(y)$, $\widehat{f}$ equals the product of the Fourier transforms of f_1 and f_2.

Proposition 7.7. *The Fourier transform of a function f of $L_2(\mathbb{R}^3)$ satisfies*

$$\|f\|_{L_2} = \{\sum_{k=0}^{\infty} \sum_{l=0}^{\infty} |\lambda|_{kl}^2\}^{\frac{1}{2}} = \|\widehat{f}\|_{L_2}.$$

This equality is a consequence of the Fubini theorem and the equality $(2\pi)^{-1} \int_{-\pi}^{\pi} e^{i(m-n)x} \, dx = 1_{\{m=n\}}$ for every real x and for all n and m. Equation (7.3) is also true for the scalar product in $\mathbb{C}_2$.

Let $p \geq 2$, by the same argument as for Proposition 7.7, the norms $L_p(\mathbb{R}^3)$ of f is the sum $\|f\|_{L_p} = \{\sum_{k=0}^{\infty} \sum_{l=0}^{\infty} |\lambda_{kl}|^p\}^{\frac{1}{p}}$. From the expression of the inverse Fourier transform

$$\|f\|_{L_p} = \|\widehat{f}\|_{L_p}.$$

Moreover, each coefficient has the bound

$$|\lambda_{kl}|^p \leq \frac{1}{(2\pi)^2} \int_{-\pi}^{\pi} \int_{-\pi}^{\pi} |f(x, y)|^p \, dx \, dy$$

and there exist constants a and b such that

$$a\|f\|_{L_p}^p \leq \|\widehat{f}\|_{L_{p'}} \leq b\|f\|_{L_p}^p.$$

Let (X, Y) be a random variable with values in $\mathbb{R}^2$, with a joint distribution function F, the function $\psi_{XY}(s, t) = Ee^{isX+jtY}$ has derivatives with respect to (s, t) and they satisfy

$$\frac{\partial^k}{\partial s^k} \frac{\partial^l}{\partial t^l} \psi_{XY}(s, t) = i^k j^l E\{X^k Y^l e^{isX+jtY}\}$$

whereas in $\mathbb{R}^2$, the function $\varphi_{XY}(s, t) = Ee^{i(sX+tY)}$ has the derivatives

$$\frac{\partial^k}{\partial s^k} \frac{\partial^l}{\partial t^l} \varphi_{XY}(s, t) = i^{k+l} E\{X^k Y^l e^{i(sX+tY)}\}.$$

A real function f of $C_1(\mathbb{C}_2)$ has a complex derivative $f^{(1)}$ such that

$$f(t + \delta) = f(t) + \delta_x f^{(1)}(t) + i\delta_y f^{(1)}(t) + j\delta_z f^{(1)}(t) + o(\|\delta\|), \ t, \delta \in \mathbb{C}_2.$$

The Cauchy equations for the derivatives in $\mathcal{C}$ are extended to $C(\mathbb{C}_2)$ in the next proposition.

Proposition 7.8. *A function* $f(x + iy + jz) = P(x, y, z) + iQ(x, y, z) + jR(x, y, z)$ *defined from* $\mathbb{C}_2$ *to* $\mathbb{C}_2$ *is continuously differentiable at* $t = x + iy + jz$ *if and only if the real functions* P, Q *and* R *belong to* $C_1(\mathbb{R}^3)$ *and*

$$\frac{\partial P(x, y, z)}{\partial x} = \frac{\partial Q(x, y, z)}{\partial y} = \frac{\partial R(x, y, z)}{\partial z},$$

$$\frac{\partial Q(x, y, z)}{\partial x} = -\frac{\partial P(x, y, z)}{\partial y},$$

$$\frac{\partial R(x, y, z)}{\partial x} = -\frac{\partial P(x, y, z)}{\partial z}.$$

Then, its derivative at t *is*

$$
\begin{aligned}
f^{(1)}(t) &= \frac{\partial P(x, y, z)}{\partial x} + i\frac{\partial Q(x, y, z)}{\partial x} + j\frac{\partial R(x, y, z)}{\partial x} \\
&= \frac{\partial Q(x, y, z)}{\partial y} - i\{\frac{\partial P(x, y, z)}{\partial y} - j\frac{\partial R(x, y, z)}{\partial y}\} \\
&= \frac{\partial R(x, y, z)}{\partial z} + j\{i\frac{\partial Q(x, y, z)}{\partial z} - \frac{\partial P(x, y, z)}{\partial z}\}.
\end{aligned}
$$

Proof. Let $t = x + iy + jz$ and $\delta = \delta_x + i\delta_y + j\delta_z$ in $\mathbb{C}_2$, the real functions P, Q, R are defined in $\mathbb{R}^3$ and the derivative of f has the form $f^{(1)} = A + iB + jC$ where the functions A, B and C are defined from $\mathbb{R}^3$ to $\mathbb{R}^3$. There exist real functions ε_k, $k = 1, 2, 3$, defined in $\mathbb{C}_2$ and converging to zero as $\|t\| \to 0$ and such that

$$
\begin{aligned}
f(t + \delta) &= f(t) + (\delta_x, i\delta_y, j\delta_z) \\
&\quad \times \{A(x, y, z) + \varepsilon_1, iB(x, y, z) + i\varepsilon_2, jC(x, y, z) + j\varepsilon_3\}^T \\
&= f(t) + \delta_x\{A(x, y, z) + iB(x, y, z) + jC(x, y, z)\} + \delta_y\{iA(x, y, z) \\
&\quad - B(x, y, z) + ijC(x, y, z)\} + \delta_z\{jA(x, y, z) + ijB(x, y, z) \\
&\quad - C(x, y, z)\} + o(\delta_x) + o(\delta_y) + o(\delta_z),
\end{aligned}
$$

it follows that the partial derivatives of f with respect to (x, y, z) satisfy the above conditions, the equalities for $f^{(1)}$ follow. $\square$

The norm $L_2(\mathbb{C}_2)$ of $f(t)$ is

$$
\begin{aligned}
\|f^{(1)}(t)\|_2 &= \{\frac{\partial P(x, y, z)}{\partial x}\}^2 + \{\frac{\partial Q(x, y, z)}{\partial x}\}^2 + \{\frac{\partial R(x, y, z)}{\partial x}\}^2 \\
&= \{\frac{\partial Q(x, y, z)}{\partial y}\}^2 + \{\frac{\partial P(x, y, z)}{\partial y}\}^2 + \{\frac{\partial R(x, y, z)}{\partial y}\}^2 \\
&= \{\frac{\partial R(x, y, z)}{\partial z}\}^2 + \{\frac{\partial Q(x, y, z)}{\partial z}\}^2 + \{\frac{\partial P(x, y, z)}{\partial z}\}^2.
\end{aligned}
$$

Under the conditions of Proposition 7.8 and as δ tends to zero in $\mathbb{C}_2$

$$f(t + \delta) = f(t) + \delta f^{(1)}(t) + o(\|\delta\|).$$

Expansions of a $\mathbb{C}_2$-valued function are similar to the Taylor expansions in $\mathbb{R}^3$ in an orthogonal basis, via the representation of the function as $f(x + iy + jz) = P(x, y, z) + iQ(x, y, z) + jR(x, y, z)$. Let f belong to $C_n(\mathbb{C}_2)$, equivalently the real functions P, Q, R belong to $C_n(\mathbb{R}^3)$ and satisfy equalities similar to those of Proposition 7.8 for all derivatives up to n. As $\|\delta\|$ tends to zero

$$f(t + \delta) = f(t) + \sum_{k=1}^{n} \frac{\delta^k}{k!} f^{(n)}(t) + o(\|\delta\|^k). \tag{7.8}$$

The isometry between $\mathbb{R}^3$ and $\mathbb{C}_2$ extends to higher dimensions. Let p be an integer larger or equal to 3 and let $p - 1$ roots $(i_1, \ldots, i_{p-1})$ of -1, they define a complex space $\mathbb{C}_{p-1}$ isometric to $\mathbb{R}^p$ by the bijection $(x_1, \ldots, x_p) \mapsto x_1 + \sum_{k=2}^{p-1} i_{k-1} x_k$. Functions of $C_n(\mathbb{C}_{p-1})$ have expansions like (7.8) under Cauchy conditions of dimension p and order n.

7.7 Stochastic integrals

In Section 4.9, the stochastic integral of a predictable process A with respect to an adapted process M of $\mathcal{M}^2_{0,loc}(\mathbb{R}^2_+)$ is defined in $L_{2,loc}$ from the integral of A^2 with respect to the predictable compensator $< M >$ of M in rectangles $R_z = [0, z]$ and in rectangles $R_{]z,z']}$, with ordered z and z' in $\mathbb{R}^2_+$. If z and z' are not ordered, for example their components satisfy $z_1 < z_2$ and $z'_1 > z'_2$, the increment of the martingale between z and z' follows the same rule (4.24) and it is the opposite of its increment between the ordered points (z_1, z'_2) and (z'_1, z_2). By splitting a surface into rectangles with perimeter tending to zero, the stochastic integral is defined over every Borel set of $\mathbb{R}^2_+$.

The integral satisfies $\int_{R_z} \int_{R_{z_{k-1}}} \cdots \int_{R_{z_2}} dM_{s_k} \cdots dM_{s_1} = (k!)^{-1} M^k_{R_z}$ and, applying this equality, it follows that

$$\int_{R_z} \cdots \int_{R_{z_2}} A_{s_k} \cdots A_{s_1} dM_{s_1} \cdots dM_{s_k} = (k!)^{-1} \{ \int_{R_z} A_s \, dM_s \}^k.$$

With $k = 2$, if $\int_{R_z} A_s \, dM_s$ belongs to $\mathcal{M}^2_{0,loc}(\mathbb{R}^2_+)$

$$E \int_{R_z} \int_{R_{z_2}} A_{s_2} A_{s_1} \, dM_{s_1} \, dM_{s_2} = \frac{1}{2} E \{ \int_{R_z} A_s \, dM_s \}^2$$

$$= \frac{1}{2} E \int_{R_z} A_s^2 \, d < M >_s.$$

Let M_1 and M_2 be local martingales of $\mathcal{M}_{0,loc}^2(\mathbb{R}_+^2)$ and let A_1 and A_2 be predictable processes of $L_{2,loc}(< M >_1)$ and $L_{2,loc}(< M >_2)$, respectively, the process $X_z = \int_{R_z} \int_{R_{z_2}} A_{s_2} B_{s_1} \, dM_{s_1} \, dM_{s_2}$ is defined from (1.22) as the scalar product of the local martingales $A_1.M_1 = \int_{R_z} A_1 \, dM_1$ and $\int_{R_z} A_2 \, dM_2$.

Let $\mathcal{M}_{0,S,loc}^p(\mathbb{R}_+^2)$ be the space of the L_p local strong martingales with mean zero. For M belonging to $\mathcal{M}_{0,S,loc}^4(\mathbb{R}_+^2)$, let $M^{(1)}(z)$ be the martingale on $\mathbb{R}_+$ defined at fixed z_2, with respect to the marginal filtration $\mathbb{F}_1$, and let $M^{(2)}(z)$ be the martingale on $\mathbb{R}_+$ defined at fixed z_1, with respect to the marginal filtration $\mathbb{F}_2$. Let A be a $\mathbb{F}_1$ and $\mathbb{F}_2$-predictable process, belonging to $L_2(< M^{(1)} >< M^{(2)} >)$. The integral of A with respect to $M^{(1)}$, then $M^{(2)}$, is denoted $\int_{R_z} A \, dMM$ and the process A belongs to $L_2(MM)$ if it is $L_2(< M^{(1)} >< M^{(2)} >)$-integrable. Cairoli and Walsh (1975) proved that the integral $A.MM_z = \int_{R_z} A \, dMM$ belongs to $\mathcal{M}_{0,S,loc}^2$ and it is continuous if M is, and

$$< A.MM, B.MM >_z = \int_{R_*^2} AB \, d < M^{(1)} > d < M^{(2)} >,$$

$$E(A.MM_z \, M_z) = 0.$$

A Wiener process W in $[0,1]^2$ is a strong martingale with respect to its natural filtration, its mean is zero and it belongs to $L_p([0,1]^2)$ for every $p \geq 1$. Let $\mathbb{F}_W$ be the filtration generated by a Wiener process W in $\mathbb{R}_+^2$, Wong and Zakaï (1974) established that every M of $\mathcal{M}_{S,loc}^2(\mathbb{R}_+^2, \mathbb{F}_W)$ has an integral representation

$$M_z = M_0 + \varphi.W_z + \psi.WW_z, \ z \in \mathbb{R}_+^2,$$

where φ is a function of $L_2(W)$ and ψ is a function of $L_2(WW)$, and $E(M_z - M_0)^2 = \int_{R_z} \varphi_z^2 \, d < M >_z + \int_{R_z} \varphi_z^2 \, d < M^{(1)} >_z \, d < M^{(2)} >_z$.

By a change of variable, for every $z' < z$ in $\mathbb{R}_+^2$, the variations of M in $R_{z',z}$ have the representation

$$M_{R_{z',z}} = M_{z'} + \varphi.W_{R_{z',z}} + \psi.WW_{R_{z',z}}.$$

A Poisson point process with parameter λ has the Laplace transform

$$L_{N_{R_z}}(u) = \exp\{\lambda |z_1| \, |z_2| (e^u - 1)\}$$

and a martingale with respect to the natural filtration of the process has a similar representation.

Let us consider integrals in open balls with radius a, $B_a(z)$, z in $\mathbb{R}^2$. By the isometry between $\mathbb{R}^2$ and $\mathbb{C}$, every z' of $B_a(z)$ is written $z' = z + \rho e^{i\theta}$

with $0 < r < a$ and θ in $[-\pi, \pi]$, and the integral of the Brownian motion W in $B_a(z)$ satisfies $E\{\int_{B_a(z)} dW_s\}^2 = \int_{B_a(z)} ds = 0$, z in $\mathbb{C}$, so it is a.s. zero. For every left-continuous function f with right-hand limits in $L_2(B_a(z))$, the integral $\int_{B_a(z)} f(s) dW_s$ is defined as a linear map such that

$$E\{\int_{B_a(z)} f(s) dW_s\}^2 = \frac{1}{2\pi} \int_{-\pi}^{\pi} \int_0^r f^2(z + re^{i\theta}) e^{i\theta} dr\, d\theta, \ z \in \mathbb{C},$$

and this defines $\int_{B_a(z)} f(s) dW_s$ like in $L_2(\mathbb{R}^2)$. A Wiener process W in $\mathbb{R}^2$ also has an integral $\int_{B_a(z)} dW_s = 0$ a.s., and for every z in $\mathbb{R}^2$ or in $\mathbb{C}$

$$E\{\int_{B_a(z)} f(s) dW_s\}^2 = \int_{B_a(z)} f^2(s) ds - \{\int_{B_a(z)} f(s) ds\}^2.$$

A Poisson point process with parameter λ has the Laplace transform $L_{N_{B_a(z)}}(u) = \exp\{\lambda \pi a^2 (e^u - 1)\}$ in balls, for every z in $\mathbb{C}$. For every function f of $L_2(\mathbb{R}^2)$, $E\{\int_{B_a(z)} f(s) dN_s\}^2 = \lambda \int_{B_a(z)} f^2(s) ds$ or it equals $E \int_{B_a(z)} f^2(s) d\Lambda(s)$ if N has a cumulative intensity Λ.

Let A be a subset of $\mathbb{R}^2$ and let f be a left-continuous Borel function with right-hand limits in $L_2(A)$, for the Lebesgue measure μ. The function f is the limit as ε tends to zero of a sum $S_\varepsilon = \sum_{i=1}^{n_\varepsilon} \zeta_i 1_{B_r(x_i)}$ where $\zeta_i = f(x_i)$ and n_ε is the number of balls with radius $r = r_\varepsilon$ belonging to the interval $I_2(\varepsilon) =]\frac{\varepsilon}{\sqrt{2}}, \varepsilon]$, centered at points x_i such that $|x_i - x_j| = \varepsilon$ and defining the coverage of A by r_ε-balls in an ε-net in A (Fig. 6.2). Then $n_\varepsilon = O(\varepsilon^{-2}|A|)$, the integral

$$\int_A f(z) dz = \lim_{\varepsilon \to 0} \pi r_\varepsilon^2 \sum_{i=1}^{n_\varepsilon} f(x_i)$$

is finite and

$$\int_A f^2(z) dz = \lim_{\varepsilon \to 0} \sum_{i=1}^{n_\varepsilon} \sum_{j=1}^{n_\varepsilon} \zeta_i \zeta_j \mu(B_r(x_i) \cap B_r(x_j))$$

$$= \lim_{\varepsilon \to 0} \frac{r_\varepsilon^2}{2} (\theta_\varepsilon - \sin \theta_\varepsilon) \sum_{i=1}^{n_\varepsilon} \sum_{x_j \in \mathcal{V}(x_i)} f(x_i) f(x_j),$$

where $\theta_\varepsilon = 2\arccos \frac{\varepsilon}{2r_\varepsilon}$, hence $\cos \frac{1}{2}\theta_\varepsilon$ belongs to the interval $[\frac{1}{\sqrt{2}}, \frac{1}{2}]$, so $\lim_{\varepsilon \to 0}(\theta_\varepsilon - \sin \theta_\varepsilon)$ is bounded, and this sum converges as ε tends to zero. Its limit also equals $\int_A f^2(re^{i\theta}) dr\, d\theta$.

The integrals $\int_A f_z dW_z$ and $\int_A f_z dW_z^{1)} dW_z^{2)}$ with respect to the Brownian motion are defined as the linear map with L_2-norms satisfying

$$E(\int_A f_z dW_z)^2 = \int_A f_z^2 dz = E(\int_A f_z dW_z^{1)} dW_z^{2)})^2.$$

The Brownian processes have therefore representations in the balls, with the complex and the real parametrizations. Similar representations hold for integrals with respect to the Wiener process and with respect to the martingale $N - \Lambda$, for a Poisson processes N with a cumulative intensity Λ.

Appendix A

Probability

A.1 Definitions and convergences in probability spaces

Let $(\Omega, \mathcal{A}, P)$ be a probability space and let $(A_n)_{n \geq 0}$ be a sequence of measurable sets. They are independent if and only if for every finite subset $\mathcal{K}$ of $\mathbb{N}$, $P(\cap_{k \in \mathcal{K}} A_k) = \prod_{k \in \mathcal{K}} P(A_k)$.

Lemma A.1 (Borel-Cantelli's lemma). *A sequence of measurable sets* $(A_n)_{n \geq 0}$ *satisfies* $\limsup_{n \to \infty} A_n = \emptyset$ *a.s. if and only if* $\sum_{n \leq 0} P(A_n)$ *is finite. If the sets* A_n *are independent and* $\sum_{n \leq 0} P(A_n)$ *is infinite, then* $\limsup_{n \to \infty} A_n = \Omega$ *a.s.*

Let $(X_n)_{n \geq 0}$ be a sequence of real random variables on $(\Omega, \mathcal{A}, P)$, the σ-algebra generated by X_n is generated by the sets $\{X_n < x\}$, x in $\bar{\mathbb{R}}$. The variables are independent if and only if for every finite subset $\mathcal{K}$ of $\mathbb{N}$ and every sequence $(x_k)_{k \in \mathcal{K}}$ of $\bar{\mathbb{R}}$, $P(\cap_{k \in \mathcal{K}} \{X_k < x_k\}) = \prod_{k \in \mathcal{K}} P(X_k < x_k)$. An equivalent characterization of the independence of the sequence $(X_n)_{n \geq 0}$ is the equality $E \prod_{k \in \mathcal{K}} f_k(X_k) = E \prod_{k \in \mathcal{K}} f_k(X_k)$, for every sequence of measurable functions $f_k : \mathbb{R} \mapsto \mathbb{R}$ and for every finite subset $\mathcal{K}$ of $\mathbb{N}$. The distribution of a vector of independent random variables is the product of their distributions and their characteristic function (Fourier transform) is the product of the marginal characteristic functions.

A sequence of random variables $(X_n)_{n \geq 0}$ converges a.s. to a variable X if $P(\limsup_n X_n = \liminf_n X_n < \infty) = 1$ and it is equivalent to the existence of a convergent series $(\varepsilon_n)_{n \geq 0}$ such that $\sum_{n \geq 0} P(|X_n - X| > \varepsilon)$ converges. The convergence properties of random variables are not equivalent and their relationships depend on the integrability properties of the variables. For all conjugate integers p and p', there exist constants $k_1 \leq p$ and $k_2 \geq (p')^{-1}$

such that for every function f and for every random variable X

$$k_1\{E|f(X)|^p\}^{\frac{1}{p}} \leq \{E|f(X)|^{p'}\}^{\frac{1}{p'}} \leq k_2\{E|f(X)|^p\}^{\frac{1}{p}}$$

and the convergences in mean are equivalent. The a.s. convergence of a sequence of random variables implies its convergence L_p, for every $p \leq 1$, and the convergence in a space L_p implies the convergence in probability. Conversely, every sequence of variables converging in probability to a variable X has a sub-sequence that converges a.s. to X. The weak convergence of a sequence of random variables is equivalent to the convergence of their characteristic functions to the characteristic function of a variable X.

A random variable X belongs to L_1 if and only if $\sum_{n \geq 0} P(|X| \geq n)$ is finite. Let $(X_n)_{n \geq 0}$ be a sequence of random variables of L_p, $p \leq 1$. The variable sequence $(X_n)_{n \geq 1}$ is *equi-integrable* if

$$\lim_{a \to \infty} \sup_{n \geq 1} E(|X_n|1_{|X_n|>a}) < \infty.$$

If there exists a variable X of $L_1(P)$ such that $|X_n| \leq |X|$ a.s. for every integer n, then $(X_n)_{n \geq 1}$ is equi-integrable. The $L_p(P)$ integrability for $p > 1$ is a sufficient condition for the equi-integrability and the following assertions are equivalent as n tends to infinity (Neveu, 1970)
1. $(X_n)_{n \geq 1}$ is equi-integrable and converges in probability to a variable X,
2. $(X_n)_{n \geq 1}$ converges in $L_1(P)$ to a variable X of $L_1(P)$.

The limiting behaviour of the supremum of weighted normal variables is deduced from the 0-1 law of the Borel-Cantelli Lemma A.1.

Proposition A.1. *Let $(a_i)_{i=1,\ldots,n}$ be a decreasing sequence of positive real numbers and let $(X_n)_{n \geq 1}$ be a sequence of independent normal variables on $(\Omega, \mathcal{A}, P)$, then*

$$P(\sup_{n \geq 1} |a_n X_n| < \infty) = 1 \text{ if } \lim_{n \to \infty} a_n \sqrt{\log n} < \infty,$$

$$= 0 \text{ if } \lim_{n \to \infty} a_n \sqrt{\log n} = \infty.$$

On a probability space $(\Omega, \mathcal{A}, P)$ let $(\mathcal{F}_n)_{n \geq 0}$, with a discrete filtration on $(\Omega, \mathcal{A}, P)$. A sequence of variables $(X_n)_{n \geq 0}$ is *adapted* if X_n is $\mathcal{F}_n$-measurable for every integer n. It follows that for a discrete stopping T, $X_T 1_{\{T \leq n\}}$ is $\mathcal{F}_n$-measurable. A *stopping time* T is a random variable such that the event $\{T \leq n\}$ is $\mathcal{F}_n$-measurable, then $\{T > n\}$ is also $\mathcal{F}_n$-measurable. It is predictable if $\{T \leq n\}$ is $\mathcal{F}_{n-1}$-measurable.

A sequence of adapted random variables $(X_n)_{n \geq 0}$ is a martingale with respect to a filtration $(\mathcal{F}_n)_{n \geq 0}$ if $E(X_{n+1}|\mathcal{F}_n) = X_n$ for every integer n. A well-known example of martingale defined from a stopped sequence is the Snell envelope. It is defined for an adapted sequence $(Z_n)_{n=0,...,N}$ with respect to a filtration $(\mathcal{F}_n)_{n=0,...,N}$ as $U_N = Z_N$ and $U_n = \max\{Z_n, E(U_{n+1}|\mathcal{F}_n)\}$, for every n belonging to $\{0, \ldots, N-1\}$. Then $\nu_0 = \inf\{n : U_n = Z_n\}$ is a stopping time and the sequence $(U_n \wedge \nu_0)_{n=0,...,N}$ is a martingale, with respect to the filtration $(\mathcal{F}_n)_{n=0,...,N}$.

Every positive supermartingale $(X_n)_{n\geq 0}$ converges a.s. to a limit X_∞ and $E(X_\infty|\mathcal{F}_n) = X_n$ for every integer n. A submartingale $(X_n)_{n\geq 0}$ such that $\sup_{n\in\mathbb{N}} EX_n^+$ is finite converges a.s. to a limit in L_1. A martingale of L_1 converges a.s. to a limit in L_1.

Theorem A.1 (Neveu, 1972). *Let $(X_n)_{n\geq 0}$ be a martingale such that $\sup_{n\in\mathbb{N}} E|X_n|\,(\log|X_n|)^+$ is finite, then $E \sup_{n\in\mathbb{N}}|X_n|$ is finite.*

With a continuous filtration $(\mathcal{F}_t)_{t\geq 0}$, a *stopping time* T is a random variable such that the event $\{T \leq t\}$ is $\mathcal{F}_t$-measurable, then $\{T \geq t\}$ is $\mathcal{F}_{t-}$-measurable, the time variable T is a predictable stopping time if $\{T \leq t\}$ is $\mathcal{F}_{t-}$-measurable. An adapted process X satisfies: X_t is $\mathcal{F}_t$-measurable for every real t and $X_T 1_{\{T\leq t\}}$ is $\mathcal{F}_t$-measurable for every real t and for every stopping time T.

Let $(X_n)_{n\geq 1}$ be a sequence of independent random variables of L_2 and let $S_n = \sum_{i=1}^n X_i$. Wald's equalities for an integrable stopping time T are

$$ES_T = E(X_1)E(T), \text{ if } X_1 \in L_1, T \in L_1,$$
$$ES_T^p = E(X_1^p)E(T), \text{ if } X_1 \in L_p, T \in L_1, \ p \geq 2. \tag{A.1}$$

It is proved recursively for centered variables writing

$$ES_{T\wedge n}^2 = E(S_{T\wedge(n-1)} + X_n 1_{\{T\geq n\}})^2$$
$$= E(S_{T\wedge(n-1)}^2 + X_n^2 1_{\{T\geq n\}})$$
$$= E(X_1^2)\{\sum_{k=1}^n P(T \geq k)\} = E(X_1^2)\,E(T)$$

since $\{T \geq n\}$ is $\mathcal{F}_{n-1}$-mesurable, and the result is deduced for $p = 2$. It is generalized to an integer $p > 2$ by induction.

From the strong law of large numbers, $n^{-1}S_n$ converges a.s. to EX_1 if the variables belong to $L_2(P)$. It converges in probability to EX_1 if they belong to $L_1(P)$. The criteria of integrability for S_n and for the variables X_i, or their maxima are related, by the Bürkholder-Davis-Gundy inequality and by other properties. Let $(a_i)_{i=1,...,n}$ be a sequence of positive real numbers and $A_n = \sum_{i=1}^n a_i$, then

$$P(|S_n| > A_n) \leq \sum_{i=1}^n P(|X_i| > a_i) \leq \sum_{i=1}^n \frac{E|X_i|}{a_i}.$$

Proposition A.2. *Let $(X_n)_{n\geq 0}$ be a sequence of independent normal variables and let $\alpha = (\alpha_n)_{n\geq 0}$ be a real sequence such that $\|\alpha\|_2 = (\sum_{n\geq 0} \alpha_n^2)^{\frac{1}{2}}$ is finite, then $\sum_{n\geq 0} \alpha_n X_n$ is a normal variable with variance $\|\alpha\|_2^2$.*

The law of the iterated logarithm for sums of independent Gaussian variables can be written like in Section 4.8.

Theorem A.2. *Let $(X_k)_{k\geq 1}$ be a sequence of independent normal variables, let $(\alpha_n)_{n\geq 1}$ be a sequence of real numbers and let $A_n = \sum_{n\geq 1} \alpha_n^2$. For very real function $h > 0$ on $\mathbb{R}$ such that $\sum_{n=1}^{\infty} h_n^{-1}$ is finite, the variable $Y_n = \sum_{n\geq 1} \alpha_n X_n$ satisfies*

$$\limsup_{n\to\infty} \frac{Y_n}{\sqrt{2A_n \log h_n}} \leq 1, \qquad \liminf_{n\to\infty} \frac{Y_n}{\sqrt{2A_n \log h_n}} \geq -1, \quad \text{a.s.}$$

Proof. The sum of the probabilities of the sets $\{Y_n > \sqrt{2A_n \log n}\}$ is bounded using the Laplace transform of the Gaussian variable Y_n

$$\sum_{n\geq 2} P(Y_n > \sqrt{2A_n \log h_n}) = \sum_{n\geq 2} \exp\{-\frac{2A_n \log h_n}{2 var Y_n}\} = \sum_{n\geq 2} h_n^{-1}$$

and it is finite. The result follows from the 0-1 law (Lemma A.1).　　□

Chow and Lai (1973) related the behaviour of $\|\alpha\|_2$ to other properties of a sum of weighted variables $\sum_{n\geq 0} \alpha_n X_n$, in particular for $\alpha \geq 2$ there exists a constant B_α such that

$$\sup_{n\geq 1} E|\sum_{i=1}^{n} \alpha_i X_i|^\alpha \leq B_\alpha (\sum_{n\geq 0} \alpha_n^2)^\alpha E|X_1|^\alpha.$$

Let $(X_k)_{k\geq 1}$ be a sequence of i.i.d. centered random variables. For every $\alpha \geq 1$, the next statements are equivalent
(1) the variables X_i belong to $L_\alpha(P)$,
(2) $\lim_{n\to\infty} n^{-\frac{1}{\alpha}} X_n = 0$ a.s.,
(3) there exists a sequence of real numbers $(\alpha_n)_{n\geq 0}$ such that $\sum_{n\geq 0} \alpha_n^2$ is finite and $\lim_{n\to\infty} n^{-\frac{1}{\alpha}} \alpha_{n-i} X_i = 0$ a.s.

Theorem A.3 (Chow and Lai, 1973). *Let $(X_k)_{k\geq 1}$ be a sequence of i.i.d. centered random variables. The following statements are equivalent*
(1) $E e^{t|X_1|} < \infty$, for every real t,
(2) $\lim_{n\to\infty} \frac{X_n}{\log n} = 0$ a.s.,
(3) there exists a sequence of real numbers $(\alpha_n)_{n\geq 0}$ such that $\sum_{n\geq 0} \alpha_n^2$ is finite and $\lim_{n\to\infty} \alpha_{n-i} \frac{X_i}{\log n} = 0$ a.s.

The variable $M_n = \max_{i=1,\ldots,n} X_i$, with independent uniform variables X_i on the interval $[0,1]$ has first moments
$$EM_n = \frac{n}{n+1}, \quad EM_n^p = \frac{n}{n+p}, \quad p \geq 2.$$
For every A in $]0,1[$, $F(A)$ belongs to $]0,1[$ and $P(M_n < A) = F^n(A)$ tends to zero as n tends to infinity therefore $\limsup_{n\to\infty} M_n = 1$ a.s., similarly $\liminf_{n\to\infty} \min_{i=1,\ldots,n} X_i = 0$ a.s. The maximum M_n of n independent variables X_i with a common distribution function F on $\mathbb{R}$ have the mean $EM_n = n \int_{\mathbb{R}^n} x_n 1_{x_n = \max\{x_1,\ldots,x_n\}} \prod_{i=1,\ldots,n} dF(x_i) = nE\{X_1 F^{n-1}(X_1)\}$ and higher moments $EM_n^p = nE\{X_1^p F^{n-1}(X_1)\}$, its extrema satisfy $\limsup_{n\to\infty} M_n = +\infty$ a.s. and $\liminf_{n\to\infty} \min_{i=1,\ldots,n} X_i = -\infty$ a.s. More generally, $\limsup_{n\to\infty} M_n$ and $\liminf_{n\to\infty} \min_{i=1,\ldots,n} X_i$ are a.s. equal to the maximum and, respectively, the minimum of the support of the distribution of the variables X_i.

Bennet and Hoeffding inequalities. Let $(X_i)_{i=1,\ldots,n}$ be independent variables with respective values in $[a,b]$, then the Bennet inequality is modified as follows. For every $t > 0$ and every integer n
$$P(S_n - ES_n \geq t) \leq \exp\left\{-n\phi\left(\frac{t}{n(b-a)^2}\right)\right\}$$
where $\phi(x) = (1+x)\log(1+x) - x$. In Hoeffding's inequality, the bound is exponential.

Theorem A.4 (Hoeffding's inequality). *Let $(X_i)_{i=1,\ldots,n}$ be independent variables with respective values in $[a_i, b_i]$, then for every $t > 0$*
$$P(n^{-\frac{1}{2}}(S_n - ES_n) \geq t) \leq \exp\left\{-\frac{2t^2}{\sum_{i=1}^n (b_i - a_i)^2}\right\}. \tag{A.2}$$
It is proved along the same arguments as Chernov's inequality. The following bound is weaker and easily proved. For every $t > 0$
$$P(n^{-\frac{1}{2}}(S_n - ES_n) \geq t) \leq \exp\left\{-\frac{t^2}{2\sum_{i=1}^n (b_i - a_i)^2}\right\}. \tag{A.3}$$
Proof. For every integer n, the Laplace transform L of $S_n - ES_n$ satisfies $\log L(0) = 0$, $L'(0) = 0$ for the centered variables and $L'(0) \leq B_n^2$, with the constants $B_n^2 = \sum_{i=1}^n (b_i - a_i)^2$. Therefore
$$\log P(n^{-\frac{1}{2}}(S_n - ES_n) > t) < \liminf_{\lambda \to 0}\{\log L(\lambda) - \lambda t\}$$
$$= \inf_\lambda\left\{\frac{1}{2}\lambda^2 b_n^2 - \lambda t\right\} = -\frac{t^2}{2B_n^2}$$
which yields (A.3). $\qquad\square$

The sum S_n has the moments $E(S_n - ES_n)^k \leq \sum_{i=1}^{n}(b_i - a_i)^k \leq \|a - b\|_{\infty}^{k}$ for every $k \geq 2$, from Bennett's inequality this implies

$$P(S_n - ES_n \geq t) \leq \exp\{-n\phi(\frac{t}{n\|a - b\|_{\infty}})\}.$$

Bernstein inequality for a sum S_n of n independent and centered variables X_i in a metric space such that $\|X_i\|_{\infty} \leq M$ for every $i \geq 1$ is

$$P(S_n \geq t) \leq \exp\{-\frac{t^2}{2\sigma_n + \frac{2}{3}Mt}\}, t > 0,$$

where $varS_n = \sigma_n^2$ (Giné, 1974). It is a slight modification of inequality (A.3) above.

From Freedman (1975), a martingale $S_n = \sum_{i=1}^{n} X_i$ with quadratic variations $T_n = \sum_{i=1}^{n} E(X_i^2|\mathcal{F}_{i-1})$ and such that $|X_n| \leq 1$ satisfies the inequality

$$P(S_n > a, T_n \leq b, n \geq 1) \leq (\frac{b}{a+b})^{a+b} \leq \exp\{-\frac{a^2}{2(a+b)}\}$$

for all $a > 0$ and $b > 0$.

A.2   Boundary-crossing probabilities

Robbins and Siegmund (1970) proved many uniform boundary-crossing probabilities for the Brownian motion W. In particular, let $f_{\mathcal{N}}$ be the normal density and $F_{\mathcal{N}}$ be its distribution function, for all $a > 0$ and b in $\mathbb{R}$

$$P(\max_{t \geq 1}(W_t - at) \geq b) = P(\max_{0 \leq t \leq 1}(W_t - bt) \geq a)$$

$$= 1 - F_{\mathcal{N}}(b + a) + e^{-2ab}F_{\mathcal{N}}(b - a),$$

$$P(\max_{t \geq 1}t^{-1}W_t \geq a) = P(\max_{0 \leq t \leq 1}W_t \geq a) = \{1 - F_{\mathcal{N}}(a)\}$$

and non uniform inequalities such as

$$P(\exists t : |W_t| \geq \sqrt{t}\alpha^{-1}\{\log t^{1-\gamma} + \alpha(a)\}) = 1 - F_{\mathcal{N}}(a)$$

$$+f_{\mathcal{N}}(a)\frac{\int_0^{\infty} F_{\mathcal{N}}(a - y)y^{-\gamma}\, dy}{\int_0^{\infty} f_{\mathcal{N}}(a - y)y^{-\gamma}\, dy},$$

$$\alpha(x) = x^2 + 2\log\int_0^{\infty} f_{\mathcal{N}}(y - x)y^{-\gamma}\, dy,$$

where $\gamma < 1$, $\alpha(x) \sim_{x \to \infty} x^2$ and α^{-1} is the inverse function of α.

Jain, Jogdeo and Stout (1975, Theorem 5.2) proved the a.s. convergence for a martingale sequence $S_n = \sum_{i=1}^{n} X_i$ such that the martingale differences X_i are stationary and ergodic and $EX_i^2 = 1$: Let $\varphi > 0$ be an increasing function and $I(\varphi) = \int_1^{\infty} t^{-1}\varphi(t)e^{-\frac{1}{2}\varphi^2(t)}\, dt$, then

$$P\{S_n > \frac{\sqrt{n}}{\varphi(n)}\,\text{a.s.}\} = 0 \quad \text{if } I(\varphi) < \infty,$$
$$= 1 \quad \text{if } I(\varphi) = \infty.$$

Other level crossing problems than those presented in Section 4.8 were reviewed by Blake and Lindsey (1973). For a Brownian motion B and with the Lebesgues measure μ, the number of crossings of zero is replaced by the duration over this threshold $\mu\{s \in [0, \tau]; B(s) > 0\}$. From Erdos and Kac (1947)

$$P\{\mu\{s \in [0, \tau] : B(s) > 0\} < t\} = 2\pi^{-1}\arcsin(\tau^{-1}t)^{\frac{1}{2}}$$

and this result has been extended by other authors.

Let $N_c = \min\{k : k^{-\frac{1}{2}}S_k > c\}$, in the Binomial case $EN_c < \infty$ if and only if $c < 1$ (Blakewell and Freeman, 1964) and in the general case of i.i.d. variables X_i with $EX_i = 0$ and $varX_i = \sigma^2$, the necessary and sufficient condition is $c^2 < \sigma^2$ (Chow, Robbins and Teicher, 1965). If $\sigma^2 = 1$, $E(N_c^2) < \infty$ if and only if $c < \sqrt{3 - \sqrt{6}}$ (Chow and Teicher, 1966).

For centered Gaussian processes with a stationary covariance function function $r(t)$, the number $M(t, u)$ of values of t such that $X_t = u$ has a mean given by the Rice formula (Dudley, 1973)

$$EM(t, u) = \frac{T}{\pi}(-\frac{r''(0)}{r(0)})^{\frac{1}{2}}\exp\{-\frac{u^2}{2r(0)}\}.$$

A.3 Distances between probabilities

Let P and Q be probabilities on a measurable space $(\mathcal{X}, \mathcal{B})$, with the Borel σ-algebra, and let $\mathcal{F}$ be the set of positive functions f on $(\mathcal{X}, \mathcal{B})$ such that f and $\frac{1}{f}$ are bounded. Kullback's information of P with respect to Q is defined as

$$I_Q(P) = \sup_{f \in \mathcal{F}}\{\int_{\mathcal{X}} \log f\, dP - \log \int_{\mathcal{X}} f\, dQ\}.$$

Theorem A.5. *Kullback's information $I_Q(P)$ is finite if and only if P is absolutely continuous with respect to Q and $g = \frac{dP}{dQ}$ belongs to $L_1(P)$, then $I_Q(P)$ is a lower semi-continuous real function satisfying*

$$I_Q(P) = \int_\mathcal{X} \log(\frac{dP}{dQ})\, dP$$

and there exists a continuous function $\phi : \mathbb{R} \mapsto [0,2]$ such that $\phi(0) = 0$ and

$$\|P - Q\|_1 \leq \sup_{A \in \mathcal{B}} |P(A) - Q(A)| \leq \phi(I_Q(P)).$$

It follows that the convergence of a sequence of probabilities $(P_n)_n$ to a limiting probability P is satisfied if $\lim_{n\to\infty} I_Q(P_n) = I_Q(P)$ for some dominating probability Q, which is equivalent to $\lim_{n\to\infty} \|P_n - P\| = 0$. The distance $\|P - Q\|_1$ defined above is the supremum over the balls and the convergence for this metric implies the convergence in probability, it is equivalent to the convergence in total variation in a probability space endowed with the Borel σ-algebra. Other relationships between probability metrics are reviewed by Gibbs and Su (2002).

The Hausdorff metric on a separable probability space $(\Omega, \mathcal{A}, P)$ is $\rho(A, B) = P(A \Delta B)$ where $A \Delta B = (A \cup B) \setminus (A \cap B)$, where A and B are measurable sets. It is extended to sub-σ fields $\mathcal{B}$ and $\mathcal{C}$ of $\mathcal{F}$

$$\delta(\mathcal{B}, \mathcal{C}) = \max\{\sup_{B \in \mathcal{B}} \inf_{C \in \mathcal{C}} P(B \Delta C), \sup_{C \in \mathcal{C}} \inf_{B \in \mathcal{B}} P(B \Delta C)\}.$$

Let (X, δ) be a metric space, with the Hausdorff metric δ.

Lemma A.2 (Rogge's lemma (1974)). *Let $\mathcal{A}$ and $\mathcal{B}$ be sub-σ-algebras of $\mathcal{F}$, then for every $\mathcal{B}$-measurable function $f : X \to [0,1]$*

$$\|P^\mathcal{A} f - f\|_1 \leq 2\delta(\mathcal{A}, \mathcal{B})\{1 - \delta(\mathcal{A}, \mathcal{B})\},$$
$$\|P^\mathcal{A} f - f\|_2 \leq [\delta(\mathcal{A}, \mathcal{B})\{1 - \delta(\mathcal{A}, \mathcal{B})\}]^{\frac{1}{2}}.$$

Let Φ be the set of all $\mathcal{F}$-measurables functions $f : X \to [0,1]$. It follows that for every sub-σ-algebras $\mathcal{A}$ and $\mathcal{B}$ of $\mathcal{F}$

$$\sup_{f \in \Phi} \|P^\mathcal{A} f - P^\mathcal{B} f\|_2 \leq [2\delta(\mathcal{A}, \mathcal{B})\{1 - \delta(\mathcal{A}, \mathcal{B})\}]^{\frac{1}{2}}.$$

It follows that

$$\delta(\mathcal{A}, \mathcal{B}) \leq \sup_{f \in \Phi} \|P^\mathcal{A} f - P^\mathcal{B} f\|_1$$

and $\sup_{f \in \Phi} \|P^{\mathcal{F}_n} f - P^{\mathcal{F}_\infty} f\|_1$ tends to zero for every sequence of sub-σ-algebras of $\mathcal{F}$ such that $\delta(\mathcal{F}_n, \mathcal{F}_\infty) \to 0$ as n tends to infinity. This convergence was applied to the equi-convergence of martingales (Boylan, 1971).

Let $(\mathbb{X}, \mathcal{X}, d)$ be a metric space with the Borel σ-algebra and let $\mathcal{P}(\mathbb{X})$ be a space of probability measures on $(\mathbb{X}, \mathcal{X})$. The Prohorov distance between probabilities P and Q of $\mathcal{P}(\mathbb{X})$ is defined by

$$\Pi(P, Q) = \inf\{\varepsilon > 0; P(A) < Q(A^\varepsilon) + \varepsilon, \; Q(A) < P(A^\varepsilon) + \varepsilon, \; \forall A \in \mathcal{X}\}$$

where the ε-neighbourhood of a subset A of $\mathbb{X}$ is

$$A^\varepsilon = \{x \in \mathbb{X}; \inf_{y \in \mathbb{X}} d(x, y) < \varepsilon\}, \; \varepsilon > 0.$$

Equivalently

$$\Pi(P, Q) = \inf\{\varepsilon > 0; |P(A) - Q(B)| < \varepsilon, A, B \in \mathcal{X} \text{ s.t. } d(A, B) < \varepsilon\}.$$

Let $(C[0, 1], \mathcal{C}, \|\cdot\|)$ be the space of continuous real functions defined on $[0, 1]$, provided with the norm uniform $\|x\| = \sup_{t \in [0,1]} |x(t)|$, for every continuous function x of $[0, 1]$, and the Borel σ-algebra. Let $X = (X_t)_{t \in [0,1]}$ and $Y = (Y_t)_{t \in [0,1]}$ be continuous processes indexed by $[0, 1]$, the Prohorov distance between the paths of the processes is defined with respect to sets of the Borel-σ-algebra $\mathcal{B}$ on $\mathbb{R}$

$$\Pi(X, Y) = \inf\{\varepsilon > 0; |P(X_t \in A) - P(Y_t \in B)| < \varepsilon, A, B \in \mathcal{B} : d(A, B) < \varepsilon\}.$$

On $C[0, 1]$, it is equivalent to the L_1 distance $\sup_{t \in [0,1]} E|X_t - Y_t|$.

Proposition A.3. *For every function f of $C_b(\mathbb{R})$ and let X and Y be processes with paths in $C_b(\mathbb{R})$. For every $\varepsilon > 0$, there exists $\eta > 0$ such that $\Pi(X, Y) < \eta$ implies $\sup_{t \in [0,1]} E|f(X_t) - f(Y_t)| < \varepsilon$ and $\Pi(f(X), f(Y)) < \varepsilon$.*
Let f be a Lipschitz function of $C_b(\mathbb{R})$, then the Prohorov distances $\Pi(X, Y)$ and $\Pi(f(X), f(Y))$ are equivalent.

A.4   Expansions in $L_2(\mathbb{R})$

Fourier transform

Lemma A.3. *In the expansion* $f(x) = \frac{a_0}{2} + \sum_{k=1}^{\infty} \{a_k \cos(kx) + b_k \sin(kx)\}$, *the coefficients are defined as*

$$a_k = \frac{1}{\pi} \int_{-\pi}^{\pi} f(x) \cos(kx)\, dx,$$

$$b_k = \frac{2}{\pi} \int_{-\pi}^{\pi} f(x) \sin(kx)\, dx,$$

$$a_0 = \frac{1}{2\pi} \int_{-\pi}^{\pi} f(x)\, dx.$$

Proof.   The integral $I(f) = \frac{1}{\pi} \int_{-\pi}^{\pi} f(x) \cos(nx)\, dx$, $n \neq 0$ integer, develops as a sum of integrals with coefficients a_k and b_k

$$I_{1n}(f) = \sum_{k=1}^{\infty} \frac{a_k}{2\pi} \int_{-\pi}^{\pi} \cos(kx) \cos(nx)\, dx$$

$$= \sum_{k=1, k\neq n}^{\infty} \frac{a_k}{2\pi} \int_{-\pi}^{\pi} \{\cos(k+n)x + \cos(k-n)x\}\, dx$$

$$+ \frac{a_n}{2\pi} \int_{-\pi}^{\pi} \{1 + \cos(2nx)\}\, dx = a_n,$$

where the other terms of the sum are zero by periodicity of the sine function. Similarly

$$I_{2n}(f) = \sum_{k=1}^{\infty} \frac{b_k}{2\pi} \int_{-\pi}^{\pi} \sin(kx) \sin(nx)\, dx$$

$$= \sum_{k=1, k\neq n}^{\infty} \frac{b_k}{2\pi} \int_{-\pi}^{\pi} \{\cos(k+n)x - \cos(k-n)x\}\, dx$$

$$+ \frac{b_n}{2\pi} \int_{-\pi}^{\pi} \{\cos(2nx) - 1\}\, dx = b_n,$$

the expression of a_0 is a consequence of the periodicity of the trigonometric functions and

$$I_{3n}(f) = \sum_{k=1}^{\infty} \int_{-\pi}^{\pi} \cos(kx) \sin(nx)\, dx$$

$$= \sum_{k=1, k\neq n}^{\infty} \int_{-\pi}^{\pi} \{\sin(k+n)x + \sin(k-n)x\}\, dx + \int_{-\pi}^{\pi} \sin(2nx)\, dx = 0. \qquad \square$$

The coefficients of the Fourier transform and the trigonometric series are decreasing as $o(k)$ when k tends to infinity.

Proposition A.4. *The set of functions e_n defined by*

$$e_n(t) = \exp\{\frac{2\pi}{T}nt\}, \ t \in [0, T]$$

is an orthogonal basis of $L_2[0, T]$ and every periodic function of $L_2[0, T]$ is the uniform limit of a series of trigonometric polynomials.

Giné (1974) proved a Lipschitz property for the Fourier transform.

Proposition A.5. *Let $f(t) = \sum_{n=-\infty}^{\infty} a_n e^{2\pi i n t}$, t in $[0, 1]$, with coefficients such that $\sum_{n \neq 0} |a_n| |\log n|^\alpha$ is finite, then $|f(s) - f(t)| \leq C_\varepsilon |\log |s - t||^{-\alpha}$ on $[0, 1]$, with the constant*

$$C_\varepsilon = 2\pi \sum_{|n| \leq e^\alpha} \{(\frac{n}{\varepsilon})^\alpha |n| |a_n| + |a_n| (\log |n|)^\alpha\}.$$

Hermite polynomials

Let $f_{\mathcal{N}}$ be the normal density and let $H_0 = 1$ and H_k, $k \geq 1$, be the Hermite polynomials defined by

$$H_k(t) = (-1)^k \frac{d^k e^{-\frac{t^2}{2}}}{dt^k} e^{\frac{t^2}{2}}, \ k \geq 1.$$

The recursive equation $H_{k+1} = xH_k - H_k'$, for $k \geq 2$, provides an algorithm to calculate their expression. We obtain

$$H_1(x) = x,$$
$$H_2(x) = x^2 - 1,$$
$$H_3(x) = x^3 - 3x,$$
$$H_4(x) = x^4 - 6x^2 + 3,$$
$$H_5(x) = x^5 - 10x^3 + 15x,$$
$$H_6(x) = x^6 - 15x^4 + 45x^2 - 15,$$
$$H_7(x) = x^7 - 21x^5 - 105x^3 - 105x, \text{ etc.}$$

They have the form

$$H_{2k}(x) = \sum_{j=0}^{k-1} h_{2j} x^{2j} + x^{2k},$$

$$H_{2k+1}(x) = \sum_{j=0}^{k-1} b_{2j+1} x^{2j+1} + x^{2k+1}.$$

and all coefficients are calculated iteratively: for every $k \geq 2$, the coefficient of x in $H_{2k-1}(x)$ is the even moment $m_{2k} = 3.5\ldots(2k-1)$ of the normal distribution and this is the constant of $H_{2k}(x)$. The difference $\alpha_{k-2} - \alpha_{k-1}$ of the coefficients α_{k-2} of x^{k-2} in H_k and α_{k-1} of x^{k-1} in H_{k+1} is equal to k, etc. Their norms are calculated from the moments m_{2k} of the normal distribution

$$c_k = \{\int_{\mathbb{R}} H_k^2(x) f_{\mathcal{N}}(x)\, dx\}^{\frac{1}{2}} = (k!)^{\frac{1}{2}}.$$

Since $m_{2k} = \{2^k(k)!\}^{-1}(2k)!$, the square of the constant terms of the normalized polynomial $h_{2k}(x)$ is $C_{2k}^k 2^{-2k}$ and it tends to zero as a power of $\frac{1}{2}$ as k increases and their sum converge.

The functions H_k satisfy

$$\sum_{k \geq 0} H_k(t) \frac{w^k}{k!} = e^{-\frac{(t+w)^2}{2}} e^{\frac{t^2}{2}} = e^{-\frac{w^2+2wt}{2}}.$$

From the derivatives of

$$e^{-\frac{t^2}{2}} = \frac{1}{\sqrt{\pi}} \int_{\mathbb{R}} e^{-\frac{x^2}{2} + ixt}\, dx,$$

the Hermite polynomials are also expressed as

$$H_k(t) = \frac{1}{\sqrt{\pi}} \int_{\mathbb{R}} (-ix)^k e^{-\frac{(x-it)^2}{2}}\, dx.$$

The polynomials are sometimes defined by similar expressions where $\exp\{-\frac{x^2}{2}\}$ of the normal density is replaced by $\exp\{-x^2\}$, the recursive equation becomes $H_{k+1} = 2xH_k - H_k'$ but these equations generate non orthogonal polynomials. For instance, $H_1(x) = 2x$, $H_2(x) = 2(2x^2 - 1)$, $H_3(x) = x^3 - \frac{3}{2}x$ and $E\{H_1(X)H_3(X)\} = 2$.

Bibliography

Adler, R. J. and Taylor, J. E. (2007). *Random and Geometry* (Springer, Berlin).

Alzer, H. (1987). The central limit theorem for empirical processes on Vapnik-Chervonenkis classes, *Ann. Probab.* **15**, pp. 178–203.

Alzer, H. (1990a). Inequalities for the arithmetic, geometric and harmonic means, *Bull. London Math. Soc.* **22**, pp. 362–366.

Alzer, H. (1990b). An inequality for the exponential function, *Arch. Math.* **55**, pp. 462–464.

Assouad, P. (1981). Sur les classes de Vapnik-Chervonenkis, *C. R. Acad. Sci. Paris, I* **292**, pp. 921–924.

Barnett, N. S. and Dragomir, S. S. (2001). A perturbed trapezoid inequality in terms of the third derivative and applications, *Ineq. Theor. Appl.* **5**, pp. 1–11.

Barnett, N. S. and Dragomir, S. S. (2002). A perturbed trapezoid inequality in terms of the fourth derivative, *J. Appl. Math. Comput.* **9**, pp. 45–60.

Beckner, W. (1975). Inequalities in Fourier analysis, *Ann. Math. USA* **102**, pp. 159–182.

Bennet, G. (1962). Probability inequalities for sums of independent random variables, *Amer. Statist. Assoc.* **57**, pp. 33–45.

Berkes, I. and Philipp, W. (1979). Approximation theorems for independent and weakly dependent random vectors, *Ann. Probab.*, pp. 29–54.

Bickel, P. J., Klassen, C. A., Ritov, Y. and Wellner, J. A. (1993). *Efficient and adaptive estimation in semiparametric models* (Johns Hopkins University Press, Baltimore).

Blake, I. F. and Lindsey, W. C. (1973). Level crossing problems for random processes, *IEEE Trans. Inf. Theor.* **19**, pp. 295–315.

Boylan, E. S. (1971). Equi-convergence of martingales, *Ann. Math. Statist.* **42**, pp. 552–559.

Bradley, R. E., d'Antonio, L. A. and Sandifer, C. E. (2007). *Euler at 300: an appreciation* (Math. Ass. Amer., Washington).

Breiman, L. (1968). *Probability* (Addison-Wesley, Reading, Massachusetts).

Bürkholder, D. L. (1973). Distribution function inequalities for martingales, *Ann. Prob.* **1**, pp. 19–42.

Bürkholder, D. L., Davis, B. J. and Gundy, R. F. (1972). Convex functions of operators on martingales, *Proceedings of the sixth Berkeley Symposium on Mathematical Statistics and Probability, Vol. 2-3*, pp. 789–806.

Byerly, E. B. (1893). *An elementary treatise on Fourier series and spherical, cylindrical, and ellipsoidal harmonics, with applications to problems in mathematical physics* (Ginn and Company, Boston, New York, Chicago London).

Cairoli, R. and Walsh, J. B. (1975). Stochastic integrals in the plane, *Acta. Math.* **134**, pp. 111–183.

Carlson, B. C. (1966). Some inequalities for hypergeometric functions, *Proc. Amer. Math. Soc.* **17**, pp. 32–39.

Cauchy, A. L. (1821). *Cours d'Analyse de l'Ecole Royale Polytechnique, I. Analyse Algébrique* (Editions Jacques Gabay, Sceaux).

Cauchy, A. L. (1833). *Résumés analytiques* (Imprimerie royale, Turin).

Chow, Y. S. and Lai, T. L. (1973). Limit behavior of weighted sums of independent random variables, *Ann. Probab.* **5**, pp. 810–824.

Chow, Y. S., Robbins, H. and Teicher, H. (1965). Moments of randomly stopped sums, *Ann. Math. Statist.* **36**, pp. 789–799.

Chow, Y. S. and Teicher, H. (1966). On second moments of stopping rules, *Ann. Math. Statist.* **37**, pp. 388–392.

Christofides, T. S. and Serfling, R. (1990). Maximal inequalities for multidimensionally indexed submartingale arrays, *Ann. Probab.* **18**, pp. 630–641.

Cox, D. R. (1960). *Point Processes* (Chapman and Hall, Londres).

Csörgö, M., Kolmós, J., Major, P. and Tusnády, G. (1974). On the empirical process when parameters are estimated, *Transactions of the seventh Prague conference on informtion theorey, statistical decision functions, random processes. Academia, Prague* **B**, pp. 87–97.

Cunnigham, F. and Grossman, N. (1971). On Young's inequality, *Amer. Math. Month.* **78**, pp. 781–783.

Dehling, H. (1983). Limit theorems for sums of weakly dependant Banach space valued random variable, *Z. Wahrsch. verw. Geb.* **63**, pp. 393–432.

Dembo, A. and Zeitouni, O. (2009). *Large Deviation Techniques and Applications, 3rd ed. Stochastic Modelling and Applied Probability 38* (Springer, Berlin Heidelberg).

den Hollander, F. (2008). *Large Deviations* (Amer. Math. Soc., London).

Deuschel, J.-D. and Stroock, D. W. (1984). *Large Deviations* (Aademic Press, London).

Doob, J. L. (1953). *Stochastic Process* (Wiley, New York).

Doob, J. L. (1975). Stochastic process measurability conditions, *Ann. Instit. J. Fourier* **25**, pp. 163–176.

Dragomir, S. S. and Rassias, T. M. (2002). *Ostrowski Type Inequalities and Applications in Numerical Integration* (Kluwer Acaemic Publisher, Dordrecht).

Dragomir, S. S. and Sofo, A. (2000). An integral inequality for twice differentiable mappings and applications, *Tamkang J. Math.* **31**, pp. 257–266.

Dudley, R. M. (1973). Sample functions of the Gaussian process, *Ann. Prob.* **1**, pp. 66–103.

Dudley, R. M. (1974). Metric entropy and the central limit theorem in *C(s)*, *Ann. Inst. Fourier* **24**, pp. 40–60.

Dudley, R. M. (1984). *A course on empirical processes. Ecole d'été de Probabilité de St Flour. Lecture Notes in Math.* **1097** (Springer-Verlag, Berlin).

Dunham, W. (1999). *Euler. The Master of Us All* (Math. Ass. Amer., Washington).

Durbin, J. (1971). Boundary-crossing probabilities for the Brownian motion and Poisson processes and techniques for computing the power of the Kolmogorov-Smirnov test, *J. Appl. Prob.* **8**, pp. 431–453.

Durrett, R. (2010). *Probability: Theory and Examples (4th edition)* (Cambridge University Press, New York).

Erdös, P. and Kac, M. (1947). On the number of positive sums of independent random variables, *Bull. Amer. Math. Soc.* **53**, pp. 1011–1020.

Feller, W. (1966). *An Introduction to Probability Theory and its Applications, Vol. 2 (second ed.)* (Wiley, London).

Fernholz, L. T. (1983). *Von Mises calculus for statistical functionals. Lecture Notes in Statistics, 19* (Springer-Verlag, Berlin).

Fourier, J. (1822). *Théorie analytique de la chaleur* (Firmin Didot, Paris).

Fournier, J. J. F. (1977). Sharpness in Young's inequality for convolution, *Pacific. J. Math.* **72**, pp. 383–397.

Freedman, D. A. (1973). Another note on the Borel-Cantelli lemma and the strong law, with the Poisson approximation as a by-product, *Ann. Probab.* **1**, pp. 910–925.

Freedman, D. A. (1975). On tail probabilities for martingales, *Ann. Probab.* **3**, pp. 100–118.

Geetor, R. K. and Sharpe, M. J. (1979). Excursions of Brownian motion and Bessel processes, *Z. Wahrsch. verw Geb.* **47**, pp. 83–106.

Gibbs, A. L. and Su, F. E. (2002). On choosing and bounding probability metrics, *Instit. Statist. Rev.* **70**, pp. 419–435.

Hardy, G. H., Littlewood, J. E. and Pólya, G. (1952). *Inequalities, 2nd ed.* (Cambridge University Press, Cambridge).

Hutton, C. (1811). *Mathematical tables* (Rivington, Wilkie *et al.*, London).

Itô, K. and McKean, H. P. (1996). *Diffusion processes and their sample paths, 2nd ed.* (Springer, Berlin-Heidelberg-New York).

Kac, M. (1947). Random walks and the theory of Brownian motion, *Amer. Math. Month.* **54**, pp. 369–391.

Kaijser, S., Nikolova, L., Perrson, L.-E. and Wedestig, A. (2005). Hardy-type inequalities via convexity, *Math. Inequal. Appl.* **8**, pp. 403–417.

Kaijser, S., Perrson, L.-E. and Öberg, A. (2002). On Carleman and Knopp's inequalities, *J. Approx. Theory* **117**, pp. 140–151.

Karlin, S. and McGregor, J. (1965). Occupation time law for birth and death processes, *Trans. Amer. Math. Soc.* **88**, pp. 249–272.

Keilson, J. and Steutel, F. W. (1974). Mixtures of distributions, moment inequalities and measures of exponentiality and normality, *Ann. Probab.* **2**, pp. 112–130.

Kiefer, J. (1961). On large deviations of the empiric d.f. of vector chance variables and a law of the iterated logarithm, *Pacific J. Math.* **11**, pp. 649–660.

Kolmos, J., Major, P. and Tusnady, G. (1975). An approximation of partial sums of independent rv's, and the sample df. I, *Z. Wahrsch. verw. Geb.* **32**, pp. 111–131.

Kolmos, J., Major, P. and Tusnady, G. (1976). An approximation of partial sums of independent rv's, and the sample df. II, *Z. Wahrsch. verw. Geb.* **34**, pp. 33–58.

Krishnapur, M. (2003). *Probability Theory* (Lecture Notes, Berkely university).

Kufner, A. and Persson, L.-E. (2003). *Weighted inequalities of the Hardy type* (World Scientific Publishing, Singapore).

Lagrange, J.-L. (1826). *Traité de la résolution des équations numériques de tous les degrés* (Bachelier, Paris).

Lamperti, J. (1958). An occupation time theorem for a class of stochastic processes, *Trans. Amer. Math. Soc.* **88**, pp. 380–387.

Legendre, A. M. (1805). *Nouvelles méthodes pour la détermination des orbites des comètes* (Firmin Didot, Paris).

Lenglart, E. (1977). Relation de domination entre deux processus, *Ann. Inst. H. Poincaré* **13**, pp. 171–179.

Lepingle, D. (1978). Sur le comportement asymptotique des martingales locales, *Séminaire Probab.* **12**, pp. 148 161.

Lohwater, A. (1982). *Introduction to Inequalities* (Unpublished).

Luor, D.-C. (2009). A variant of a general inequality of the Hardy-Knopp type, *J. Ineq. Pure Appl. Math.* **3**, pp. 1–9.

Massart, P. (1986). Rates of convergence in the central limit theorem for empirical processes, *Ann. Instit. Henri Poincaré* **22**, pp. 381–423.

Maurey, B. (2004). Inégalités de Brunn-Minkovski-Lusternik, et d'autres inégalités géométriques et fonctionnelles, *Séminaire Bourbaki* **928**, pp. 1–19.

Meyer, P.-A. (1969). Les inégalités de Bürkholder en théorie des martingales, d'après Gundy, *Sém. Probab. Strasbourg* **3**, pp. 163–174.

Mona, P. (1994). Remarques sur les inégalités de Bürkholder-Davis-Gundy, *Sém. Probab. Strasbourg* **28**, pp. 92–97.

Mulholland, H. P. (1950). On generalizations of Minkowski's inequality in the form of a triangle inequality, *Proc. London Math. Soc.* **51**, pp. 294–307.

Neveu, J. (1970). *Bases mathématiques du calcul des probabilités* (Masson, Paris).

Neveu, J. (1972). *Martingales à temps discret* (Masson, Paris).

Pachpatte, B. G. (2005). *Mathematical Inequalities* (North-Holland, Elsevier).

Pechtl, A. (1999). Distributions of occupation times of Brownian motion with drift, *J. Appl. Math. Decis. Sci.* **3**, pp. 41–62.

Pitt, L. D. (1977). A Gaussian correlation inequality for symmetric convex sets, *Ann. Probab.* **5**, pp. 470–474.

Pollard, D. (1981). A central limit theorem for empirical processes, *J. Aust. Math. Soc; A* **33**, pp. 235–248.

Pollard, D. (1982). A central limit theorem for k-means clustering, *Ann. Probab.* **10**, pp. 919–926.

Pollard, D. (1984). *Convergence of Stochastic Processes* (Springer, New York).

Pons, O. (1986). A test of independence between two censored survival times, *Scand. J. Statist.* **13**, pp. 173–185.

Pons, O. (2008). *Statistique de processus de renouvellement et markoviens* (Hermès Science Lavoisier, London and Paris).

Pons, O. and Turckheim, E. (1989). Méthode de von-Mises, Hadamard différentiabilité et bootstrap dans un modèle non-paramétrique sur un espace métrique, *C. R. Acad. Sc. Paris, Ser. I* **308**, pp. 369–372.

Pratelli, M. (1975). Deux inégalités concernant les opérateurs de Bürkholder sur les martingales, *Ann. Probab.* **3**, pp. 365–370.

Revuz, D. and Yor, M. (1986). *Continuous martingales and Brownian motion* (Springer, Berlin).

Robbins, H. and Siegmund, D. (1970). Boundary-crossing probabilities for the Wiener process and sample sums, *Ann. Math. Statist.* **41**, pp. 1410–1429.

Rogge, L. (1974). Uniform inequalities for conditional expectations, *Ann. Probab.* **2**, pp. 486–489.

Sahoo, P. K. and Riedel, T. (1998). *Mean value theorems and functional equations* (World Scientific Publishing, Singapore-London).

Sandifer, C. E. (2007). *The Early Mathematics of Leonhard Euler* (Math. Ass. Amer., Washington).

Schladitz, K. and Baddeley, A. J. (2000). Uniform inequalities for conditional expectations, *Scand. J. Statis.* **27**, pp. 657–671.

Serfling, R. J. (1968). Contributions to central limit theory for dependent variables, *Ann. Math.Statist.* **39**, pp. 1158–1175.

Shorack, G. R. and Wellner, J. A. (1986). *Empirical processes and applications to statistics* (Wiley, New York).

Sjölin, P. (1995). A remark on the Hausdorff-Young inequality, *Proc. Am. Math. Soc.* **123**, pp. 3085–3088.

Slepian, D. (1962). The one-sided barrier problem for Gaussian noise, *Bell System Techn. J.* **1**, pp. 463–501.

Stout, W. F. (1970). The Hartman-Wintner law of the iterated logarithm for martingales, *Ann. Math. Statist.* **41**, pp. 2158–2160.

Strassen, V. (1967). Almost sure behaviour of sums of independent random variables and martingales, *Proceed. 5th Berkeley Symp. Mathematical Statistics and Probability, Vol. 2* , pp. 315–343.

Talagrand, M. (1987). Donsker classes and random geometry, *Ann. Probab.* **15**, pp. 1327–1338.

Talagrand, M. (1994). Sharper bounds for Gaussian and empirical processes, *Ann. Probab.* **22**, pp. 28–76.

Talagrand, M. (1995). Concentration of measures and isoperimetric inequalities in product spaces, *Pub. Math. IHES* **81**, pp. 73–205.

Thorin, O. (1970). Some comments on the Sparre Andersen model in the risk theory, *Scand. Actuar. J.* **1**, pp. 29–50.

Van der Vaart, A. and Wellner, J. A. (1995). *Weak Convergence and Empirical Processes* (Springer, New York).

Inequalities in Analysis and Probability

Varadhan, S. R. S. (1984). *Large Deviations and Applications* (SIAM, Philadel-
    phia).
Walsh, J. B. (1974). Stochastic integrals in the plane, *Proc. Intern. Congress
    Math.* **Sect. 11**, pp. 189–194.
Weisz, F. (1994). One-parameter martingale inequalities, *Ann. Univ. Sci. Bu-
    dapest, Sect. Comp.* **14**, pp. 249–278.
Wong, E. and Zakai, M. (1974). Martingaales and stochastic integrals for pro-
    cesses with multidimensional parameter, *Z. Wahrsch. verw. Geb.* **29**, pp.
    109–122.

Index

Arcsine law, 20, 120, 155

Bennet inequality
  continuous martingale, 106
  discrete martingale, 103
Bennett inequality, 25, 108
  functional inequality, 132
Bernstein inequality, 206
Berry-Essen incquality, 12
Biased length variable, 148
  Bienaymé-Tchebychev, 149
Bienaymé-Chebychev inequality, 10
  maximum variable, 12
Birnbaum-Marshal inequality, 100
Boundary crossing, 120
  Brownian motion, 206
Brownian bridge, 21
  Doob inequality, 21
Brownian motion, 19, 111
  arcsine law, 114
  duration out of zero, 20
  Laplace transform, 24, 112
  moments inequality, 111
  stopping time, 19
  transformed, 113
    exponential inequality, 138
    modulus of continuity, 138
Bürkholder inequality, 17
Bürkholder-Davis-Gundy inequality
  functional inequality, 129
  independent variable, 92
  martingale, 99

Carlson inequality, 80
Cauchy convex inequality
  arithmetic mean, 65
Cauchy distribution, 67
Cauchy equations, 184, 192, 196
Cauchy inequality
  arithmetic mean, 64
Chernov's theorem, 23
Complex polynomial, 182
Complex series, 181
Convexity, 6
  Carleman inequality, 46
  Cauchy inequality, 6
  Hadamard inequality, 6
  Minkowski, 48
Current time, 121, 158

Dependent variables, 98, 151
  Chernov theorem, 99
  mixing coefficient, 97
  moment inequality, 98
Differentiability
  functional, 142
Diffusion process, 116
  exponential solution, 117
  polynomial solution, 118
  ruin model, 160
Distances
  Hausdorff, 208
  probability, 207
  Prohorov, 209

Distribution mixture
  continuous, 71
Distribution mixture, 70
  continuous mixture, 166

Empirical process, 27, 127, 131
  Donsker class, 129
  exponential inequality, 115
  functional inequality, 133
  tightness, 176
Entropy, 88
  functional, 128
  metric, 127
Ergodic process, 151
Exponential distribution, 166

Fisher information, 90
Fourier series, 179
Fourier transform, 12, 194
  convergence, 210
  Hölder's inequality, 5
  operator, 186
  Parseval equality, 186
Functional equations, 74
Functional means, 83

Generating function, 11
Geometric mean, 66

Hadamard inequality, 76
Hardy inequality, 38
  convex transform, 43
  integral, 13
  multidimensional, 54
  series, 13
Hermite polynomials
  coefficients, 211
Hermite transform
  partial sums, 190
Hilbert inequality, 36
Hoeffding inequality, 205
  functional inequality, 133
holomorph function
  norm of derivatives, 193
Hölder inequality, 36

Inequalities
  functional, 127
  moments, 33
  norms, 91
  series, 32

Kolmogorov inequality, 101, 162
Kullback-Leibler information, 90

Lévy inequality, 20, 120
Laplace transform, 11
  convex function, 52
large deviations, 26
Law of iterated logarithm, 122
Lenglart inequality
  continuous martingale, 106
  discrete martingale, 103
Level-crossing probability, 120
Logarithmic mean, 80
lower bound, 92

Martingale, 162
  d-dimensional, 18
  Bennett's inequality, 102
  Chernov's theorem, 106
  continuous index, 19, 104
    predictable quadratic
    variations, 21
  convergence, 203
  discrete index, 15
    predictable quadratic
    variations, 16
  functional inequality, 135
  Kolmogorov inequality, 16
  Lenglart, 22
  Lepingle, 21
  maximum variable, 17, 101
  second order moments, 125
  spatial, 124
Maximum variable
  convergence, 202
  moment inequality, 42
Mean residual time, 72
Mean value theorem, 14, 75, 77
Median, 68
Mode, 68

Neveu inequality, 103
Norm
  Cauchy inequality, 3
  equivalences, 3
  geometric equality, 2
  inequality, 2
  Minkowski inequality, 5
  norm ℓ_p, 2
Norm L_p
  Cauchy-Schwarz inequality, 3, 10
  Hölder inequality, 4
  Jensen inequality, 5
  Kinchin inequality, 4
  random variable, 9

Orthonormal basis, 211
  Fourier transform, 183
  Hermite polynomials, 183, 211
Ostrowski inequality, 14

Partial sums, 61
Point process
  functional inequality, 140
Poisson process, 108, 120, 125
  Chernov theorem, 110
  Laplace transform, 110
  Lenglart inequality, 110
  moments inequality, 110
  spatial, 170
  spatial moments, 172
Probability of ruin, 156
  diffusion process, 160
  infinite ruin time, 157

Regression
  functional, 146

functional Kolmogorov inequality,
  148

scalar product, 2
Slepian inequalities, 154
Snell envelope, 203
Spatial inequalities, 169
spatial process, 124, 171
  empirical process, 176
  moments, 176
  Poisson process, 176
Stationarity
  covariance function, 153
Stationary process, 154
  Gaussian process, 154
Stochastic convergences, 202
Stochastic order, 154
Stopping time, 16, 19, 202
  Brownian motion, 19
  Doob's theorem, 17
submartingale, 15
Sum of weighted variables, 94, 204
supermartingale, 15, 103

Talagrand's inequality, 24

unimodal density, 70

Vapnik-Chervonenkis, 28

waiting time, 109
  residual waiting time, 121
Wald equalities, 203

Young's inequality, 86